Nonlinear Analysis and Applications

PURE AND APPLIED MATHEMATICS

A Program of Monographs, Textbooks, and Lecture Notes

Contributions to *Lecture Notes in Pure and Applied Mathematics* are reproduced by direct photography of the author's typewritten manuscript. Potential authors are advised to submit preliminary manuscripts for review purposes. After acceptance, the author is responsible for preparing the final manuscript in camera-ready form, suitable for direct reproduction. Marcel Dekker, Inc. will furnish instructions to authors and special typing paper. Sample pages are reviewed and returned with our suggestions to assure quality control and the most attractive rendering of your manuscript. The publisher will also be happy to supervise and assist in all stages of the preparation of your camera-ready manuscript.

LECTURE NOTES

IN PURE AND APPLIED MATHEMATICS

1. *N. Jacobson*, Exceptional Lie Algebras
2. *L.-Å. Lindahl and F. Poulsen*, Thin Sets in Harmonic Analysis
3. *I. Satake*, Classification Theory of Semi-Simple Algebraic Groups
4. *F. Hirzebruch, W. D. Newmann, and S. S. Koh*, Differentiable Manifolds and Quadratic Forms (out of print)
5. *I. Chavel*, Riemannian Symmetric Spaces of Rank One (out of print)
6. *R. B. Burckel*, Characterization of C(X) Among Its Subalgebras
7. *B. R. McDonald, A. R. Magid, and K. C. Smith*, Ring Theory: Proceedings of the Oklahoma Conference
8. *Y.-T. Siu*, Techniques of Extension of Analytic Objects
9. *S. R. Caradus, W. E. Pfaffenberger, and B. Yood*, Calkin Algebras and Algebras of Operators on Banach Spaces
10. *E. O. Roxin, P.-T. Liu, and R. L. Sternberg*, Differential Games and Control Theory
11. *M. Orzech and C. Small*, The Brauer Group of Commutative Rings
12. *S. Thomeier*, Topology and Its Applications
13. *J. M. López and K. A. Ross*, Sidon Sets
14. *W. W. Comfort and S. Negrepontis*, Continuous Pseudometrics
15. *K. McKennon and J. M. Robertson*, Locally Convex Spaces
16. *M. Carmeli and S. Malin*, Representations of the Rotation and Lorentz Groups: An Introduction
17. *G. B. Seligman*, Rational Methods in Lie Algebras
18. *D. G. de Figueiredo*, Functional Analysis: Proceedings of the Brazilian Mathematical Society Symposium
19. *L. Cesari, R. Kannan, and J. D. Schuur*, Nonlinear Functional Analysis and Differential Equations: Proceedings of the Michigan State University Conference
20. *J. J. Schäffer*, Geometry of Spheres in Normed Spaces
21. *K. Yano and M. Kon*, Anti-Invariant Submanifolds
22. *W. V. Vasconcelos*, The Rings of Dimension Two
23. *R. E. Chandler*, Hausdorff Compactifications
24. *S. P. Franklin and B. V. S. Thomas*, Topology: Proceedings of the Memphis State University Conference
25. *S. K. Jain*, Ring Theory: Proceedings of the Ohio University Conference
26. *B. R. McDonald and R. A. Morris*, Ring Theory II: Proceedings of the Second Oklahoma Conference
27. *R. B. Mura and A. Rhemtulla*, Orderable Groups
28. *J. R. Graef*, Stability of Dynamical Systems: Theory and Applications
29. *H.-C. Wang*, Homogeneous Banach Algebras
30. *E. O. Roxin, P.-T. Liu, and R. L. Sternberg*, Differential Games and Control Theory II
31. *R. D. Porter*, Introduction to Fibre Bundles
32. *M. Altman*, Contractors and Contractor Directions Theory and Applications
33. *J. S. Golan*, Decomposition and Dimension in Module Categories
34. *G. Fairweather*, Finite Element Galerkin Methods for Differential Equations
35. *J. D. Sally*, Numbers of Generators of Ideals in Local Rings
36. *S. S. Miller*, Complex Analysis: Proceedings of the S.U.N.Y. Brockport Conference
37. *R. Gordon*, Representation Theory of Algebras: Proceedings of the Philadelphia Conference
38. *M. Goto and F. D. Grosshans*, Semisimple Lie Algebras
39. *A. I. Arruda, N. C. A. da Costa, and R. Chuaqui*, Mathematical Logic: Proceedings of the First Brazilian Conference

40. *F. Van Oystaeyen*, Ring Theory: Proceedings of the 1977 Antwerp Conference
41. *F. Van Oystaeyen and A. Verschoren*, Reflectors and Localization: Application to Sheaf Theory
42. *M. Satyanarayana*, Positively Ordered Semigroups
43. *D. L. Russell*, Mathematics of Finite-Dimensional Control Systems
44. *P.-T. Liu and E. Roxin*, Differential Games and Control Theory III: Proceedings of the Third Kingston Conference, Part A
45. *A. Geramita and J. Seberry*, Orthogonal Designs: Quadratic Forms and Hadamard Matrices
46. *J. Cigler, V. Losert, and P. Michor*, Banach Modules and Functors on Categories of Banach Spaces
47. *P.-T. Liu and J. G. Sutinen*, Control Theory in Mathematical Economics: Proceedings of the Third Kingston Conference, Part B
48. *C. Byrnes*, Partial Differential Equations and Geometry
49. *G. Klambauer*, Problems and Propositions in Analysis
50. *J. Knopfmacher*, Analytic Arithmetic of Algebraic Function Fields
51. *F. Van Oystaeyen*, Ring Theory: Proceedings of the 1978 Antwerp Conference
52. *B. Kedem*, Binary Time Series
53. *J. Barros-Neto and R. A. Artino*, Hypoelliptic Boundary-Value Problems
54. *R. L. Sternberg, A. J. Kalinowski, and J. S. Papadakis*, Nonlinear Partial Differential Equations in Engineering and Applied Science
55. *B. R. McDonald*, Ring Theory and Algebra III: Proceedings of The Third Oklahoma Conference
56. *J. S. Golan*, Structure Sheaves over a Noncommutative Ring
57. *T. V. Narayana, J. G. Williams, and R. M. Mathsen*, Combinatorics, Representation Theory and Statistical Methods in Groups: YOUNG DAY Proceedings
58. *T. A. Burton*, Modeling and Differential Equations in Biology
59. *K. H. Kim and F. W. Roush*, Introduction to Mathematical Consensus Theory
60. *J. Banas and K. Goebel*, Measures of Noncompactness in Banach Spaces
61. *O. A. Nielsen*, Direct Integral Theory
62. *J. E. Smith, G. O. Kenny, and R. N. Ball*, Ordered Groups: Proceedings of the Boise State Conference
63. *J. Cronin*, Mathematics of Cell Electrophysiology
64. *J. W. Brewer*, Power Series Over Commutative Rings
65. *P. K. Kamthan and M. Gupta*, Sequence Spaces and Series
66. *T. G. McLaughlin*, Regressive Sets and the Theory of Isols
67. *T. L. Herdman, S. M. Rankin, III, and H. W. Stech*, Integral and Functional Differential Equations
68. *R. Draper*, Commutative Algebra: Analytic Methods
69. *W. G. McKay and J. Patera*, Tables of Dimensions, Indices, and Branching Rules for Representations of Simple Lie Algebras
70. *R. L. Devaney and Z. H. Nitecki*, Classical Mechanics and Dynamical Systems
71. *J. Van Geel*, Places and Valuations in Noncommutative Ring Theory
72. *C. Faith*, Injective Modules and Injective Quotient Rings
73. *A. Fiacco*, Mathematical Programming with Data Perturbations I
74. *P. Schultz, C. Praeger, and R. Sullivan*, Algebraic Structures and Applications: Proceedings of the First Western Australian Conference on Algebra
75. *L. Bican, T. Kepka, and P. Němec*, Rings, Modules, and Preradicals
76. *D. C. Kay and M. Breen*, Convexity and Related Combinatorial Geometry: Proceedings of the Second University of Oklahoma Conference

77. *P. Fletcher and W. F. Lindgren*, Quasi-Uniform Spaces
78. *C.-C. Yang*, Factorization Theory of Meromorphic Functions
79. *O. Taussky*, Ternary Quadratic Forms and Norms
80. *S. P. Singh and J. H. Burry*, Nonlinear Analysis and Applications

Other Volumes in Preparation

Nonlinear Analysis and Applications

edited by
S. P. Singh and J. H. Burry
Department Of Mathematics and Statistics
Memorial University of Newfoundland
St. John's, Newfoundland
Canada

MARCEL DEKKER, INC. New York and Basel

Library of Congress Cataloging in Publication Data
Main entry under title:

Nonlinear analysis and applications.

(Lecture notes in pure and applied mathematics ; v.80)
Proceedings of an International Conference on Nonlinear Analysis and Applications, held at Memorial University of Newfoundland in St. John's from June 1-3, 1981 and jointly orgainzed by Memorial University of Newfoundland and the University of Sherbrooke.
Includes index.
1. Mathematical analysis--Congresses.
2. Nonlinear theories--Congresses. I. Singh, Sankatha Prasad, . II. Burry, J. H. (John H.), . III. International Conference on Nonlinear Analysis and Applications (1981 : Memorial University of Newfoundland) IV. Memorial University of Newfoundland. V. Université de Sherbrooke. VI. Series.
QA299.6.N663 1982 519.4 82-14875

MARCEL DEKKER, INC.
270 Madison Avenue, New York, New York 10016

Current printing (last digit):
10 9 8 7 6 5 4 3 2 1

PRINTED IN THE UNITED STATES OF AMERICA

PREFACE

An International Conference on Nonlinear Analysis and Applications was held at Memorial University of Newfoundland in St. John's from June 1-3, 1981. The Conference was a joint endeavour organized by Memorial and the University of Sherbrooke in Sherbrooke, Quebec.

The Conference dealt with various aspects of Nonlinear Analysis and Applications including Fixed Point Theory, Approximation Theory, Summability Theory and other areas of modern applied mathematics. The aim was to bring together mathematicians working in these areas to stimulate further interest and interaction for the future.

This volume consists of the proceedings of the Conference. It includes papers that were delivered as invited one-hour and half-hour addresses and papers by those who intended to attend, but could not come. The Conference was supported by Memorial University of Newfoundland (particularly, the Dean of the School of Graduate Studies and the Department of Mathematics and Statistics), the University of Sherbrooke and by various individual operating research grants.

The Conference was opened and the participants welcomed by Dr. Leslie Harris, President of Memorial University, who was introduced to the participants by Dr. John Burry, Head of the Department of Mathematics and Statistics.

The editors are most pleased to express thanks to each member of the Department who willingly helped with the planning and organizing of the Conference. A special note of thanks goes to Professor Gilles Fournier, University of Sherbrooke, who helped enthusiastically from the beginning until the end; to Mrs. Mary Pike, Harold Johnson and G. S. Singh for their valuable assistance in the actual running of the Conference and to Mr.

G. S. Singh, again, for his fine job of proofreading of these proceedings. Finally, our thanks go to Ms. R. Genge for her careful and excellent typing of the entire manuscript, and to the staff of Marcel Dekker, Inc., for undertaking publication of the proceedings.

S. P. Singh

J. H. Burry

CONTENTS

APPLICATIONS

APPROXIMATION THEORY AND RELATED RESULTS

LIST OF CONTRIBUTORS

1. Al-Bassam, M. A., University of Kuwait, Kuwait.
2. Altman, M., University of Louisiana, Baton Rouge, Louisiana, U. S. A.
3. Borwein, D., University of Western Ontario, London, Ontario, Canada.
4. Burbea, J., University of Pittsburgh, Pittsburgh, PA, U.S.A.
5. Cheney, E. W., University of Texas at Austin, Austin, Texas, U.S.A.
6. Dube, L. S., Vanier College, Montreal, Quebec, Canada.
7. Dubinsky, E., Clarkson College of Technology, Potsdam, New York, U.S.A.
8. Dunham, C. B., University of Western Ontario, London, Ontario, Canada.
9. Edelstein, M., Dalhousie University, Halifax, Nova Scotia, Canada.
10. Fisher, B., Leicester University, Leicester, United Kingdom.
11. Fournier, G., Universite de Sherbrooke, Sherbrooke, Quebec, Canada.
12. Gaines, R. E., Colorado State University, Fort Collins, Colorado, U.S.A.
13. Grace, S. R., University of Saskatchewan, Saskatoon, Saskatchewan, Canada.
14. Groetsch, C. W., University of Cincinnati, Cincinnati, Ohio, U.S.A.
15. Guay, M. D., University of Southern Maine, Portland, Maine, U.S.A.
16. Guénard, F.*, Ecole Normale Superieure de l'Enseignment, Cachan, France.
17. Holland, A. S. B.†, University of Calgary, Calgary, Alberta, Canada.
18. Jakimovski, A., Tel-Aviv University, Ramat-Aviv, Tel-Aviv, Israel.
19. Jaworowski, J., Indiana University, Bloomington, Indiana, U.S.A.
20. Kiang, M. T., St. Mary's University, Halifax, Nova Scotia, Canada.
21. Kirk, W. A., University of Iowa, Iowa City, Iowa, U.S.A.
22. Lakshmikantham, V., University of Texas at Arlington, Arlington, Texas, U.S.A.
23. Lalli, B. S., University of Saskatchewan, Saskatoon, Saskatchewan, Canada.

---

*Current affiliation: Centre Pédigogigue Régional de l'Académie de Créteil, Créteil, Cedex, France.
†Deceased.

24. Lancaster, P., University of Calgary, Calgary, Alberta, Canada.

25. Mohapatra, R. N., American University of Beirut, Beirut, Lebanon.

26. Naimpally, S. A., Lakehead University, Thunder Bay, Ontario, Canada.

27. Pai, D. V., Indian Institute of Technology, Bombay, India.

28. Peterson, J. K., Michigan Technological University, Houghton, Michigan, U.S.A.

29. Respess, J. R., University of Texas at Austin, Austin, Texas, U.S.A.

30. Rhoades, B. E., Indiana University, Bloomington, Indiana, U.S.A.

31. Saff, E. B., University of South Florida, Tampa, Florida, U.S.A.

32. Sahney, B. N.*, University of Calgary, Calgary, Alberta, Canada.

33. Schirmer, H., Carleton University, Ottawa, Ontario, Canada.

34. Sehgal, V. M., University of Wyoming, Laramie, Wyoming, U.S.A.

35. Sharma, A., University of Alberta, Edmonton, Alberta, Canada.

36. Sine, R., University of Rhode Island, Kingston, Rhode Island, U.S.A.

37. Singh, K. L.,† Lakehead University, Thunder Bay, Ontario, Canada.

38. Singh, S. P., Memorial University of Newfoundland, St. John's, Newfoundland, Canada.

39. Srivastava, P., Allahabad University, Allahabad, U.P., India.

40. Srivastava, S. C., K. A. Degree College, Allabahad, U.P., India.

41. Thomeier, S., Memorial University of Newfoundland, St. John's, Newfoundland, Canada.

42. Turett, B. J., Oakland University, Rochester, Michigan, U.S.A.

43. van Dulst, D., University of Amsterdam, Amsterdam, The Netherlands.

44. Varga, R. S., Kent State University, Kent, Ohio, U.S.A.

45. Veeramani, P., Indian Institute of Technology, Bombay, India.

46. Watson, B., Memorial University of Newfoundland, St. John's, Newfoundland, Canada.

47. Whitfield, J. H. M., Lakehead University, Thunder Bay, Ontario, Canada.

48. Wong, C. S., University of Windsor, Windsor, Ontario, Canada.

49. Yadav, B. S., University of Delhi, Delhi, India.

---

* Deceased.

†Current affiliation: University of Minnesota - Duluth, Duluth, Minnesota, U.S.A.

Nonlinear Analysis and Applications

MAIN ADDRESSES

RECENT DEVELOPMENT OF THE THEORY OF CONTRACTORS AND CONTRACTOR DIRECTIONS AND ITS APPLICATIONS

M. Altman

University of Louisiana
Baton Rouge, Louisiana
U.S.A.

ABSTRACT

This paper is a survey of recent development of the theory of Contractors and Contractor Directions, which is basically designed for solving equations in abstract spaces, and of its applications to classical and modern analysis. The theory which is presented in the research monograph: Contractors and Contractor Directions - Theory and Applications, Marcel Dekker, 1977 (304 pages), is characterized by its generality, unifying property, capacity for applications, potential for further development, and flexibility concerning adaptation of methods related to some other branches of mathematics. Since, the publication of the monograph, both the theory and its applications, have undergone intensive development. General existence and surjectivity theorems have been obtained for new classes of equations: with closed operators as well as with operators without continuity conditions. The most surprising result is that the theory of contractor directions provides a tool for solving general quasi-linear evolution equations of "hyperbolic" type in nonreflexive Banach spaces whereas no other method is available. Kato's theory applies to reflexive Banach spaces.

A new concept has been introduced: generalized contractor directions. Among other developments: self-accelerating iterative methods of contractor directions, the diagonal method of contractor directions which is an iterative method of solving equations in locally convex metric spaces with application to optimization problems in such spaces. More progress is expected along with applications: to nonlinear integral equations, to nonlinear evolutions equations and partial differential equations of hyperbolic type, to eigenvalue problems and bifurcation theory, to nonlinear programming and general control theory. The conclusion of the program will be the second volume of the monograph mentioned above.

Introduction

Since the completion of the manuscript of the research monograph "Contractors and Contractor Directions - Theory and Applications" (see A[135]) which is an outgrowth of research during 1971-1975, both the theory and its applications have undergone intensive development. Let us recall that two new concepts are basic for the entire theory: the concept of contractors and the concept of contractor directions.

Let $P : X \to Y$ be a nonlinear mapping and let $\Gamma(x) : Y \to X$ be a bounded linear operator associated with $x \in X$, where X and Y are Banach spaces. Then Γ is a contractor for P with majorant function Q if

$$\|P(x + \Gamma(x)y) - Px - y\| \leq Q(\|y\|) \tag{1.1}$$

is satisfied for $x \in X$ and $y \in Y$ defined by the problem. If $Q(s) = qs$ for some $0 < q < 1$, then Q is a linear majorant function. If $Px = x - Fx$ and F is a contraction of X, then the identity operator $I = \Gamma(x)$ for all $x \in X$, is a contractor for P. This observation justifies the term "contractor".

While searching for a class of nonlinear majorant functions, we discovered a new chapter in the classical theory of infinite series: Series of iterator of a positive function (A[123]). The following simple example shows the usefulness of the discovery: Let

$$s_{n+1} = s_n(1 - s_n^{\alpha}(1 + \alpha)^{-1}) \text{ with } 0 < \alpha < 1 \text{ and } 0 < s_1 < 1.$$

Then the convergence of the series $\sum_{n=1}^{\infty} s_n$ follows immediately from our new criterion while none of the criterions given in the classical monograph by Knopp [1] seems to be working in this case.

The following is another nontrivial example of a series of iterates of a positive function

$$s_{n+1} = s_n - s_n^{1+\alpha}(1 - s_n)^{1-\alpha}, \quad n = 1,2,\ldots,$$

where $0 < \alpha < 1$ and $0 < s_1 < 1$. The convergence of the series $\sum_{n=1}^{\infty} s_n$ also follows from our theory, and cannot be obtained by classical criterions.

Series of iterates of a positive function constitute a nice class of majorant functions for contractors (see A[117], A[119]). The contractor method is based on the following general iterative procedure

$$x_{n+1} = x_n - \Gamma(x_n)Px_n, \quad n = 0,1,\ldots \tag{1.2}$$

for approximate solutions of equation $Px = 0$. It is the basis for a unified theory for a very large class of iterative methods, including the most important ones: (a) the method of successive approximations; (b) the Newton-Kantorovich method for operator equations; (c) the following method which is called the Newton-Altman method (see A[24]) for nonlinear functionals F defined on a Banach space X

$$x_{n+1} = x_n - \frac{Fx_n}{F'(x_n)y_n} y_n, \quad n = 0,1,\ldots, \tag{1.3}$$

where F' is the Fréchet derivative of F and $F'(x_n)y_n$ with $\|y_n\| = 1$ is approximately equal to $\|F'(x_n)\|$; (d), the method of steepest descent and other gradient type methods. Both existence and convergence theorems are obtained (see A[135]) and the contractor method also reveals the character of the convergence and yields error estimates. Entirely new techniques are there developed, for instance, the method of series of iterates of a positive function mentioned above.

In a recent series of investigations Pták ([1] - [4]) has proposed a new method of estimating the convergence of iterative processes and proved a very general closed graph theorem with many important applications, e.g., to Newton's method and its modification due to Jurgen Moser, to factorization in Banach algebras and selection theorems. In all these investigations a crucial role is played by a convergent series of iterates of a positive function. Therefore, our results concerning the convergence of series of iterates of a positive function can also be applied in all situations investigated by Pták.

2. Applications Of The Contractor Methods To The Quantum Field Theory

Yoshimura [1] has applied the contractor method to the Quantum Field theory. He tried almost all existence and iterative methods of functional analysis, and he arrived at the conclusion: "... the most powerful method so far available is Altman's method of contractor with nonlinear majorant." (see ibid. pp. 261, 264, 267). Yoshimura [2] has also used the contractor method in his paper "Applications of Altman's contractor techniques to nonlinear integral equations arising in Green's function formalizm of Klauder's augmented Quantum Field theory."

3. Applications Of The Contractor Method To Random Nonlinear Equations In Probability Theory And To Stochastic Integral Equations

Padgett and Lee [1] have applied the contractor method to random nonlinear equations in Probability Theory.

Lee [1] has used the contractor method in his Ph.D. dissertation.

Kuo [1] has applied the contractor method to stochastic integral equations in order to prove an existence and uniqueness theorem.

4. Contractor Directions

The second basic concept employed in the general theory is the concept of contractor directions. Given $0 < q < 1$, then $\Gamma_x(P) = \Gamma_x(P, q)$ is a set of contractor directions for P at x, i.e., $y \in \Gamma_x(P)$ if there exist $0 < \varepsilon = \varepsilon(x, y) \leq 1$ and $h = h(x, y) \in X$ such that

$$\|P(x + \varepsilon h) - Px - \varepsilon y\| \leq q\varepsilon\|y\| . \tag{4.1}$$

The element h is called the strategic direction at x. In general, the strategic direction is supposed to satisfy the following condition

$$\|h\| \leq B(g(x) \cdot \|y\|) , \tag{4.2}$$

where g is an arbitrary positive functional on X which is bounded on closed bounded sets, and B is a continuous increasing function such that $B(0) = 0$, $B(s) > 0$ for $s > 0$, and

$$\int_0^1 s^{-1} B(s)ds < \infty . \tag{4.3}$$

The class of such functions B is denoted by $\mathbb{B}$. The method yields a unified approach to solvability theorems. In case of an operator with closed range restriction (5) is not required. The function B plays an important part in the theory of contractor directions, and does not seem to be introduced artificially. In fact, it is shown in A[127] that in the case of the Newton-Kantorovich iterative method, the function B is given by $B(s) = (1/2K)^{1/2}\, s^{1/2}$, where K is the Lipschitz constant for the Fréchet derivative P'.

5. Weak Contractor Directions And Weak Directional Contractions For An Operator With Closed Range

DEFINITION. (A[142]). Let X be an abstract set and $P : X \to Y$ a mapping of X into a (real or complex) Banach space Y. Given an upper semicontinuous function q such that $0 < q(s) < s$ for $s > 0$, we define sets $\Gamma_x^*(P) \subset Y$ of weak contractor directions for P at $x \in X$ as follows: $y \in \Gamma_x^*(P)$ if there exist a positive $\varepsilon = \varepsilon(x, y) \leq 1$ and an element $\bar{x} \in X$ such that

$$\|P\bar{x} - Px - \varepsilon y\| \leq \varepsilon q(\|y\|) .$$

If $q(\|y\|) = q\|y\|$ with $0 < q < 1$, then obviously $\Gamma_x^*(P) = \Gamma_x(P)$.

THEOREM. (A[142]). If $\Gamma_x^*(P) = Y$ for all $x \in X$, then $P(X) = Y$.

This theorem generalizes Theorem 1.2 of A[122] and recent "normal solvability" theorems of Pohozhayev, Browder, Krasnoselskii-Zabreiko, and others.

DEFINITION. Let X be a Banach space and q an upper semicontinuous function such that $0 < q(s) < s$ for $s > 0$. A mapping $F : D(F) \subset X \to X$ is called a weak directional contraction if for each $x \in D(F)$ and $y \in X$, there exists a positive $\varepsilon = \varepsilon(x, y) \leq 1$ such that $x + \varepsilon y \in D(F)$ and

$$\|F(x + \varepsilon y) - Fx\| \leq \varepsilon q(\|y\|) . \tag{5.1}$$

THEOREM. (A[142]). Let $F : D(F) \subset X \to X$ be a weak directional contraction. If the range of $I - F$ is closed in X, where I is the identity mapping, then P is a mapping onto X.

A Fixed Point Theorem For Weak Directional Contractions In the Narrow Sense

DEFINITION. Let $W \subset X$ be a convex set. A mapping $F : W \to W$ is called a weak directional contraction in the narrow sense if relation (5.1) is satisfied for all $x \in W$ and $y = Fx - x$.

THEOREM. Let $F : W \to W$ be a weak directional contraction in the narrow sense. If the range $(I - F)(W)$ is closed in X, then F has a fixed point in W.

If $F : W \to W$ is a contraction of a closed convex set W of a Banach space X, then $(I - F)(W)$ is closed in X. Therefore, the above theorem is an extension of the Banach contraction principle, and generalizes some known results.

6. Closed Operators With Nonclosed Ranges

A general existence principle for closed operators with arbitrary ranges has been established in A[126]. Local and global existence theorems are proved there, and surjectivity theorems are proved for a new class of operators: Cauchy mappings (If $\{x_n\}$ is bounded whenever $\{Px_n\}$ is a Cauchy sequence, then P is called a Cauchy mapping). By utilizing some of the techniques of the theory of monotone operators, a new aspect of the theory of contractor directions is developed in A[132] resulting in applications to nonlinear equations of Hammerstein type and to quasilinear elliptic systems in divergence form. The crucial idea exploited there is that so called (B,g)-accretive operators are Cauchy mappings for which restriction (4.2) is satisfied.

An application of the method of contractor directions to nonlinear programming in Banach space resulted in a generalization of Lusternik's theorems on existence of tangent directions and on constraint minimum problems in Banach space, see A[128].

The theory of contractor directions yields also a method of reducing nonlinear problems to linearized ones. In this conjunction extensive studies of solutions of linearized equations are needed. Since the linearized equation depends upon a variable vector of a Banach space, so does the solution. Therefore, a priori estimate is needed in general so as to comply with requirement (4.2). Some results have been obtained for partial differential equations, see A[125] and for evolution equations, see A[129].

The general theory of contractor directions can also be developed for finding roots of a nonlinear functional on a Banach space. The applicability of the generalized Newton-Kantorovich method (see A[30], Bartle [1]) can be broadened. New results are also expected for the Newton-Altman method (see (1.3)). Let us mention that the Newton-Kantorovich method is of no use to the problem of finding roots of a nonlinear functional on a Banach space.

It follows from Sections 5 and 6 and from the monograph (A[135]) that the method of contractor directions unifies and generalizes recent surjectivity theorems of Pohozhayev, Browder, Krasnoselskii-Zabreiko, and others as well as the implicit function theorem of the type studied by Nash, Moser and others.

7. Existence Theorems For Operator Equations Without Continuity Condition

In his interesting paper Ekeland [1] has given a survey of the development of the idea which is the crux of the proof of a well-known Bishop-Phelps lemma. A number of diverse results including Ekeland's theorem and results obtained by Browder and others led to a very general principle on ordered sets by Brézis and Browder [1]. A reformulation of the Brézis-Browder principle is also given by Ekeland [1] who provides a concise proof due to Crandall. A reformulation of Ekeland's theorem yields the Browder-Caristi-Kirk fixed point theorem. A further generalization of the Brézis-Browder principle is given in A[144]. A generalization of the

Browder-Caristi-Kirk fixed point theorem follows then immediately. The generalization of the Brézis-Browder principle is such that makes its proof very short and straightforward.

The proofs of the result of Sections 6 and 7 for the method of contractor directions are based on a transfinite induction argument. However, it turns out that local and global existence theorems can also be proved for the method of contractor directions by using Ekeland's theorem instead of the transfinite induction. Then the hypothesis that the operator in question is closed should be replaced by other assumptions. For example, the following theorem holds.

THEOREM. (A[141]). Let $P : W \to Y$ be a nonlinear mapping where W is a closed set of X and let B be a stricIty increasing function with $B(0) = 0$ and such that

$$0 < k = \inf\{f(s, \varepsilon) \mid s > 0,\ 0 < \varepsilon \leq 1\}\ ,$$

where

$$f(s, \varepsilon) = \{1 - B[\exp(-(1 - q)\varepsilon) \cdot s/B(s)]/\varepsilon\}\ .$$

Suppose that $-Px \in \Gamma_x(P) = \Gamma_x(P, q)$ for all $x \in W$ and with strategic directions $h = h(x)$ satisfying the relation

$$\|h\| \leq B(\|Px\|) \quad \text{for all} \quad x \in W.$$

If the functional $x \to \|Px\|$ is lower semicontinuous, then equation $Px = 0$ has a solution x in W.

It is easily seen that, for instance, the functions $B(s) = Cs$ or $B(s) = Cs^{1/2}$, where $C > 0$ is a constant, satisfy the above conditions.

Let us observe that no continuity condition is required.

More results are given in A[141] including global existence theorems and fixed point theorems as well as generalizations of some known results.

At this point, let us mention that the application of the transfinite induction argument to the method of contractor directions provides some information on the whereabout of the solution of the equation (see A[126]) whereas in the cases in which the Ekeland type argument is applied to the method of contractor directions, no such information is available.

8. Self-accelerating Iterative Methods Of Contractor Directions

Since the method of contractor directions gives only a tool to prove existence theorems for solutions of general equations, there is a need in a constructive version of the method in order to find existing solutions in a practical way. The first attempt has already been made in A[126] and A[127]. Among other things, the Newton-Kantorovich method with small steps is investigated there, and convergence theorems are proved under less restrictive hypotheses. As a consequence, the initial approximate solution is not required to be as good as in the case with step size equal to one. However, the method automatically turns into the classical Newton-Kantorovich method, i.e., with step size equal to one, as soon as the approximate solution improves sufficiently. A general theory of self-accelerating iterative methods based upon the concept of contractor directions is developed in A[139].

The Fundamental Lemma For Iterative Methods of Contractor Directions

Let $P : D(P) \subset X \to Y$ and $U_o = D(P) \cap S$, where S is an open ball with center $x_o \in D(P)$ and radius r. We assume that $-Px \in \Gamma_x(P) = \Gamma_x(P, q)$ for all $x \in U_o$, i.e., relation (4.1) is satisfied with $y = -Px$, and the strategic directions $h = h(x)$ satisfy

$$||h|| \le B(||Px||) \text{ for all } x \in U_o,$$

where $B \in \mathbb{B}$ (see (4.3)), and

$$r \ge (1 - q)^{-1} \int_o^a s^{-1} B(s) ds \quad \text{with} \quad a \ge ||Px_o|| \exp(1 - q) .$$

LEMMA. (A[139]). Let

$$x_{n+1} = x_n + \varepsilon_n h_n, \ n = 0,1,\dots, \tag{8.1}$$

where $0 < \varepsilon_n \le 1$ and h_n are such that

$$||P(x_n + \varepsilon_n h_n) - (1 - \varepsilon_n)Px_n|| \le q\varepsilon_n ||Px_n|| \tag{8.2}$$

and

$$\|h_n\| \leq B(\|Px_n\|) . \tag{8.3}$$

Then $\{Px_n\}$ is a Cauchy sequence, $\{x_n\}$ lies in U_o and converges to some x^* such that

$$\|x_n - x^*\| \leq (1 - q)^{-1} \int_b^{b_n} s^{-1}B(s)ds \quad \text{for } n = 0,1,\ldots, \tag{8.4}$$

where

$$b_n = \|Px_o\| \exp((1 - q)(1 - t_n)),$$

$$b = \|Px_o\| \exp((1 - q)(1 - T)), \quad t_o = 0,$$

$$t_n = \sum_{i=0}^{n-1} \varepsilon_i, \quad \text{and} \quad T = \sum_{i=0}^{\infty} \varepsilon_i .$$

By using this new technique one can construct general iterative procedures for the method of contractor directions. The choice of $\{\varepsilon_n\}$ in (8.1) can be made as follows: put

$$\Phi(\varepsilon, x, h) = \|P(x + \varepsilon h) - (1 - \varepsilon)Px\|/\varepsilon .$$

Then put $\varepsilon_n = 1$ if $\Phi(1, x_n, h_n) \leq q\|Px_n\|$, otherwise, under very general conditions, there exists $0 < \varepsilon_n < 1$ such that

$$\beta q\|Px_n\| \leq \Phi(\varepsilon_n, X_n, h_n) \leq q\|Px_n\| . \tag{8.5}$$

where $0 < \beta, q < 1$ are properly chosen. (In practice, we put $\varepsilon = 1/2$, $\varepsilon = 1/4$, ..., until (8.5) is satisfied). Condition (8.5) implies $Px_n \to 0$ as $n \to \infty$.

Based upon the technique mentioned above are several iterative procedures proposed for the method of contractor directions (see A[139]). Included there is also a procedure for non-differentiable operators.

9. A Self-Accelerating Damped Newton-Kantorovich Method With Approximate Solutions Of The Linearized Equations For Operators With Continuous Fréchet Derivatives

The Newton-Kantorovich method is based on the exact solution of a linear equation with the Fréchet derivative at each step. This is practically impossible to achieve, and therefore, what is in practical use and is supposed to be the Newton-Kantorovich method is in fact a different one.

Suppose that P' is continuous and $U = D(P) \cap \overline{S}(x_o, r)$, where $\overline{S}$ is the closure of S. Then the iterative procedure (8.1) with h_n satisfying (8.3) and, for some $0 < \overline{q} < 1$,

$$\|P'(x_n)h_n + Px_n\| \leq \overline{q}\,\|Px_n\|, \tag{9.1}$$

and ε_n satisfying (8.5) or $\varepsilon_n = 1$ as above, where $\overline{q} < q < 1$ and $\overline{q}/q < \beta < 1$ are arbitrary. Then $\{x_n\}$ converges to a solution $x^* \in U$ of equation $Px = 0$, and the error estimate (8.4) holds.

Krasnoselskii and Rutickii [1] have proved the convergence of the process (8.1) with $\varepsilon_n = 1$ for all n, under certain conditions including that the Fréchet derivative P' is Hölder continuous. However, condition (9.1) is there replaced by the following

$$\|h_n(x_n) - h\| \leq \overline{q}\|P'(x_n)^{-1}Px_n\| \quad \text{with} \quad 0 < \overline{q} < 1, \tag{9.2}$$

where $h_n(x_n)$ is the approximate solution and H is the exact solution of the linear equation

$$P'(x_n)h + Px_n = 0. \tag{9.3}$$

But (9.2) is not verifiable in practice.

As a consequence of the new technique, we obtain the following interesting property of the Newton-Kantorovich method.

Under Kantorovich's hypotheses (see Kantorovich and Akilov [1]), the method (8.1), where h_n is the exact solution of the equation

$$P'(x_n)h + Px_n = 0,$$

is convergent to a solution of equation $Px = 0$, whenever $\sum_{n=0}^{\infty} \varepsilon_n = \infty$ with $0 < \varepsilon_n \leq 1$. Kantorovich has proved the convergence for $\varepsilon_n = \varepsilon < 1$ for all n.

Hence, we infer that by putting $\varepsilon_n = 1$ for all n, we artificially limit the capacity of Newton's method.

Another interesting feature results from the new technique for the Banach contraction principle for F: Let

$$x_{n+1} = (1 - \varepsilon_n)x_n + \varepsilon_n Fx_n (n = 0,1,\ldots) \quad \text{with} \quad 0 < \varepsilon_n \leq 1$$

where $x_0 \in X$ is arbitrary. Then $x_n \to x^* = Fx^*$ whenever $\sum_{n=o}^{\infty} \varepsilon_n = \infty$. Moreover,

$$\|x_n - x^*\| \leq (1 - q)^{-1}\|x_o - Fx_o\| \exp((1 - q)(1 - t_n)),$$

where $t_o = 0$, $t_n = \sum_{i=o}^{n-1} \varepsilon_i$, and $0 < q < 1$ is the Lipschitz constant for F.

This is another extension of Banachs contraction principle, and follows from the new technique without using Banach's theorem.

Notice that in general case of (8.1), by taking ε_n larger, we get a faster convergence. If $\varepsilon_n = 1$ is admissible, then it will appear in (8.1). Hence, the procedure (8.1), where h_n satisfies (9.3), turns into the Newton-Kantorovich method automatically if conditions improve.

10. The Diagonal Method Of Contractor Directions

Let X, Y be vector spaces, each endowed with an increasing sequence of seminorms to form a complete locally convex metric space. Let $P : D(P) \subset X \to Y$ be a nonlinear mapping. The diagonal method of contractor directions is an extension of the method of contractor directions to the class of spaces mentioned above and is based upon the following notion.

DEFINITION. (A[136]). Given $0 < q < 1$, $\Gamma_x(P, n, q) = \Gamma_x(P, n) \subset Y$ is a set of contractor directions at x for P of order n, i.e., $y \in \Gamma_x(P, n)$ if there exist a positive $\varepsilon = \varepsilon(x, y) \leq 1$ and $h \in X$ such that

$$\|P(x + \varepsilon h) - Px - \varepsilon y\|_i \leq q\, \varepsilon \|y\|_i \quad \text{for} \quad 0 \leq i \leq n,$$

where $x + \varepsilon h \in D(P)$.

A theory of iterative algorithms for the diagonal method of contractor directions is developed in A[136], where existence and convergence theorems are obtained as well as an inverse function theorem for Fréchet spaces. As an example, the following theorem holds.

THEOREM. (A[136]). Suppose that for each $x \in U_o$, neighborhood of x_o, there exists $h(x)$ such that

$$P'(x)h(x) = -Px, \tag{10.1}$$

and for each $k = 0,1,\ldots,$ there exist $B_k \subset \mathbf{B}$ and an integer $p(k) \geq 0$, $p(o) \geq 0$, such that

$$\|h(x)\|_k \leq B_k(\|Px\|_{k+p(k)}) \quad \text{for} \quad k = 0,1,\ldots \quad . \tag{10.2}$$

There exist positive constants $C_o, C_1, \ldots,$ such that

$$\max_{0 \leq t \leq 1} \|P''(x + t\, h)(h, h)\|_i \leq C_i \|Px\|_i \tag{10.3}$$

for all $i = 0, 1$ and $x \in U_o$, where $P'(x)h = Px$.

$$\sum_{n=o}^{\infty} 1/\max_{0 \leq i \leq n} C_i = \infty. \tag{10.4}$$

Then the sequence $\{x_n\}$, where

$$x_{n+1} = x_n + \varepsilon_n h_n \quad (n = 0,1,\ldots)$$

with $h_n = h(x_n)$ satisfying (10.1), (10.2) and proper choice of ε_n, converges to a solution of the equation $Px = 0$. By using the diagonal method of contractor directions, the existence of tangent directions as well as the Lagrange multipliers theorem is proved for locally convex metric spaces in A[137]. This generalizes Lusternik's result for Banach spaces. The diagonal method of contractor directions also plays a crucial part in the theory of nonlinear programming in locally convex metric spaces, in A(145]. Some results obtained there are based upon A[137] and A[133]. More results can be obtained based upon A[148].

10*. An Existence Theorem For Equations In Locally Convex Metric Spaces

Based on the diagonal method of contractor directions, by using an extension to locally convex metric spaces of the fundamental lemma for iterative methods of contractor directions the following existence theorem can be proved.

THEOREM. (A[148]). Suppose that $-Px \in \Gamma_x(P, n)$, for $n = 1,2,\ldots,$ $x \in U_0$ (a neighborhood of x_0 defined in a certain way) with $h(x)$ satisfying (10.2). Then $Px = 0$ as a solution $x \in U_0$.

11*. An Application Of Contractor Directions To Quasi-linear Evolution Equations in Nonreflexive Banach Spaces

Introduction

As a result of our previous investigations a new approach is presented for solving abstract quasi-linear equations in non-reflexive Banach spaces. This approach is also based on the theory of contractor directions ([135]) and is not based directly on the C_0-semi-groups theory. The general ideas developed in A[146] are exploited here and lead to a rather surprising result. It turns out that by imposing an extra Lipschitz condition on the family $\{A(t, x)\}$ which appears in the evolution equation under consideration, one can eliminate the restriction which requires the existence of a family of isomorphisms $\{S(t)\}$ (see Kato [3]) or substitutional assumptions (see Graff [1], Altman [146, 147]).

1. Let X, Y, Z be Banach spaces, and denote by $\|\cdot\|_X, \|\cdot\|_Y, \|\cdot\|_Z$ the norms in $X, Y, Z,$ respectively. We assume that the embedding $Z \subset Y \subset X$ is continuous and Z is dense in X. Let W_0 be an open ball in Y with center $x_0 \in Z$ and radius $r > 0$, and denote by W its closure in Y, and let Z_0 be an open ball in Z with the same center x_0 and radius $R > 0$. Given $0 < b$, denote by $C(0, b; X)$, $C(0, b; Y)$, $C(0, b; Z)$ the spaces of all continuous functions $x = x(t)$, $y = y(t)$, $z = z(t)$ defined on the interval $[0, b]$ with values in X,

Y, Z and norms

$$\|x\|_{C(X)} = \sup_t [\|x(t)\|_X : 0 \le t \le b],$$

$$\|y\|_C = \sup_t [\|y(t)\|_Y : 0 \le t \le b],$$

$$\|z\|_D = \sup_t [\|z(t)\|_Z : 0 \le t \le b],$$

respectively. Denote by $C^1(0, b; X)$ the vector space of all continuously differentiable functions from $[a, b]$ into X. In what follows Z may be incomplete.

We assume that there exists a vector space $V \subset Z$ which is dense in $Z(\|\cdot\|_Z)$ and has the following property: Corresponding to each $t \in [0, b]$ and x contained in the norm $\|\cdot\|_Y$-closure of $V_0 = V \cap Z_0 \cap W_0$, the mapping $A(t, x) : Y \to X$ is linear and continuous. We may assume that $\|\cdot\|_Z \ge \|\cdot\|_Y$ and $\|x_0\|_Z < R$.

Consider the Cauchy problem for the quasi-linear equation

$$Px(t) \equiv dx/dt + A(t,x)x - f(t,x) = 0,\ 0 \le t \le b,\ x(0) = x_0. \tag{1.1}$$

We make the following assumptions.

(A_1) $f(t, v) \in Z$ for all $(t, v) \in [0, b] \times V_0$, and $f(t, x) \in Y$ for all $t \in [0, b]$ and x contained in the $\|\cdot\|_Y$-closure of V_0. If $(t, x) \in [0, b] \times V_0$, then $A(t, x)v \in Z$ for all $v \in V$. We assume that $x_0 \in V_0$ and P_0 is such that

$$\|A(t, x_0)x_0 - f(t, x_0)\|_Y \le P_0 \quad \text{for all} \quad 0 \le t \le b. \tag{1.2}$$

(A_2) Let $x \in C(0, b; V_0(\|\cdot\|_Z)) \cap C^1(0, b; X)$ with $x(0) = x_0$, and $y \in L^1(0, b; V(\|\cdot\|_Z)) \cap C(0, b; X)$. Then the equation

$$dz/dt + A(t, x)z - y(t) = 0,\ 0 \le t \le b,\ x(0) = 0, \tag{1.3}$$

has a solution $z \in C(0, b; V(\|\cdot\|_Z)) \cap C^1(0, b; X)$ such that

$$\|z\|_D \le bC\|Y\|_D , \tag{1.4}$$

where C is a global constant, and if $z = x + h$, then h is obviously a solution of the equation

$$dh/dt + A(t, x)h + [dx/dt + A(t, x)x - y] = 0, \qquad (1.5)$$

$$0 \leq t \leq b, \; h(0) = 0,$$

and we assume

$$\|h\|_C \leq bC\|dx/dt + A(\cdot, x)x - y\|_C, \qquad (1.6)$$

whenever $\|dx/dt + A(\cdot, x)x - y\|_C \leq 2P_0$.

(A_3) There exists a constant K_2 such that

$$\|[A(t, u) - A(t, v)]y\|_X \leq K_2\|u - v\|_Y\|Y\|_Y, \qquad (1.7)$$

for all $t \in [0, b]$ and u, v contained in the $\|\cdot\|_Y$-closure of V_0, and $y \in Y$.

(A_4) There exists a constant K_3 such that

$$\|f(t, u) - f(t, v)\|_Y \leq K_3\|u - v\|_Y \qquad (1.8)$$

for all $t \in [0, b]$ and u, v contained in the $\|\cdot\|_Y$-closure of V_0.

We also assume that $f(\cdot, x)$ can be approximated as follows. Let

$$x \in C(0, b; V_0(\|\cdot\|_Z)) \cap C^1(0, b; X).$$

Then for arbitrary $\omega > 0$, there exists a function $Y_\omega \in L^1(0, b; V(\|\cdot\|_Z)) \cap C(0, b; X)$ such that

$$\|f(\cdot, x) - Y_\omega\|_D < \omega, \qquad (1.9)$$

and we also assume that d_0 is such that

$$\|f(\cdot, v)\|_D \leq d_0, \qquad (1.10)$$

for all $v \in V_0$. Finally, we assume the following extra Lipschitz condition.

(A_5) There exists a constant K_1 such that

$$\| [A(t, v + \varepsilon h) - A(t, v)]u\|_Y \le \varepsilon K_1 \|h\|_Y \|u\|_Z , \tag{1.11}$$

for all $t \in [0, b]$, $v \in V_0$, $h, u \in V$, and $0 < \varepsilon \le 1$ such that $v + \varepsilon h \in V_0$.

THEOREM 1.1. In addition to the hypotheses (A_1 - A_5), suppose that the radius r of the open ball $W_0 \subset Y$ with center $x_0 \in V_0$ satisfies the following relation

$$r > \bar{r} = (1 - q)^{-1}\bar{C}\, P_0 \exp(1 - q),\ bC < \bar{C},$$

where P_0, C, are determined by (1.2), (1.6), respectively and $0 < b' \le b$ and q are such that $\bar{q} = b'C_1(K_4 + K_3) < q < 1$ and $C < C_1$. Then there exists a function $x \in C(0, b'; W(\|\cdot\|_Y)) \cap C^1(0, b'; X)$ satisfying equation 1.1 for all $t \in [0, b']$.

REMARK. Kato [3] writes: "We do not know whether or not the reflexivity can be eliminated, possibly, be adding other assumptions." The above theorem gives a positive answer to the question.

2. Kato's [3] theory of evolution operators for solving abstract linear evolution equations can be used to obtain sufficient conditions for assumptions (A_2) to be satisfied.

(i) In what follows Z is assumed to be complete and $x_0 = 0$. Suppose that the family of linear bounded operators $\{A(t, x)\} \subset B(Z, X)$ satisfies Kato's [3] hypotheses so that the evolution operators $\{U(t, s; x)\}$ exist and the solution of equation (1.3) is given by the formula

$$\text{(S)}\quad z(t) = \int_0^t U(t, s; x)y(s)ds.$$

In addition, we assume that

$$\|U(\cdot, \cdot,; x)\|_{\infty,Z} \le M, \tag{2.1}$$

for some constant M and all $x \in Z_0$, using Kato's [3] notation. Then using Kato's estimate we obtain

$$\|z\|_D = \|z\|_{\infty,Z} \leq M \int_0^b \|y(t)\|_Z\,dt. \tag{2.2}$$

Hence, we obtain relation (1.4).

(ii) Suppose that we replace in (i) Z by Y and Z_0 by W_0 and we assume that similar assumptions are made with regard the family $\{A(t, x)\} \subset B(Y, X)$.

(iii) In addition, we assume that if

$$x, z \in C(0, b; V_0(\|\cdot\|_Z)) \cap C^1(0, b; X),$$

then the function $A(\cdot, x)z \in L^1(0, b; V(\|\cdot\|_Y))$.

(iv) We also assume that for each $v \in V_0$, the function $t \to A(t, v)$ is continuous in the $B(Y, X)$ norm. This implies, by virtue of (1.7), that $A(\cdot, x)z \in C(0, b; X)$ and we can apply Kato's estimate to obtain relation (1.6).

Along with Kato's theory of evolution operators Remark 1.1 has to be applied, and Banach spaces $Z \subset Y$ should be chosen so as to satisfy assmuption (A_5). The remaining hypotheses are without change.

Under the above hypotheses the proof of Theorem 1.1 carries over with minor modifications indicated in Remark 1.1 (see A[149]).

Remark 1.2 also applies here.

EXAMPLE. The following example is given by Kato [3]. Consider the quasi-linear equation

$$u_t + a(u)u_x = b(u)u, \quad t > 0, \quad -\infty < x < \infty,$$

where $u = u(t, x)$ is real-valued and $u_t = \partial u/\partial t$, $u_x = \partial u/\partial x$. The real-valued functions a and b are assumed to be sufficiently smooth. We choose

$$X = H^0(-\infty, \infty), \quad Y = H^{s_1}(-\infty, \infty), \quad Z = H^{s_2}(-\infty, \infty),$$

$s_2 > s_1$, where H^s denotes the Sobolev space of order s of the L^2-type, so that $H^0 = L^2$. The vector space $V \subset Z$ consists of sufficiently smooth functions of $H^{s_2}(-\infty, \infty)$.

We put $D = d/dx$, and

$$A(t, y) = A(y)D, \; f(t, y) = f(y) = f(y) = yb(y).$$

Kato [3] has shown that his hypotheses are satisfied, which imply the existence of the evolution operator $U(t, s; x)$ for solving equations (1.3) by the formula (S). This yields the estimate given by (2.2) to obtain (1.4). Kato [3] also noticed that $Y = H^s(-\infty, \infty)$ may be chosen with any integer $s \geq 2$. Hence, Kato's hypotheses are also satisfied for $s_2 > s_1 \geq 2$, so that the estimate given by (2.2) can be applied to the solution of equation (1.5) yielding (1.6) because assumption (iii) can obviously be satisfied for $A(t, y) = A(y) = a(y)D$. It is easily seen that $s_2 > s_1$ can be chosen so as to satisfy the extra Lipschitz condition (1.11) in (A_5). The remaining assumptions can be verified without difficulty.

REMARK 2.1. Theorem 1.1 yields only existence of a solution of the quasi-linear evolution equation. However, it is easily seen that a uniqueness theorem can be obtained if we assume the following stronger Lipschitz conditions.

$$\|[A(t, u) - A(t, v)]y\|_X \leq K_1\|u - v\|_X\|y\|_Y, \tag{2.3}$$

and

$$\|f(t, u) - f(t, v)\|_X \leq K_2\|u - v\|_X , \tag{2.4}$$

for some $K_1, K_2 > 0$ and all u, v in the $\|\cdot\|_Y$-closure of V_0 and all $t \in [0, b]$ and $y \in Y$. In fact, suppose that $\bar{x}$ and x are solutions of equation (1.1), then we get

$$d(\bar{x}-x)/dt + A(t,x)(\bar{x}-x) = f(t,\bar{x}) - f(t,x) - [A(t, \bar{x}) - A(t,x)]\bar{x}, \; 0 \leq t \leq b', \; \bar{x}(0) - x(0) = 0. \tag{2.5}$$

Hence, the uniqueness follows if b' is small.

Recently, Graff [1] proved a general uniqueness theorem for the nonlinear Cauchy problem which implies uniqueness of the solution constructed in Theorem 1.1 under more general conditions than the above remark.

12*. A Simplified Approach To Linear Evolution Equations

Kato's [1, 2] technique combined with the introduction of an extra inherently admissible subspace yields a simplified approach to linear evolution equations. Let $Z \subset Y \subset X$ be Banach spaces with Y, Z dense in X, Y, respectively, and continuous embedding. For each $t \in [0, b]$ and $y \in Y$, we have $A(t)y \in X$ and $A(t) : Y \to X$ is linear and continuous. Consider the Cauchy problem for the linear equation

$$Lx(t) \equiv dx/dt + A(t)x - f(t) = 0, \; 0 \leq t \leq b, \; x(0) = y. \tag{1}$$

Denote by $B(Y, X)$ the space of all bounded linear operators on Y to X with norm $\|\cdot\|_{Y,X}$, and by $G(X)$ the set of all negative infinitesimal generators of C_0-semigroups.

DEFINITION. $Z(Z \subset Y \subset X)$ is inherently A-admissible if Y is A-admissible and the semigroup of class C_0 on Y which is formed by $\{e^{-tA}\}$, leaves Z invariant and in turn forms a semigroup of class C_0 on Z.

We make the following assumptions.

(i) $\{A(t)\}$ is quasi-stable with stability index $\{M_X, \beta_X\}$ (see Kato [2], Dorroh and Graff [1]).

(ii) Y is $A(t)$-admissible for each $t \in [0, b]$. If $\tilde{A}(t) \in G(Y)$ is the part of $A(t)$ in Y, then $\{\tilde{A}(t)\}$ is quasi-stable in Y with index $\{M_Y, \beta_Y\}$. Moreover, Z is $\tilde{A}(t)$-admissible (or inherently admissible), and if $\tilde{\tilde{A}}(t) \in G(Z)$ is the part of $\tilde{A}(t)$ in Z, then $\{\tilde{\tilde{A}}(t)\}$ is (inherently) quasi-stable in Z with index $\{M_Z, \beta_Z\}$.

(iii) $A(t) \in B(Y, X)$ for each $t \in [0, b]$, and the map $t \to A(t)$ is norm-continuous in $B(Y, X)$. $Z \subset D(\tilde{A}(t))$ for each $t \in [0, b]$, and $A(t) \in B(Z, Y)$. In addition, the map $t \to A(t)$ is norm-continuous in $B(Z, Y)$.

THEOREM 1. Let conditions (i), (ii) and (iii) be satisfied. Then there exists a unique evolution operator $U = \{U(t, s)\}$ such that $U(t, s) \in B(X)$ which is defined on the triangle $\Delta : b \geq t \geq s \geq 0$, and has the following properties.

(a) For each $x \in X$, $U(\cdot, \cdot)x$ is jointly continuous from Δ into X, with $U(s, s) = 1$, and $\sup \|U(t, s)\|_X < \infty$.

(b) $U(t, s)U(s, r) = U(t, r)$.

(c) $U(t, s)Y \subset Y$, and $\sup \|U(t, s)\|_Y < \infty$. $U(\cdot, \cdot)Y$ is jointly continuous from Δ into Y, for each $y \in Y$.

(d_1) $(d/ds)U(t, s)y = U(t, s)A(s)y$, $y \in Y$, $0 \leq s \leq t \leq b$.

(d_2) $(d/dt)U(t, s)y = A(t)u(t, s)y$, $y \in Y$, $0 \leq s \leq t \leq b$.

THEOREM 2. Let the assumption of Theorem 1 be satisfied. Then with $y \in Y$ and $f(\cdot)$ continuous in Y,

$$u(t) = U(t, 0y + \int_0^t U(t, s)f(s)ds$$

is the solution of (1.1), and

$$\|u\|_{\infty,Y} \leq \|U\|_{\infty,Y}(\|Y\|_Y + \|f\|_{1,Y}),$$

$$\|du/dt\|_{\infty,X} \leq \|f\|_{\infty,X} + \|A\|_{\infty,Y,X}(\|Y\|_Y + \|f\|_{1,Y}),$$

where

$$\|f\|_{\infty,X} = \sup_{0\leq t\leq b} \|f(t)\|_X,$$

$$\|f\|_{1,Y} = \int_0^b \|f(t)\|_Y dt, \quad \|U\|_{\infty,Y} = \sup\|U(t, s)\|_Y,$$

$$\|A\|_{\infty,Y,X} = \sup_{0\leq t\leq b} \|A(t)\|_{Y,X}.$$

13*. A New Approach To Quasi-linear Evolution Equations In Nonreflexive Banach Spaces

Using the same notation as in Sections 11* and 12*, we put $V = Z$ and

$V_0 = Z_0 \cap W_0$, and make the following assumptions.

(A_0) Corresponding to each $(t, v) \in [0, b] \times V_0$, $A(t, v)Z \subset Y$ holds, and $A(t, v) : Y \to X$, $A(t, v) : Z \to Y$ are linear and continuous. In addition, for each $v \in V_0$, the function $t \to A(t, v)$ is continuous in the $B(Y, X)$ norm as well as in the $B(Z, Y)$ norm.

(A_2') For each $x \in C(0, b; V_0(\|\cdot\|_Z)) \cap C^1(0, b; X)$, the family $\{A(t, x(t))\}$ satisfies conditions (i) and (ii) of A[151], i.e., Y is admissible and Z is inherently admissible. Moreover, the family $\{A(t, x(t))\}$ is quasi-stable in Y and Z with stability indices $\{M_X, \beta_X\}$, $\{M_Y, \beta_Y\}$ and $\{M_Z, \beta_Z\}$ independent of the function x.

THEOREM. Let condition (A_0) be satisfied. Then Theorem 1.1, 11* holds with (A_2) replaced by (A'_2).

The proof of this theorem is based on a refinement of Theorem 1.1, 11*, and on an application of the simplified approach to linear evolution equation by utilizing only the approximate evolution operators $\{U_n(t, s; x)\}$ constructed for $\{A(t, x(t))\}$ with $x \in C(0, b; V_0(\|\cdot\|_Z)) \cap C^1(0, b; X)$, $x(0) = x_0 \in V_0$. This interesting feature is due to the flexibility of the method of contractor directions.

The remark after Theorem 1.1, 11*, concerning uniqueness is also valid.

14*. An Application Of The Theory Of Contractor Directions To The Theory Of Dynamically Generated Mass

Recently, Yoshimura [3] has applied the theory of contractor directions to Pagels' theory of dynamically generated mass. It seems to be worthy to notice that on page 11 of his paper he came to the following conclusion. "Then one of the best theorems available, also due to Altman[15], is Theorem 4".

15*. Generalized Contractor Directions

Let X_i, $Y_i (i = 1, \ldots, n)$ be Banach spaces and $P_i : D \subset X_1 \times \ldots \times X_n \to Y_i$ be nonlinear operators. Then $y = \{y_i\} \in Y_1 \times \ldots \times Y_n$ is a generalized contractor direction for $P = \{p_i\}$ at $x = \{x_i\} \in X_1 \times \ldots \times X_n$, i.e., $y \in \Gamma_x(P)$, if there exist $0 < \varepsilon \le 1$ and $h = \{h_i\}$ (called strategic direction) such that

$$\|P_i(x_1 + \varepsilon h_1, \ldots, x_n + \varepsilon h_n) - P_i(x_1, \ldots, x_n) - \varepsilon y_i\|_i$$

$$\le \varepsilon \sum_{k=1}^{n} a_{ik} \|y_k\|_k ,$$

where $\{a_{ik}\}$ is a fixed nonnegative $n \times n$ real matrix such that there exist $0 < q < 1$ and $\{r_i\}$ $(1 \le i \le n)$ with

(a) $\sum_{k=1}^{n} a_{ik} r_k \le q r_i$, $i = 1, \ldots, n$. Conditions for a matrix to satisfy (a) are given by Matkowski [1, 2].

THEOREM. If P is closed on D and $-Px \in \Gamma_x(P)$, for each $x \in D$, with $h = \{h_i(x)\}$ such that $\|h_i\|_i \le B_i(\|P_i x\|_i)$, for $B_i \in \mathbf{B}$, $i = 1, \ldots, n$, (see p. 17), then $Px = 0$ has a solution $x \in D$.

16*. Generalized Directional Contractions

Let $X = X_1 \times \ldots \times X_n$ and $F_i : X \to X_i$ be nonlinear. Then $F = \{F_i\}$ is a generalized directional contraction if for arbitrary $x, y \in X$, there exists $0 < \varepsilon \le 1$ such that

$$\|F_i(x + \varepsilon y) - F_i x\|_i \le \varepsilon \sum_{k=1}^{n} a_{ik} \|y_k\|_k , \quad (i = 1, \ldots, n),$$

where $\{a_{ik}\}$ is a fixed matrix with property (a) given in 15*. If y is replaced by $y = \{F_i x - x_i\}$, then F is called a directional contraction in the narrow sense.

A Fixed Point Theorem

THEOREM. A closed generalized directional contraction F in the narrow sense has a fixed point, i.e., $F_i x = x_i$, for some $x = \{x_i\}$.

The above fixed point theorem generalizes the fixed point theorem of Matkowski [1, 2], Pavaloiu [1], and Rus [1].

17*. Another Way Of Proving Existence Theorems In The Theory Of Contractor Directions

Recently Ray and Cramer [1] have directly applied the Brézis-Browder principle on ordered set by introducting a specialy partial ordering in order to prove their basic results concerning the theory of contractor directions as well as other results related to the theory of normal solvability as developed by F. E. Browder and others. This is the first proof which used partial ordering and gives the same information concerning the whereabout of the solution as can be obtained by the transfinite induction argument. The only difference is that the hypotheses hold in a closed ball instead of an open one as in the latter argument.

18*. A New Contribution To The Contractor Concepts

A new contribution to the contractor concept has recently been made by K. Balakrishna Reddy and P. V. Subramanyam [1]. In their paper "Altman's contractors and Matkowski's fixed point theorem", they made a very interesting observation that the apparently disconnected theorems of Matkowski [1, 2] and Altman [135] can be unified on the basis of a contractor concept extended to a product of Banach spaces. A special matrix introduced by Matkowski serves as a majorant function for the extended contractor concept. In this way, they have obtained a number of new results. Their main result generalizes Matkowski's fixed point theorem (Theorem 1.4, Matkowski [1], see also Matkowski [2]) and Altman's [135], Theorem 5.1, p. 13. Other results include a further extension of Altman's generalization of Krasnoselskii's fixed point theorem. Also based on the contractor theory which they developed is an application to the solution of a system of evolution equations in Banach spaces, besides an implicit function theorem.

19*. More Applications Of The Theory Of Contractors And Contractor Directions

More applications are expected, e.g., to:

(a) the eigenvalue problem

(b) the bifurcation theory

(c) the development of some gradient type methods

(d) the development of some optimization techniques

(e) the abstract control theory

(f) the development of the Newton-Altman method for finding roots of nonlinear functionals on Banach spaces

(g) some problems in Banach algebras

(h) nonlinear evolution equations with application to nonlinear partial defferential equation of hyperbolic type.

20*. The Methodology Of Proofs In The Theory Of Contractor Directions

Four different methods of proofs are now available in the theory of contractor directions:

1. The transfinite induction argument.
2. An argument based on the Brézis-Browder principle on ordered sets and its generalization.
3. The method of iterations.
4. The Caristi-Kirk fixed point argument.

Among more recent contributions, let us mention:

Padgett and Rao [1] have applied the contractor concept in their studies of existence and stability for generalized McShane systems.

Lee and Padgett [1] have applied the concept of integral contractors (see A[135]) to study random nonlinear integral equations in population growth problems. This approach allows to obtain very general existence theorems for the above equations.

Padgett and Rao [2] have also applied the concept of the integral contractor to obtain more general results in the theory of stochastic integral equations.

Ray and Walker [1] have applied their interesting extension of Ekeland's theorem to nonlinear mapping theorems via contractor directions.

REFERENCES

M. Altman, Publications: [24, 30, 122, 123], [125-129], [132, 133], [135-137], [139, 141, 142, 144, 145]. A[ ] = M. Altman [ ].

[1] R. G. Bartle, "Newton's method in Banach spaces", Proc. Amer. Math. Soc. 6 (1955), 827-831.

[1] H. Brézis and F. E. Browder, "A general principle on ordered sets in nonlinear functional analysis", Advances Math., 21 (1976), 355-364.

[1] F. E. Browder, "On the Fredholm alternative for nonlinear operators", Bull. Amer. Math. Soc. 76 (1970), 993-998.

[1] D. Downing and W. A. Kirk, "A generalization of Caristi's theorem with application to nonlinear mapping theory", Pacif. J. Math., 69 (1977), 339-346.

[1] I. Ekeland, "Nonconvex minimization problems", Bull. Amer. Math. Soc. (N.S.) 1 (1979), 443-474.

[1] L. V. Kantorovich and G. P. Akilov, "Functional Analysis in Normed Spaces", Pergammon Press, New York, 1964.

[1] Konrad Knopp, "Theory and Applications of Infinite Series", Blackie and Son, London and Glasgow, 1944.

[1] M. A. Krasnoselŝkii and Ya. B. Rutickii, "Some approximate methods of solving nonlinear operator equations based on linearization", Soviet Math. Doklady, Vol. 2, No. 6 (1961).

[1] H. H. Kuo, "Stochastic integral contractor", J. Integral. Equ. 1 (1979), 35-46.

[1] A. C. H. Lee, "Random contractors and random nonlinear operator equations", Ph.D. Dissertation, Univ. South Carolina, Columbia, 1977.

[1] L. A. Lusternik and V. I. Sobolev, "Elements of Functional Analysis", Moscow 1951.

[1] V. Ptâk, "A theorem on the closed graph type", Manuscripta Math. 13 (1974), 109-130.

[2] V. Ptâk, "A quatitative refinement of the closed graph theorem", Czech. Math. J. 99 (1974), 503-506.

[3] V. Ptâk, "The rate of convergence of Newton's process", Numer. Math. 25 (1976), 279-285.

[4] V. Ptâk, "A modification of Newton's method", Casopis Pest. Mat. 101 (1976), 188-194.

[1] T. Yoshimura, "Descending problem in Green's functional approach to Quantum Field theory", Commun. Math. Phys. 40 (1975), 259-272.

[2] T. Yoshimura, "Applications of Altman's contractor techniques to nonlinear integral equations arising in Green's function formalizm of Klauder's augmented Quantum Field theory", Dept. Math., King's College, London WC2R2LS, England, Preprint.

[1] W. J. Padgett and A. C. H. Lee, "Random contractors and the solution of random nonlinear equations", J. Nonlinear Analysis 1 (1977), 175-185. M. Altman [146-152].

[1] J. R. Dorroh and R. A. Graff, "Integral equations in Banach spaces, a general approach to the linear Cauchy problem, and applications to the nonlinear problem", J. Integral Equations 1 (1979), 309-359.

[1] R. A. Graff, "A functional analytic approach to existence and uniqueness of solutions of some nonlinear Cauchy problems", 1979, (preprint).

[1] T. Kato, "Linear evolution equations of "hyperbolic" type", J. Fac. Sci. Univ. of Tokyo, Sec. I, 17 (1970), 241-258.

[2] T. Kato, "Linear evolution equations of "hyperbolic" type II", J. Math. Soc. Japan 25 (1973), 648-666.

[3] T. Kato, "Quasi-linear equations of evolution with applications to partial differential equations", Lecture Notes in Mathematics, 448 (1975), 25-70, Springer-Verlag.

[1] A. C. H. Lee and W. J. Padgett, "A random nonlinear integral equation in population growth problems", J. Integral Equations 2 (1980), 1-9.

[1] W. J. Padgett and A. N. V. Rao, "Existence and stability of solutions of generalized McShane systems", J. Integral Equations 1 (1979), 47-59.

[2] W. J. Padgett and A. N. V. Rao, "Solution of a stochastic integral equation using integral contractors", Information and Control 31 (1979), to appear.

[1] J. Matkowski, "Some inequalities and a generalization of Banach's principle", Bull. Acad. Polon. Sci., 21 (1973), 322-324.

[2] J. Matkowski, "Integrable solutions of functional equations", Dissertationes Math. (Rozprawy Mat.) CXXVII (1975), 1-68.

[1] K. Balakrishna Reddy and P. V. Subrahmanyam, "Altman's contractors and Matkowski's fixed point theorem", (preprint).

[1] I. Pavaloiu, "La résolution des systèmes d'équations opérationelles ā ℓ'aide des methodes iteratives", Mathematics (Cluj), 11 (34), (1969), 137-141.

[1] I. A. Rus, "Asupra punctelor fixe ale aplicatiilor definite per un produs catezian II: spatii metrice", St. Cerc. Mat. 24 (1972), 897-904.

[3] T. Yoshimura, "On the operator theoretical aspect of dynamical mass generation", preprint.

[1] W. O. Ray and W. T. Cramer, "Solvability of nonlinear operator equations", Pacif. J. Math., to appear.

[1] W. O. Ray and A. M. Walker, "Mapping theorems for Gateaux differentiable and accretive operators", J. Nonlin. Analys., to appear.

[1] M. Turinici, "Mapping theorems via contractor directions in metrizable locally convex spaces", Bull. Acad. Sci. Polon., to appear.

M. Altman:

24. "Concerning approximate solutions of nonlinear functional equations", Bull. Acad. Sci. Polon. 5 (1957), 461-465.

30. "On the generalization of Newton's method", ibid. 789-795.

122. "Contractor directions, directional contractors and directional contractions for solving equations", Pacif. J. Math., Vol. 62, No. 1, (1976), 1-8.

123. "Series of iterates", Collqui. Math. 38 (1977), 305-317.

124. "General solvability theorems", Pacif. J. Math. 68 (1977).

125. "An application of the method of contractor directions to partial differential equations", J. Differential Equations 28 (1978), 406-436.

126. "An existence principle in nonlinear functional analysis", J. Nonlinear Analysis 2 (1978), 765-771.

127. "A strategy theory of solving equations in: Functional Analysis Methods in Numerical Analysis", Lecture Notes in Mathematics, Springer-Verlag, Edit. M. Z. Nashed, Vol. 701, (1979).

128. "An application of the method of contractor directions in nonlinear programming", Numer. Funct. Anal. Opt. 1 (1979), 647-663.

129. "Contractor directions and evolution equations", Nonlin. Analys., 3 (1979), 325-336.

130. "Principe general de l'extremum", to appear.

131. "Contractor directions and monotone operators", J. Integral Equation, 1, (1979), 17-33.

132. "Monotone ε-mappings", Bull. Acad. Polon. Sci., 27 (1979), 793-796.

133. "A general exrremum principle for optimization problems", Nonlin. Analys., 3 (1979), 891-895.

134. "A stronger fixed point theorem for contraction mappings", to appear.

135. "Contractors and Contractor Directions - Theory and Applications", Marcel Dekker (1977) (Research Monograph, 304 pages).

136. "An inverse function theorem for Fréchet spaces", to appear.

137. "The Lagrange multipliers theorem for locally convex metric spaces", to appear.

138. "A perturbation property of Banach algebras", to appear.

139. "Iterative methods of contractor directions", Nonlin. Analys., 4 (1980), 761-772.

140. "Iterative methods for the eigenvalue problem", to appear.

141. "Existence theorems for nonlinear operator equations without continuity condition", Nonlin. Analys., 4 (1980), 437-449.

142. "Weak contractor directions and weak directional contractions", to appear.

143. "A common fixed point theorem for two mappings", to appear.

144. "A generalization of the Brézis-Browder principle on ordered sets", J. Nonlin. Analys. (1981).

145. "An application of the diagonal method of contractor directions to nonlinear programming in locally convex metric spaces", to appear.

146. "Quasi-linear evolution equations in non-reflexive Banach spaces I", J. Nonlin. Analys., to appear.

147. "Quasi-linear evolution equations in non-reflexive Banach spaces II", J. Nonlin. Analys., to appear.

148. "An existence theorem for equations in locally convex metric spaces", to appear.

149. "Quasi-linear evolution equations in non-reflexive Banach spaces", J. Integral Equations, 4 (1980).

150. "Generalized contractor directions", to appear.

151. "A simplified approach to linear evolution equations", to appear.

152. "A new approach to quasi-linear evolution equations in non-reflexive Banach spaces", to appear.

153. "The method of contractor directions for nonlinear elliptic equations", J. Nonlin. Analys. (1981).

154. "An application of the method of contractor directions to the Boltzmann equation for neutron transport in L^1-space", to appear.

155. "A new technique for solving general nonlinear evolution equations in Banach spaces", to appear.

GENERALIZATION OF THE HAUSDORFF MOMENT PROBLEM

D. Borwein

University of Western Ontario
London, Ontario
Canada

A. Jakimovski

Tel-Aviv University
Ramat-Aviv, Tel-Aviv
Israel

ABSTRACT*

Let $\{k_n\}$ be a sequence of positive integers, $\{\ell_n\}$ a sequence of distinct numbers such that $\ell_0 \geq 0$, $k_0 = 1$ if $\ell_0 = 0$, $\ell_n > 0$ for $n \geq 1$, $\ell_n \to \infty$ and $\sum_{n=1}^{\infty} k_n/\ell_n = \infty$, and let $\{u_n^{(r)}\}$ $(r = 0,1,\ldots,k_n-1,\ n = 0,1,\ldots)$ be a sequence of real numbers. We discuss the problem of establishing necessary and sufficient conditions for there to be a function α satisfying

$$(1) \qquad (-1)^r u_n^{(r)} = \int_0^1 t^{\ell_n} \log^r t \, d\alpha(t) \text{ for } r = 0,1,\ldots,k_n-1,\ n = 0,1,\ldots$$

and certain additional conditions. The case $k_n = 1$ for $n \geq 0$, $0 = \ell_0 < \ell_1 < \ell_2 < \ldots$ of the problem is the version of the classical moment problem considered originally by F. Hausdorff - the above formulation

---

*For the complete text see Canadian Journal of Mathematics 33(1981), No. 4, 946-960.

is a natural generalization thereof. An alternative formulation of the problem is to express it as the "infinite Hermite interpolation problem" of establishing necessary and sufficient conditions for a function F to be a Mellin transform of the form $F(z) = \int_0^1 t^z \, d\alpha(t)$ and to satisfy $F^{(r)}(\ell_n) = (-1)^r u_n^{(r)}$ for $r = 0,1,\ldots,k_n-1$, $n = 0,1,\ldots$.

Considerable simplification is obtained by adoption of the following notation. Construct a monotonic sequence $\{\lambda_s\}$ from $\{\ell_n\}$ by repeating each ℓ_n k_n times. Then

$$0 \leq \lambda_0 \leq \lambda_1 \leq \lambda_2 \leq \cdots \leq \lambda_n, \ \lambda_1 > 0, \ \lambda_n \to \infty, \ \sum_{n=1}^{\infty} \frac{1}{\lambda_n} = \infty.$$

For each s there is an integer n(s) such that $\lambda_s = \ell_{n(s)}$. Let $m_s = k_{n(s)}$ and construct a sequence $\{\mu_s^{(r)}\}$ $(r = 0,1,\ldots,m_s-1,\ s = 0,1,\ldots)$ from $\{u_n^{(r)}\}$ by setting $\mu_s^{(r)} = u_{n(s)}^{(r)}$. Then m_s is the multiplicity of λ_s, i.e., it is the number of indices j for which $\lambda_j = \lambda_s$; and $\mu_j^{(r)} = \mu_s^{(r)}$ whenever $\lambda_j = \lambda_s$. Formula (1) can be written in the equivalent form

$$(2) \qquad (-1)^r \mu_s^{(r)} = \int_0^1 t^{\lambda_s} \log^r t \, d\alpha(t) \text{ for } r = 0,1,\ldots,m_s-1,\ s = 0,1,\ldots .$$

For $0 \leq k \leq s \leq n$, let $m_s(k,n)$ be the multiplicity of λ_s among $\lambda_k, \lambda_{k+1}, \ldots, \lambda_n$. By a standard result on Hermite interpolation there is a unique polynomial $P_n(z)$ of degree at most n such that

$$P_n^{(r)}(\lambda_s) = (-1)^r \mu_s^{(r)} \text{ for } r = 0,1,\ldots,m_s(0,n)-1,\ s = 0,1,\ldots,n.$$

It is known that

$$P_n(z) = \sum_{k=0}^{n} u[\lambda_k,\ldots,\lambda_n](\lambda_{k+1}-z)\ldots(\lambda_n-z)$$

where the divided difference $u[\lambda_k,\ldots,\lambda_n]$ is given by

$$u[\lambda_k,\ldots,\lambda_n] = -\frac{1}{2\pi i}\int_{C_{kn}} \frac{P_n(z)dz}{(\lambda_k-z)\ldots(\lambda_n-z)},$$

C_{kn} being a positively sensed Jordan contour enclosing $\lambda_k,\lambda_{k+1},\ldots,\lambda_n$. For $0 \le k \le n$, $0 < t \le 1$, let

$$\lambda_{nk} = \lambda_{k+1}\ldots\lambda_n u[\lambda_k,\ldots,\lambda_n],$$

$$\lambda_{nk}(t) = -\lambda_{k+1}\ldots\lambda_n \frac{1}{2\pi i}\int_{C_{kn}} \frac{t^z dz}{(\lambda_k-z)\ldots(\lambda_n-z)},\quad \lambda_{nk}(0) = \lambda_{nk}(0+),$$

with the convention that products such as $\lambda_{k+1}\ldots\lambda_n = 1$ when $k = n$.

Let

$$D_0 = (1+\lambda_0)d_0 = 1,\quad D_n = \left(1+\frac{1}{\lambda_1}\right)\ldots\left(1+\frac{1}{\lambda_n}\right) = (1+\lambda_n)d_n \text{ for } n \ge 1.$$

Let θ be an even continuous convex function such that $\theta(u)/u \to 0$ as $u \to 0$ and $\theta(u)/u \to \infty$ as $u \to \infty$. Associated with this function is the Orlicz class L_θ of all functions x Lebesgue integrable over [0,1] for which

$$\int_0^1 \theta(x(t))dt < \infty.$$

Let L_∞ be the space of measurable functions x on [0,1] with finite norm

$$\|x\|_\infty = \text{ess.sup}_{0<t<1} |x(t)|.$$

Let

$$M_\theta(n) = \sum_{k=0}^{n} \frac{d_k}{D_n}\,\theta\left(\frac{D_n}{d_k}\lambda_{nk}\right),$$

$$M_1(n) = \sum_{k=0}^{n} |\lambda_{nk}|,$$

$$M_\infty(n) = \max_{0\le k\le n} |\lambda_{nk}|\,\frac{D_n}{d_k},$$

and let

$$M_\theta = \sup_{n\ge 0} M_\theta(n),\ M_1 = \sup_{n\ge 0} M_1(n),\ M_\infty = \sup_{n\ge 0} M_\infty(n).$$

Let BV be the space of normalized functions α of bounded variation on $[0,1]$, i.e., $\alpha(0) = 0$, $2\alpha(t) = \alpha(t+) + \alpha(t-)$ for $0 < t < 1$.

The following two theorems set out the main results.

THEOREM 1. A necessary and sufficient condition for there to be a function

(i) $\alpha \in BV$ satisfying (1) is that $M_1 < \infty$;

(ii) $\beta \in L_\infty$ satisfying

$$(3)\qquad (-1)^r u_n^{(r)} = \int_0^1 t^{\ell_n} \log^r t\, \beta(t)dt \text{ for } r = 0,1,\ldots,k_n-1,\ n = 0,1,\ldots$$

is that $M_\infty < \infty$;

(iii) $\beta \in L_\theta$ satisfying (3) is that $M_\theta < \infty$.

Furthermore

(iv) if (1) is satisfied by a function $\alpha \in BV$, then

$$M_1 = \int_0^1 |d\alpha(t)| - \delta|\alpha(0+)| \text{ where } \delta = 0 \text{ when } \ell_0 = 0,\ \delta = 1 \text{ when } \ell_0 > 0;$$

moreover α is unique when $\ell_0 = 0$, and when $\ell_0 > 0$ it differs by a constant, over the interval $0 < t \leq 1$, from any other function in BV satisfying (1);

(v) if (3) is satisfied by a function $\beta \in L_\infty$, then β is essentially unique and $M_\infty = \|\beta\|_\infty$;

(vi) if (3) is satisfied by a function $\beta \in L_\theta$, then β is essentially unique and

$$M_\theta = \int_0^1 \theta(\beta(t))dt.$$

THEOREM 2. For $n = 0,1,\ldots,$

$$M_1(n) \leq M_1(n+1),\ M_\infty(n) \leq M_\infty(n+1),\ M_\theta(n) \leq M_\theta(n+1);$$

and

$$\lim_{n\to\infty} M_1(n) = M_1,\ \lim_{n\to\infty} M_\infty(n) = M_\infty,\ \lim_{n\to\infty} M_\theta(n) = M_\theta.$$

ON ROTATIONAL PATCHES OF VORTICITY

J. Burbea

University of Pittsburgh
Pittsburgh, Pennsylvania
U.S.A.

Abstract

An analytical method for the determination of the m-waves of Kelvin and their angular velocities is presented. These m-fold symmetric solutions are uniformly rotating and constitute a generalization of the rotating ellipse of Kirchoff which corresponds to $m = 2$.

1. Introduction

We consider the following problem: A patch, or a plane region, D of uniform vorticity ω_o is moving in a fluid at rest at infinity. The problem is to determine for which shapes of the boundary of the patch D, steady rotation is possible, and, in that case to also determine the velocity Ω of this rotation. A circular patch is an obvious solution with an arbitrary angular velocity Ω, and, in 1880, Kelvin showed that small perturbations to the circle proportional to $\cos m\theta$ also rotated uniformly, with angular velocity (see [6, p. 231]):

$$\Omega_m = \frac{\omega_o}{2} \frac{m-1}{m}, \quad m \geq 2. \tag{1.1}$$

Another solution is the well-known elliptical vortex of Kirchoff (see [6, p. 232]) with the angular velocity

$$\Omega_E = \omega_o \frac{Q}{(1+Q)^2} ; \quad Q \geq 1, \tag{1.2}$$

where Q is the axis-ratio of the ellipse E. This rotating ellipse of Kirchoff is the finite amplitude form of the $m = 2$ wave of Kelvin.

The problem of finding the finite amplitude form of the $m \geq 3$ waves of Kelvin was attacked numerically by Deem and Zabusky [5], where a "contour-dynamics" algorithm was employed.

The analysis leading to (1.2) by Kirchoff depend heavily, and crucially, on the use of elliptic coordinates and skillfully matching, on the elliptic boundary, the interior and exterior stream functions.

The main purpose of this paper is in presenting an analytical and "coordinate-free" method for the determination of the finite amplitude form of the m-waves of Kelvin and their angular velocities Ω. This will be accomplished by exploiting the conformal mapping ϕ of the exterior of the unit disk onto the exterior of the patch D. This results in a single and elegant equation of the form

$$\lambda|\phi|^2 + 2\text{ReS}(\phi) = \text{const.}, \tag{1.3}$$

where

$$\lambda = 1 - 2\Omega/\omega_o,$$

$\phi = \phi(\omega)$, ω varies in the unit circle $\Gamma = \{\omega : |\omega| = 1\}$ and

$$S(\phi) = S(\phi(\omega)) = \frac{1}{2\pi i}\int_\Gamma \overline{\phi(\tau)} \log\left(1 - \frac{\phi(\omega)}{\phi(\tau)}\right)\phi'(\tau)d\tau.$$

As one would expect, equation (1.3) is invariant under reflection, magnification and rotation. Moreover, the approximate solutions of Kelvin and the exact elliptic solution of Kirchoff, as described in (1.1) - (1.2) are, at once, deductible from (1.3). The circle $\phi_o(\omega) \equiv \omega$ is the trivial solution of (1.3) with λ, and therefore, also Ω, being an arbitrary constant.

In our recent paper [1, 2] equation (1.3) was linearized about the trivial solution $\phi_o(\omega) = \omega$, resulting in a (linear) eigenvalue problem.

The eigenvalues are the Ω_m of (1.1) and the corresponding eigenfunctions are $\omega^{-(m-1)}$, $m \geq 2$. It thus follows that the infinitesimal amplitude m-wave of Kelvin is the m-fold symmetric hypotrochoid $\phi_m(\omega) = \omega(1+b\omega^{-m})$ with $|b| < (2m - 3)^{-1}$, $m \geq 2$. Moreover, the analysis in [1, 2] leads to an improvement of (1.1) in the form of

$$\Omega_m = \frac{\omega_o}{2} \frac{m-1}{m}(1 - \frac{m}{2}|b|^2), \quad m \geq 2, \tag{1.4}$$

where for $m = 2$, the expression is exact. We note that $\phi_2(\omega)$ is <u>precisely</u> the Kirchoff solution and Ω_2 is identical with Ω_E of (1.2). Using methods from the theory of bifurcation, it was also shown in [1, 2] that for any integer $m \geq 2$ and any b, $0 < |b| < (2m - 3)^{-1}$, there exists a unique solution of (1.3) of the form

$$\phi(\omega) = \omega[1 + \sum_{n=1}^{\infty} \beta_n \omega^{-mn}] \tag{1.5}$$

where $\beta_1 \equiv \beta_1(m) \equiv b$ and $\beta_n \equiv \beta_n(m, b)$, $n \geq 2$, are functions of m and b. Here, b is a complex number but due to the invariance of the problem it can be taken as a positive number. For the same reason, the solution of (1.5) can be magnified by any $a \neq 0$.

The problem is now reduced to the determination of the coefficients β_n of the m-fold symmetric mapping ϕ of (1.5) so that (1.3) is satisfied. The case of $m = 2$ is, of course, immediate while for $m \geq 3$, such a determination is very involved. However, it is rather a fortunate circumstance that the Riemann mapping (1.5) exhibits, in effect, a perturbation away from the trivial solution of the circle. This fact will be exploited by expanding all quantities involved in powers of $\varepsilon \equiv |b|$, $0 < \varepsilon < (2m - 3)^{-1}$ and then determining the coefficients in the usual manner. In doing so we meet two completely mathematically-equivalent approaches, namely: (1) expanding each mapping coefficient β_n, $n \geq 2$, in powers of ε and then, using (1.3), determining the coefficients of ε^k, as functions of ω, of the mapping $\phi(\omega)$ in (1.5) and, (2) before expanding in powers of ε, we compare the powers of ω, $\omega \in \Gamma$, by using (1.3) and (1.5), resulting in a system of non-linear equations for the coefficients β_n and then expand these in powers of ε.

From the above mentioned two equivalent approaches we prefer, and shall use, the second approach for the following reasons:
(i) In approach (1), the multinomial expressions resulting in the ε-powers expansion of the operator $S(\phi)$ of (1.3) are very complicated and admit no closed form. On the other hand, in approach (2) the so called "Grunsky-coefficients" of the mapping emerge and they admit a reasonable closed form formula. Moreover, the Grunsky-coefficients are widely used in function theory and have been tabulated [7].
(ii) As far as the numerical algorithm is concerned, two truncations are taking place here, namely the truncation of the infinite sum in (1.5) and the truncation in expanding each mapping coefficient β_n, $n \geq 2$, in powers of $\varepsilon = |b|$. This order in performing these sequence of truncations corresponds to approach (2) while the reversed order corresponds to approach (1). In the reversed order, the resulting expansion, though convergent, may cease to represent a conformal map or, equivalently, the resulting curve may not be a simple curve and "vortex-breaking" is experienced. On the other hand, retaining the above order of truncations yields a sequence of curves all of which are simple and one can theoretically show that they indeed converge to the sought off boundary-patch.

In this paper we show how the coefficients of the mapping ϕ are determined and we compute explicitly few of them. This will result in a further improvement of the expansion of the angular velocity in (1.4). Furthermore, we numerically compute the cases of $m = 3, 4$ and compare them to the results of Deem and Zabusky [5]. We should remark here that it is very plausible that for $m \geq 3$, the exact solutions admit a closed form in terms of some standard transcendental functions. We were not able to identify these proposed functions but the approximate solution that is achieved is a true approximation, in the sense that its convergence to the exact solution is indeed improved when adding higher orders of the parameter $|b| = \varepsilon$, $0 < |b| < (2m-3)^{-1}$. The geometrical meaning of the parameter $|b|$ is that it measures the dilitation of the patch, i.e., $|b| = (Q-1)/(Q+1)$ where Q is essentially the axis-ratio of the patch D. The condition for non-breaking of the vortex-patch is $1 \leq Q < (m-1)/(m-2)$.

The condition for the stability of the vortex-patch, based on our forthcoming paper [3], is presented in [4].

There exists an entirely different method for attacking the present problem which on the surface seems to be less involved and more straightforward than ours. This method is based on representing the boundary patch in polar coordinates $r = r(\theta)$ and expanding $r(\theta)$ in the form

$$r(\theta) = \sum_{n=0}^{\infty} \delta_n \cos nm\theta, \quad (\delta_0 \equiv 1), \tag{1.6}$$

with an appropriate expansion of the velocity potential. Such a method is adopted in Su [10]. Here (see §7) we discuss Su's result and compare it to ours. However, before even discussing this method one cannot escape to notice that it must involve more effort than in ours. This is true even for the elliptic case $m = 2$. Indeed, in our method the ellipse, by (1.5), is represented by the elementary rational function $\phi(\omega) = \omega + b\omega^{-1}$ while the corresponding expansion (1.6) for $r(\theta)$ of the ellipse is infinite and very involved (see §7).

§2. The Basic Equation

A short derivation of the basic equation of (1.3) follows. To do so we employ complex notation. The patch of uniform vorticity ω_0 is assumed to be a region D whose boundary C is a closed smooth simple curve. We may assume that the origin is contained in D as its centroid. We seek a stream function $\psi = \psi(z)$ which is of class C^1 in the entire plane with the following additional properties:

(i) $\Delta\psi = \omega_0$ inside D and $\Delta\psi = 0$ in E_D, the exterior of D;

(ii) $\psi(z) = (\Omega/2)|z|^2$ for all $z \in C$ and where the bifurcation-parameter Ω is the angular velocity;

(iii) ψ and $\partial_z\psi$ are continuous across C;

(iv) $\partial_z\psi(\infty) = 0$.

The solution of items (i), (iii) and (iv) is uniquely determined, modulo an additive constant, by

$$\psi(z) = \frac{\dot{\omega}_o}{2\pi} \int_D \log|z - \zeta|\, dA(\zeta) + c. \tag{2.1}$$

In this paper all additive constants that appear are denoted by the same letter c. We need only satisfy item (ii) on the boundary with ψ as given in (2.1). This is done by transforming the area integral in (2.1) to a contour integral and thereby obtaining an equation for the boundary-patch C. There exists a variety of ways for this task to be accomplished. However, the following <u>indirect</u> approach can be shown to be the most effective one. First, compute $\partial_z \psi$ from (2.1). Next, apply Greens' formula on this $\partial_z \psi$ for z <u>inside</u> D and then integrate along the sought off streamline.

Accordingly, and bearing in mind Plemelj saltus theorem, we find that

$$\lambda|z|^2 + 2\mathrm{Re}S(z) = c, \quad z \in C, \tag{2.2}$$

where

$$\lambda = 1 - 2\Omega/\omega_o \tag{2.3}$$

and

$$S(z) = \frac{1}{2\pi i} \int_C \bar{\zeta} \log(1 - \frac{z}{\zeta})\, d\zeta, \quad z \in \bar{D} \tag{2.4}$$

Here, $\bar{D}$ denotes the closure of D and the integral is taken in the principal-value sense. The choice of the origin as the centroid of D gives

$$\int_C |z|^2 dz = 0 \tag{2.5}$$

as a constraint to the solution if (2.2)

We note that the system (2.2) - (2.5) is invariant under $z \to \bar{z}$ and $z \to az$ for any $a \in \mathbf{C} - \{0\}$ which means that the problem is invariant under reflection, magnification and rotation but, because of (2.5), it is not invariant under translation.

Equation (1.3) is obtained from (2.2) via the parametrization of the boundary by means of a canonical conformal mapping, $z = \phi(\omega)$. However, the derivation of (2.2) was based on interior points of D, the best strategy is, therefore, to exploit an exterior conformal mapping so that the "free" boundary will be "trapped" in between.

Accordingly, we consider the exterior of the unit disk $E_\Delta = \{\omega : |\omega| > 1\}$ and its boundary $\Gamma = \{\omega : |\omega| = 1\}$. Let $z = \phi(\omega)$ be the conformal mapping of E_Δ onto the exterior of the patch E_D with $\phi(\infty) = \infty$. We may assume, thanks to the invariance properties of (2.2)-(2.5), that $\phi'(\infty) = 1$. Such a mapping is of the form

$$\phi(\omega) = \omega + \sum_{n=0}^{\infty} b_n \omega^{-n}, \quad \omega \in E_\Delta. \tag{2.6}$$

We denote by $\sum$ the class of all holomorphic and univalent functions ϕ in E_Δ whose expansion is of the form (2.6). Equations (2.2), (2.4) and (2.5) can be now written as

$$\lambda|\phi|^2 + 2\mathrm{Re}S(\phi) = c, \tag{2.7}$$

$$S(\phi) = S(\phi(\omega)) = \frac{1}{2\pi i}\int_\Gamma \overline{\phi(\tau)}\log\left(1 - \frac{\phi(\omega)}{\phi(\tau)}\right)\phi'(\tau)d\tau \tag{2.8}$$

and

$$b_o\left(1 - \sum_{n=1}^{\infty} n|b_n|^2\right) = \sum_{n=1}^{\infty} n\bar{b}_n b_{n+1} + \sum_{n=1}^{\infty} nb_n\left(\sum_{k=1}^{\infty} b_k\bar{b}_{k+n}\right), \tag{2.9}$$

where $\phi = \phi(\omega)$, $\omega \in \Gamma$ and ϕ belongs to class $\sum$. Evidently, this system (2.7)-(2.9) for $\phi \in \Gamma$ is invariant under $\phi \to \bar{\phi}$ and $\phi \to \phi o e^{i\theta}$ for any $\theta \in [0, 2\pi)$.

We see that the solutions of (2.7)-(2.9) are members of the class $\sum$ but in effect, as mentioned in the introduction, they are members of the subclass $\sum_m$ described below. Let $m \geq 2$ be an integer and denote by $\sum_m$ the subclass of $\sum$ consisting of those ϕ with the expansion

$$\phi(\omega) = \omega[1 + \sum_{n=1}^{\infty} \beta_n \omega^{-mn}], \quad \omega \in E_\Delta.$$

Thus, for any $\phi \in \sum_m$, $E_D = \phi(E_\Delta)$ is m-fold symmetric which means that E_D is invariant under the rotation $z \to e^{2\pi i/m}z$. Also, for any $\phi \in \sum_m$, the constraint (2.5) or (2.9) is identically satisfied which means that the origin is indeed the centroid of the patch D. Furthermore, since if ϕ is a solution of (2.7)-(2.9), then $e^{i\theta}\phi$ is also a solution we may assume that the patch is symmetric with respect to the real axis.

At this stage we need some classical results from the theory of univalent functions as those found, for example, in the book of Pommerenke [8, pp. 56-64]. The next section is devoted to a brief review of these results.

3. Faber Polynomials

Let the function

$$\phi(\omega) = \omega + \sum_{n=0}^{\infty} b_n \omega^{-n} \tag{3.1}$$

be holomorphic in some neighborhood of ∞ so that it is unvalent in $|\omega| > R$, for some $R > 0$. For a fixed $z \in \mathbb{C}$, the function $\log[\omega^{-1}(\phi(\omega) - z)]$ is holomorphic for $|\omega| > R$ and vanishes at ∞. We write its expansion around ∞ as

$$\log \frac{\phi(\omega)-z}{\omega} = -\sum_{n=1}^{\infty} \frac{1}{n}\Phi_n(z)\omega^{-n}, \quad (|\omega| > R, \ z \in \mathbb{C}). \tag{3.2}$$

Putting $\Phi_o(z) \equiv 1$ and differentiating with respect to ω yields

$$\frac{\phi'(\omega)}{\phi(\omega)-z} = \sum_{n=0}^{\infty} \Phi_n(z)\omega^{-(n+1)}, \tag{3.3}$$

while differentiating with respect to z gives

$$\frac{1}{\phi(\omega)-z} = \sum_{n=1}^{\infty} \frac{1}{n} \Phi_n'(z)\omega^{-n}. \tag{3.4}$$

From (3.1) and (3.3) follows that $\Phi_1(z) = z - b_o$ and the recursion relation

$$\Phi_{n+1}(z) = (z - b_o)\Phi_n(z) - \sum_{k=1}^{n-1} b_{n-k} \Phi_k(z) - (n + 1)b_n;\ n \geq 1. \tag{3.5}$$

It follows that $\Phi_n(z)$ is a polynomial of degree n of the form

$$\Phi_n(z) = (z - b_o)^n - n\, b_1(z - b_o)^{n-2} + \dots;\quad n \geq 2.$$

This polynomial is called the n-th Faber polynomial of ϕ and is also defined as the principal part at infinity of the Laurent series of $[\phi^{-1}(z)]^n$. One shows

$$\Phi_0(z) = 1,\ \Phi_1(z) = z - b_o,\ \Phi_2(z) = (z - b_o)^2 - 2b_1$$

$$\Phi_3(z) = (z - b_o)^3 - 3b_1(z - b_o) - 3b_2,$$

$$\Phi_4(z) = (z - b_o)^4 - 4b_1(z - b_o)^2 - 4b_2(z - b_o) + (2b_1^{\ 2} - 4b_3).$$

We now introduce the Grunsky coefficients b_{nk} of ϕ. They are defined by

$$\log\frac{\phi(\omega)-\phi(\tau)}{\omega-\tau} = -\sum_{n,k=1}^{\infty} b_{nk}\, \omega^{-n} \tau^{-k},\quad (|\omega| > R,\ |\tau| > R). \tag{3.6}$$

Since $\phi(\omega) \neq \phi(\tau)$ (for $\omega \neq \tau$) and $\phi'(\omega) \neq 0$ in $\{|\omega| > R,\ |\tau| > R\}$, the function on the left-hand side of (3.6) is holomorphic in this domain and, thus, the above expansion is valid. Clearly,

$$b_{nk} = b_{kn},\quad (n,\ k = 1,\ 2,\ \dots).$$

Upon putting $z = \phi(\tau)$ in (3.2), we obtain

$$\log \frac{\phi(\omega)-\phi(\tau)}{\omega-\tau} = -\sum_{n=1}^{\infty} \frac{1}{n} [\Phi_n(\phi(\tau)) - \tau^n]\omega^{-n}.$$

This, together with (3.6), yields

$$\Phi_n[\phi(\omega)] = \omega^n + n \sum_{k=n}^{\infty} b_{nk}\, \omega^{-k}, \quad (|\omega| > R). \tag{3.7}$$

From $\Phi_1(z) = z - b_o$ we deduce that

$$b_{n1} = b_{1n} = b_n; \quad n \geq 1.$$

Also, from (3.6) it follows that the Grunsky coefficients are independent of b_o. Moreover, using (3.5) and (3.7) we obtain the recursion formula

$$(n+1)b_{n+1,k} = b_{n+k} + n\, b_{n,k+1} + n \sum_{m=1}^{k-1} b_{k-m}\, b_{nm} - \sum_{m=1}^{n-1} m\, b_{n-m}\, b_{mk}$$

which shows that

$$b_{22} = b_3 + \frac{1}{2} b_1^2, \; b_{23} = b_4 + b_1 b_2, \; b_{24} = b_5 + b_1 b_3 + \frac{1}{2} b_2^2$$

$$b_{33} = b_5 + b_1 b_3 + b_2^2 + \frac{1}{3} b_1^3, \; b_{34} = b_6 + b_1 b_4 + 2b_2 b_3 + b_1^2 b_2,$$

$$b_{44} = b_7 + b_1 b_5 + 2b_2 b_4 + 2b_1 b_2^2 + b_1^2 b_3 + \frac{3}{2} b_3^2 + \frac{1}{4} b_1^4.$$

The following explicit formula for b_{nm} is due to Schur [9]:

$$b_{nm} = \sum_{C_{nm}} \frac{(\gamma-1)!}{\gamma_1! \ldots \gamma_{n+m-1}!} D^{(n)}_{\gamma_1 \ldots \gamma_{n+m-1}}\, b_1^{\gamma_1} \ldots b_{n+m-1}^{\gamma_{n+m-1}}, \tag{3.8}$$

where

$$C_{nm} = \{(\gamma_1,\dots,\gamma_{n+m-1}) \in N^{n+m-1} : 2\gamma_1 + 3\gamma_2 + \dots + (n+m)\gamma_{n+m-1} = n+m\},$$

$$D^{(n)}_{\gamma_1\dots\gamma_{n+m-1}} = \left[\prod_{k=1}^{n+m-1} \left(\frac{x-x^{k+1}}{1-x}\right)^{\gamma_k}\right]_n = \sum_{0\le\alpha_j\le\gamma_j} (-1)^{\alpha} \binom{\gamma_1}{\alpha_1} \dots \binom{\gamma_{n+m-1}}{\alpha_{n+m-1}} \binom{n-1+A+\alpha}{\gamma-1}$$

and

$$\gamma = \gamma_1 + \dots + \gamma_{n+m-1},\ \alpha = \alpha_1 + \dots + \alpha_{n+m-1},\ A = 2\alpha_1 + 3\alpha_2 + \dots + (n+m)\alpha_{n+m-1}.$$

Here, N denotes the set of all non-negative integers, $\alpha_j \in N$ and $[\dots]_n$ denotes the n-th coefficient of x in the expansion of the expression in the bracket.

From (3.8) one deduces that $n\, b_{nm}$ is a polynomial of degree $\le \min(n, m)$ in $b_1, \dots, b_{n+m-1}$ with non-negative integer coefficients.

We conclude this section by mentioning the following simple example: Let $\phi(\omega) = \omega + b\omega^{-1}$. Then $b_{nm} = \delta_{nm} b^n/n$, where $\delta_{nm} = 0$ for $n \neq m$ and $\delta_{nn} = 1$. Hence

$$\Phi_n[\phi(\omega)] = \omega^n + b^n\omega^{-n};\quad (\phi(\omega) = \omega + b\omega^{-1}),$$

and

$$\Phi_n(z) = (2\sqrt{b})^n\, T_n(z/2\sqrt{b})$$

where

$$T_n(z) = 2^{-(n+1)} \cos(n \cos^{-1} z),$$

is the nth Tchebbycheff polynomial of the first kind.

4. Stream In Terms Of Faber Polynomials

We now express the stream function ψ of (2.1) in terms of the Faber polynomials of the mapping ϕ associated with the vortex D.

We first note that for $z \in \overline{D}$

$$\partial_z \psi = \frac{\omega_o}{4} [\overline{z} - \frac{1}{2\pi i} \int_C \frac{\overline{\zeta}}{\zeta - z} d\zeta]; \quad z \in \overline{D}. \tag{4.1}$$

Here

$$\frac{1}{2\pi i} \int_C \frac{\overline{\zeta}}{\zeta - z} d\zeta = \frac{1}{2\pi i} \int_\Gamma \frac{\overline{\phi(\omega)}}{\phi(\omega) - z} \phi'(\omega) d\omega .$$

We expand this integral by means of (3.3) for which we obtain that it is equal to $\sum_{n=0}^{\infty} \overline{b}_n \Phi_n(z)$. Combining this result with (4.1) yields

$$\partial_z \psi = \frac{\omega_o}{4} [\overline{z} - \sum_{n=0}^{\infty} \overline{b}_n \Phi_n(z)], \quad z \in \overline{D}.$$

This may be integrated and we obtain, using the notation of (2.2)-(2.4),

$$\psi(z) = \frac{\omega_o}{2} [\frac{1}{2} |z|^2 + \text{Re } S(z)] + c, \quad z \in \overline{D}, \tag{4.2}$$

where

$$S(z) = - \sum_{n=0}^{\infty} \overline{b}_n F_n(z) + c, \quad F_n'(z) = \Phi_n(z). \tag{4.3}$$

An alternative and useful expression for $S(z)$ may be obtained by expanding the above Cauchy integral with the aid of (3.4). This gives

$$S(z) = - \sum_{n=1}^{\infty} \frac{1}{n} [\overline{b}_{n-1} - \sum_{k=1}^{\infty} k\, b_k\, \overline{b}_{k+n}] \Phi_n(z) + c. \tag{4.4}$$

This can also be obtained from (2.8) and (3.2). Therefore, using (3.7) and (4.4) we obtain

$$S(\phi) = -\sum_{n=1}^{\infty}\frac{1}{n}[\overline{b}_{n-1} - \sum_{k=1}^{\infty} k\, b_k\, \overline{b}_{k+n}](\omega^n + n\sum_{j=1}^{\infty} b_{nj}\omega^{-j}) + c, \quad (4.5)$$

for $\omega \in \Gamma$. The above expressions are valid for $\phi \in \sum$. When $\phi \in \sum_m$, $m \geq 2$, (4.4) admits the form (apart from an additive constant)

$$S(z) = -\frac{1}{m}\sum_{n=1}^{\infty}\frac{1}{n}[\overline{b}_{nm-1} - \sum_{k=1}^{\infty}(km-1)b_{km-1}\overline{b}_{(k+n)m-1}]\Phi_{nm}(z). \quad (4.6)$$

With the aid of these expressions the system (2.7)-(2.9) may be transformed to a system of non-linear equations involving the Grunsky coefficients of $\phi \in \sum$. Indeed, by (2.6), (2.7) and (4.5) we obtain

$$b_o(1 - \sum_{n=1}^{\infty} n|b_n|^2) = \sum_{n=1}^{\infty} n\,\overline{b}_n b_{n+1} + \sum_{n=1}^{\infty} n\, b_n(\sum_{k=1}^{\infty} b_k\,\overline{b}_{k+n}) \quad (4.7)$$

$$\lambda[b_{n-1} + \sum_{k=0}^{\infty}\overline{b}_k\, b_{k+n}] = \frac{1}{n}[b_{n-1} - \sum_{k=1}^{\infty} k\,\overline{b}_k\, b_{k+n}] \quad (4.8)$$

$$+ \sum_{j=1}^{\infty} b_{nj}[\overline{b}_{j-1} - \sum_{k=1}^{\infty} k\, b_k\,\overline{b}_{k+j}], \quad n \geq 1.$$

When $\phi \in \sum_m$, $m \geq 2$, this system reduces to

$$\lambda[\beta_n + \sum_{k+1}^{\infty}\overline{\beta}_k\,\beta_{k+n}] = \frac{1}{nm}[\beta_n - \sum_{k=1}^{\infty}(km-1)\overline{\beta}_k\,\beta_{k+n}] \quad (4.9)$$

$$+ \sum_{j=1}^{\infty}\beta_{nj}[\overline{\beta}_j - \sum_{k=1}^{\infty}(km-1)\beta_k\,\overline{\beta}_{k+j}],$$

for all $n \geq 1$. Here, we have used the notation

$$\beta_n = b_{nm-1},\ \beta_{nj} = b_{nm,jm};\quad n,\ j = 1,\ 2,\ \dots\ .$$

We also used the fact, which easily follows from (3.8), that for $\phi \in \sum_m$, $m \geq 2$, $b_{n,jm} = 0$ for all n which are not multiples of m.

It is clear that $\phi(\omega) = \omega$ is the trivial solution of (4.7)-(4.8), where λ is an arbitrary real number. Another solution is the well-known elliptical vortex of Kirchoff (see Lamb [6, p. 232]). This solution is easily obtained from (4.9) as the following proposition shows:

PROPOSITION 1. The mapping $\phi \in \sum_2$ given by $\beta_n = b_{2n-1} = \delta_{n1} b$, $n \geq 1$, with $0 < |b| < 1$ solves (4.9) provided

$$\lambda = \frac{1}{2}(1 + |b|^2) \tag{4.10}$$

and therefore

$$\Omega = \frac{\omega_o}{4}(1 - |b|^2).$$

<u>Proof.</u> In this case we have $\phi = \omega[1 + b\omega^{-2}] = \omega + b\omega^{-1}$ which describes an ellipse. Equation (4.9) is now

$$\lambda\beta_n = \frac{1}{2n}\beta_n + \beta_{n1}\overline{\beta}_1;\quad n = 1,\ 2,\ \dots\ .$$

Since $\beta_{n1} = b_{2n,2} = \delta_{n1}\ b^{2n}/2n$ we see that this equation is trivially satisfied for all $n \geq 2$. The case of $n = 1$ results in (4.10).

As a complement for this proposition we shall prove:

PROPOSITION 2. The only solution of (4.7)-(4.8) among all mappings $\phi \in \sum$ of the form $\phi(\omega) = \omega + \sum_{n=0}^{N} b_n\,\omega^{-n}$, $N < \infty$, is the elliptic solution $\phi = \omega + b_1\omega^{-1}$, $|b_1| < 1$.

<u>Proof.</u> We may assume that $b_N \neq 0$ and $b_n = 0$ for $n \geq N + 1$.

Equation (4.8) is equivalent to $\lambda|\phi|^2 + 2\mathrm{Re}\ S(\phi) = \text{const.}$, with $\phi = \phi(\omega)$, $\omega \in \Gamma$. We compare the coefficients of the highest power of ω^{-1}. The coefficient of such power in $\lambda|\phi|^2$ is of the form $\lambda\ b_N\ \omega^{-(N+1)}$. Using (4.3), the corresponding term in $2\mathrm{Re}\ S(\phi)$ is contained in $-2\mathrm{Re}[\overline{b}_N F_N(\phi)]$, provided $N \geq 1$. Since $F_n' = \Phi_n$, this term is contained in

$$-\frac{2}{N+1}\ \mathrm{Re}[\overline{b}_N(\phi(\omega) - b_o)^{N+1}]$$

and thus the relevant term in $2\mathrm{Re}\ S(\phi)$ is of form $-(N+1)^{-1}\overline{b}_N b_N^{N+1}\omega^{-N(N+1)}$.

Since $b_n \neq 0$ we must have $N \leq 1$. Thus $\phi = \omega + b_o + b_1\ \omega^{-1}$. We now appeal to (4.7) to obtain $b_o(1 - |b_1|^2) = 0$ and since $|b_1| < 1$ we have $b_o = 0$. Therefore, $\phi = \omega + b_1\ \omega^{-1}$ and the proposition now follows from Proposition 1.

§5. Symmetric Vortices

We shall now assume that the given vortex is m-fold symmetric about the origin. In this case $\phi \in \sum_m$, $m \geq 2$; and

$$\phi(\omega) = \omega[1 + \sum_{n=1}^{\infty} \beta_n\ \omega^{-mn}],\ (\beta_n = b_{nm-1}). \tag{5.1}$$

The relevant equation here is that of (4.9). Thus

$$\lambda[\beta_n + \sum_{k=1}^{\infty} \overline{\beta}_k\ \beta_{k+n}] = \frac{1}{nm}[\beta_n - \sum_{k=1}^{\infty} (km-1)\overline{\beta}_k\ \beta_{k+n}] \tag{5.2}$$

$$+ \sum_{j=1}^{\infty} \beta_{nj}[\overline{\beta}_j - \sum_{k=1}^{\infty} (km-1)\beta_k\ \overline{\beta}_{k+j}],$$

for all $n \geq 1$, where λ is the bifurcation parameter

$$\lambda = 1 - 2\Omega/\omega_o \tag{5.3}$$

and β_{nj} are the Grunsky coefficients $\beta_{nj} = b_{nm,jm}$. Of course, the β_{nj} form an infinite symmetric matrix in terms of the coefficients β_k. A use of (3.8) shows that

$$\beta_{nk} = \sum_{C_{nk}} \frac{(\gamma-1)!}{\gamma_1!\cdots\gamma_{n+k}!} D^{(nm)}_{\gamma_1\cdots\gamma_{n+k}} \beta_1^{\gamma_1}\cdots\beta_{n+k}^{\gamma_{n+k}}; \quad n, k \geq 1, \tag{5.4}$$

where

$$C_{nk} = \{(\gamma_1,\ldots,\gamma_{n+k}) \in N^{n+k} : \gamma_1 + 2\gamma_2 + \ldots + (n+k)\gamma_{n+k} = n + k\},$$

$$D^{(nm)}_{\gamma_1\cdots\gamma_{n+k}} = \Big[\prod_{j=1}^{n+k} \Big(\frac{x-x^{jm}}{1-x}\Big)^{\gamma_j}\Big]_{(nm)} \tag{5.5}$$

$$= \sum_{0\leq\alpha_j\leq\gamma_j} (-1)^{\alpha} \binom{\gamma_1}{\alpha_1}\cdots\binom{\gamma_{n+k}}{\alpha_{n+k}}\binom{\alpha+m(n-A)-1}{\gamma-1}$$

and

$$\gamma = \gamma_1 + \ldots + \gamma_{n+k},\ \alpha = \alpha_1 + \ldots + \alpha_{n+k},\ A = \alpha_1 + 2\alpha_2 + \ldots + (n+k)\alpha_{n+k}.$$

From (5.4) one obtains that $nm\,\beta_{nk}$ is a polynomial of degree $n+k$ in $\beta_1, \ldots, \beta_{n+k}$ with positive integer coefficients and void of constant terms. We, of course, may assume that $\beta_1 \neq 0$. We write

$$\beta_n = \varepsilon_n b^n, \quad n \geq 1,$$

where the ε_n are real and $\varepsilon_1 = 1$ and thus $\beta_1 = b$. Using (5.4) we obtain that

$$\beta_{nk} = \varepsilon_{nk}\ b^{n+k}$$

where

$$\varepsilon_{nk} = \sum_{C_{nk}} \frac{(\gamma-1)!}{\gamma_1!\cdots\gamma_{n+k}!} D^{(nm)}_{\gamma_1\cdots\gamma_{n+k}}\ \varepsilon_1^{\gamma_1}\cdots\varepsilon_{n+k}^{\gamma_{n+k}} \tag{5.6}$$

with $D^{(nm)}_{\gamma_1\cdots\gamma_{n+k}}$ being given by (5.5). We also write

$$\varepsilon = |b| = |\beta_1| > 0,$$

and we note that $\varepsilon < (2m - 3)^{-1}$ for, otherwise, the mapping $\phi(\omega)$ becomes multiple-valued (see also [2]). With this notation equation (5.2) becomes

$$\lambda[\varepsilon_n + \sum_{k=1}^{\infty} \varepsilon_k\ \varepsilon_{k+n}\ \varepsilon^{2k}] = \frac{1}{nm}[\varepsilon_n - \sum_{k=1}^{\infty} (km - 1)\varepsilon_k\ \varepsilon_{k+n}\ \varepsilon^{2k}] \tag{5.7}$$

$$+ \sum_{j=1}^{\infty} \varepsilon_{nj}[\varepsilon_j - \sum_{k=1}^{\infty} (km - 1)\varepsilon_k\ \varepsilon_{k+j}\ \varepsilon^{2k}],$$

for all $n \geq 1$.

In order to manage equation (5.7) in an effective way we shall introduce the following quantities

$$A_n = \varepsilon_n + \sum_{k=1}^{\infty} \varepsilon_k\ \varepsilon_{k+n}\ \varepsilon^{2k}, \tag{5.8}$$

$$B_n = \varepsilon_n - \sum_{k=1}^{\infty} (km - 1)\varepsilon_k\ \varepsilon_{k+n}\ \varepsilon^{2k}, \tag{5.9}$$

and

$$C_n = \sum_{j=1}^{\infty} \varepsilon_{nj}[\varepsilon_j - \sum_{k=1}^{\infty} (km - 1)\varepsilon_k\ \varepsilon_{k+j}\ \varepsilon^{2k}]. \tag{5.10}$$

With these quantities equation (5.7) reduces to

$$A_1(B_n + nm\ C_n) = n\ A_n(B_1 + m\ C_1); \quad n \geq 2. \tag{5.11}$$

Equation (5.7) is therefore a system of equations constituting a sequence of integrability conditions in the form of (5.11). The variables ε_2, ε_3, ..., are then determined via (5.11) and the bifurcation parameter λ is given by

$$\lambda = \frac{1}{mA_1}(B_1 + m\ C_1). \tag{5.12}$$

Using (5.3), the angular velocity is then

$$\Omega = \frac{\omega_o}{2m}\ \frac{1}{A_1}\ [m(A_1 - C_1) - B_1]. \tag{5.13}$$

It is more convenient for our purposes to change the order of summation in (5.10). This yields an alternative expression for C_n in the form of

$$C_n = \sum_{k=1}^{\infty} \varepsilon_k\ \varepsilon^{2k}[\varepsilon_{kn} - \sum_{j=1}^{k-1} (jm-1)\varepsilon_j\ \varepsilon_{k-j,n}]. \tag{5.10'}$$

In view of (5.5), (5.6), (5.8), (5.9) and (5.10') we see that the solution of (5.11) must be of the form

$$\varepsilon_n = \sum_{k=0}^{\infty} \mu_n^{(k)}\ \varepsilon^{2k}, \quad n \geq 2, \tag{5.14}$$

where $\mu_n^{(k)}$ are real for all $k \geq 0$. Recalling that $\varepsilon_1 = 1$ we immediately deduce from (5.11) that $\mu_n^{(o)} = 0$ for all $n \geq 2$ and therefore,

$$\varepsilon_n = \sum_{k=1}^{\infty} \mu_n^{(k)}\ \varepsilon^{2k}, \quad n \geq 2. \tag{5.14'}$$

It remains to determine $\mu_n^{(k)}$ for a fixed $n \geq 2$ and all $k \geq 1$. This is done in the usual way be substituting (5.14') in (5.11) and comparing the corresponding powers of ε^{2k}, $k \geq 1$. In this way, $\mu_n^{(1)}$ is determined from the coefficients of ε^2, $\mu_n^{(2)}$ is determined from the coefficients of ε^4 and the $\mu_n^{(1)}$. In general, the equation for the coefficients of ε^{2k} involves $\mu_n^{(k)}$ in a linear way and the previously determined $\mu_n^{(j)}$, $1 \leq j \leq k - 1$.

We shall determine explicity $\mu_n^{(k)}$ for all $n \geq 2$ and for $k = 1, 2, 3$. In order to do so we write

$$\varepsilon_{kn} = \sum_{j=0}^{\infty} \varepsilon^{2j} \varepsilon_{kn}^{(j)}; \quad k, n \geq 1. \tag{5.15}$$

Using (5.8), (5.9) and (5.14') we deduce that, for $n \geq 2$,

$$A_1 = 1 + \mu_2^{(1)} \varepsilon^4 + \mu_2^{(2)} \varepsilon^6 + \ldots,$$

$$B_1 = 1 - (m - 1)\mu_2^{(1)} \varepsilon^4 - (m - 1)\mu_2^{(2)} \varepsilon^6 - \ldots,$$

$$A_n = \mu_n^{(1)} \varepsilon^2 + (\mu_n^{(2)} + \mu_{n+1}^{(1)}\varepsilon^4 + \mu_n^{(3)} + \mu_{n+1}^{(2)})\varepsilon^4 + (\mu_n^{(3)} + \mu_{n+1}^{(2)})\varepsilon^6 + \ldots,$$

$$B_n = \mu_n^{(1)}\varepsilon^2 + [\mu_n^{(2)} - (m-1)\mu_{n+1}^{(1)}]\varepsilon^4 + [\mu_n^{(3)} - (m-1)\mu_{n+2}^{(2)}]\varepsilon^6 + \ldots .$$

Next, using (5.10'), (5.14') and (5.15) we obtain

$$C_1 = \varepsilon_{11}^{(0)} \varepsilon^2 + \varepsilon_{11}^{(1)} \varepsilon^4 + \{\varepsilon_{11}^{(2)} + \mu_2^{(1)}[\varepsilon_{21}^{(0)} - (m-1)\varepsilon_{11}^{(0)}]\}\varepsilon^6 + \ldots,$$

$$C_n = \varepsilon_{1n}^{(0)} \varepsilon^2 + \varepsilon_{1n}^{(1)} \varepsilon^4 + \{\varepsilon_{1n}^{(2)} + \mu_2^{(1)}[\varepsilon_{2n}^{(0)} - (m-1)\varepsilon_{1n}^{(0)}]\}\varepsilon^6 + \ldots .$$

Substituting these quantities in (5.11) and equating the coefficients of

ε^{2k}, for $k = 1, 2, 3,$ to zero we obtain

$$\mu_n^{(1)} + nm\,\varepsilon_{1n}^{(0)} = n\,\mu_n^{(1)}$$

$$\mu_n^{(2)} - (m-1)\mu_{n+1}^{(1)} + nm\,\varepsilon_{1n}^{(1)} = nm\,\varepsilon_{11}^{(0)}\,\mu_n^{(1)} + n(\mu_n^{(2)} + \mu_{n+1}^{(1)})$$

$$\mu_n^{(3)} - (m-1)\mu_{n+1}^{(2)} + nm[\varepsilon_{1n}^{(2)} + \mu_2^{(1)}(\varepsilon_{2n}^{(0)} - (m-1)\varepsilon_{1n}^{(0)}] \tag{5.16}$$

$$+ \mu_2^{(1)}(\mu_n^{(1)} + nm\,\varepsilon_{1n}^{(0)})$$

$$= n\mu_n^{(1)}[m\,\varepsilon_{11}^{(1)} - (m-1)\mu_2^{(1)}] + nm(\mu_n^{(2)} + \mu_{n+1}^{(1)})\varepsilon_{11}^{(0)} + n(\mu_n^{(3)} + \mu_{n+1}^{(2)}).$$

We now compute $\varepsilon_{1n}^{(0)}$, $\varepsilon_{1n}^{(1)}$, $\varepsilon_{1n}^{(2)}$ and $\varepsilon_{2n}^{(0)}$ by appealing (5.5), (5.6), (5.14'), (5.15) and recalling that $\varepsilon_1 = 1$. To do so we first compute $\varepsilon_{1,n-1}$ for $n \geq 2$. Since $\varepsilon_{1,n-1}$ is a polynomial in $\varepsilon_1, \ldots, \varepsilon_n$ we need compute only the quadratic part of $\varepsilon_{1,n-1}$. We must have

$$\gamma_1 + 2\gamma_2 + \ldots + n\,\gamma_n = n.$$

The constant term of $\varepsilon_{1,n-1}$ is due to $(\gamma_1, \ldots, \gamma_n) = (n, 0, \ldots, 0)$, the linear term is due to $(\gamma_1, \ldots, \gamma_n) = (n - k, 0, \ldots, 1_k, 0, \ldots, 0)$ where $2 \leq k \leq n$, the pure quadratic term is due to $(\gamma_1, \ldots, \gamma_n) = (n - 2k, 0, \ldots, 2_k, 0, \ldots, 0)$ with $4 \leq 2k \leq n$ and the mixed quadratic term is due to $(\gamma_1, \ldots, \gamma_n) = (n - k - j, 0, \ldots, 1_k, 0, \ldots, 1_j, 0, \ldots, 0)$ where $2 \leq k < j \leq n$ and $k + j \leq n$. We use the table below in accordance with (5.5) and (5.6).

$(\gamma_1,\ldots,\gamma_n)$	γ	$\gamma-1$	$\frac{(\gamma-1)!}{\gamma_1!\ldots\gamma_n!}$	$D^{(m)}_{\gamma_1\ldots\gamma_n}$
$(n,0,\ldots,0)$	n	$n-1$	$1/n$	$\binom{m-1}{n-1}$
$(n-k,0,\ldots,1_k,0,\ldots,0)$	$n-k+1$	$n-k$	1	$\binom{m-1}{n-k}$
$(n-2k,0,\ldots,2_k,0,\ldots,0)$	$n-2k+2$	$n-2k+1$	$\frac{1}{2}(n-2k+1)$	$\binom{m-1}{n-2k+1}$
$(n-k-j,0,\ldots,1_k,0,\ldots,1_j,0,\ldots,0)$	$n-k-j+2$	$n-k-j+1$	$n-k-j+1$	$\binom{m-1}{n-k-j+1}$

From the last table we obtain

$$\varepsilon_{1,n-1} = \frac{1}{n}\binom{m-1}{n-1} + \sum_{k=2}^{n}\binom{m-1}{n-1}\varepsilon_k + \frac{1}{2}\sum_{k=2}^{[n/2]}(n-2k+1)\binom{m-1}{n-2k+1}\varepsilon_k^2$$

$$+\sum_{k=2}^{[(n-1)/2]}\sum_{j=k+1}^{n-k}(n-k-j+1)\binom{m-1}{n-k-j+1}\varepsilon_k\,\varepsilon_j + \ldots,\quad n\geq 2,$$

and therefore, for $n \geq 1$,

$$\varepsilon^{(0)}_{1n} = \frac{1}{n+1}\binom{m-1}{n}, \tag{5.17}$$

$$\varepsilon^{(1)}_{1n} = \sum_{k=1}^{n}\binom{m-1}{n-k}\mu^{(1)}_{k+1}, \tag{5.18}$$

$$\varepsilon^{(2)}_{1n} = \sum_{k=1}^{n}\binom{m-1}{n-k}\mu^{(2)}_{k+1} + \frac{1}{2}\sum_{k=1}^{[(n-1)/2]}(n-2k)\binom{m-1}{n-2k}[\mu^{(1)}_{k+1}]^2 \tag{5.19}$$

$$+\sum_{k=1}^{[(n-2)/2]}\sum_{j=k+1}^{n-k-1}(n-k-j)\binom{m-1}{n-k-j}\mu^{(1)}_{k+1}\,\mu^{(1)}_{j+1}.$$

In a similar fashion one finds that

$$\varepsilon^{(0)}_{2n} = \frac{1}{n+2}\binom{2m-1}{n+1} - \binom{m}{n+1},\quad n\geq 1. \tag{5.20}$$

We now return to the system of (5.16). Using (5.17)-(5.20) we deduce that, for $n \geq 2$,

$$\mu_n^{(1)} = \frac{n}{n-1} \binom{m}{n+1}, \tag{5.21}$$

$$\mu_n^{(2)} = \frac{mn}{n-1} \sum_{k=1}^{n-2} \frac{k+1}{k} \binom{m-1}{n-k} \binom{m}{k+2} + \left(\frac{n}{n-1}\right)^2 \binom{m}{2} \binom{m}{n+1} + \tag{5.22}$$

$$(m-1) \frac{n+1}{n} \binom{m}{n+2}$$

and

$$\mu_n^{(3)} = \frac{mn}{n-1} \sum_{k=1}^{n-2} \binom{m-1}{n-k} \mu_{k+1}^{(2)} + (m-1)\mu_{n+1}^{(2)} + \frac{n}{n-1} \binom{m}{2} \mu_n^{(2)} \tag{5.23}$$

$$+ \frac{1}{2} \frac{mn}{n-1} \sum_{k=1}^{[(n-1)/2]} (n-2k) \binom{m-1}{n-2k} \left[\frac{k+1}{k} \binom{m}{k+2}\right]^2$$

$$+ \frac{mn}{n-1} \sum_{k=1}^{[(n-2)/2]} \sum_{j=k+1}^{n-k-1} (n-k-j) \binom{m-1}{n-k-j} \frac{k+1}{k} \frac{j+1}{j} \binom{m}{k+2} \binom{m}{j+2}$$

$$+ 2\binom{m}{3} \frac{mn}{n-1} \left\{\frac{1}{n+2} \binom{2m-1}{n+1} - \frac{2m-1}{n+1} \binom{m-1}{n}\right\} - \frac{n+1}{n-1} \binom{m}{2} \binom{m}{n+2}.$$

From these calculations we conclude the following proposition:

PROPOSITION 3. Let $m \geq 2$ be an integer and let b be a complex number with $0 < |b| < (2m - 3)^{-1}$. Then the mapping $\phi \in \sum_m$ given by

$$\phi(\omega) = \omega[1 + \sum_{n=1}^{\infty} \beta_n \omega^{-nm}],$$

where

$$\beta_n = \varepsilon_n b^n; \quad n \geq 1,$$

with

$$\varepsilon_1 = 1, \; \varepsilon_n = \sum_{k=1}^{\infty} \mu_n^{(k)} \varepsilon^{2k}; \quad \varepsilon = |b|, \quad n \geq 2,$$

solves the system (5.2). The coefficients $\mu_n^{(k)}$; $n \geq 2$, $k \geq 1$, are uniquely determined via the scheme of (5.16). The coefficients $\mu_n^{(k)}$ for $k = 1, 2, 3$ and all $n \geq 2$ are given in (5.21)-(5.23). Especially, for $m = 2$ all $\mu_n^{(k)}$; $n \geq 2$, $k \geq 1$, are zero which results in the previously mentioned elliptical vortex of Kirchoff. The bifurcation parameter λ and the angular velocity Ω are determined via (5.12) and (5.13). Explicitly, writing up to ε^8 we have

$$\lambda = \frac{1}{m A_1} (B_1 + m C_1),$$

$$\Omega = \frac{\omega_o}{2m} \frac{1}{A_1} [m(A_1 - C_1) - B_1], \tag{5.24}$$

where

$$A_1 = 1 + \mu_2^{(1)} \varepsilon^4 + \mu_2^{(2)} \varepsilon^6 + [\mu_2^{(3)} + \mu_2^{(1)} \mu_3^{(1)}]\varepsilon^8 + \dots ,$$

$$B_1 = 1 - (m-1)\mu_2^{(1)} \varepsilon^4 - (m-1)\mu_2^{(2)} \varepsilon^6 - [(m-1)\mu_2^{(3)}$$

$$+ (2m-1)\mu_2^{(1)} \mu_3^{(1)}]\varepsilon^8 - \dots ,$$

$$C_1 = \frac{1}{2} (m-1)\varepsilon^2 + \mu_2^{(1)} \varepsilon^4 + [\mu_2^{(2)} - \frac{1}{6} (m-1)(2m-1)\mu_2^{(1)}]\varepsilon^6$$

$$+ [\mu_2^{(3)} + \mu_2^{(1)} \mu_3^{(1)} - \frac{1}{6} (m-1)(2m-1)\mu_2^{(2)} - \frac{1}{24}(m-1)(m-2)(3m-1)\mu_3^{(1)}]\varepsilon^8$$

$$+ \dots ,$$

$$\mu_2^{(1)} = 2\binom{m}{3}, \quad \mu_3^{(1)} = \frac{3}{2}\binom{m}{4},$$

$$\mu_2^{(2)} = 4\binom{m}{2}\binom{m}{3} + \frac{3}{2}(m-1)\binom{m}{4},$$

and

$$\mu_2^{(3)} = (m-1)[9\binom{m}{3}\binom{m}{3} + \frac{9}{4}\binom{m}{2}\binom{m}{4} + \frac{4}{3}(m-1)\binom{m}{5}]$$

$$+ \binom{m}{2}[8\binom{m}{2}\binom{m}{3} + 3(m-1)\binom{m}{4}] - 3\binom{m}{2}\binom{m}{4}$$

$$+ \binom{m}{3}[2\binom{2m}{4} - \frac{4}{3}\binom{m-1}{2}\binom{2m}{2}].$$

§6. Some Calculations

It is interesting to note that the present solution in the form of power series exhibits a "telescopic" character by which fewer terms appear when m is smaller. This can be seen from (5.21)-(5.23). For future applications we record here some numerical values of these coefficients in Table 1 below:

Table 1

m	n	$\mu_n^{(1)}$	$\mu_n^{(2)}$	$\mu_n^{(3)}$
2	≥ 2	0	0	0
	2	2	12	100
	3	0	9	243/2
3	4	0	0	160/3
	≥ 5	0	0	0
	2	8	201/2	3545/2
	3	3/2	315/2	8509/2
	4	0	200/3	29863/6
	5	0	15/2	21925/8
4	6	0	0	638
	7	0	0	203/4
	≥ 8	0	0	0

The actual numerical determination is carried as follows: For an integer $m \geq 2$ we fix the parameter b with $0 < b < (2m - 3)^{-1}$. The boundary of the patch is the closed curve

$$x = \cos\theta + \sum_{n=1}^{\infty} \beta_n \cos(nm-1)\theta, \quad y = \sin\theta - \sum_{n=1}^{\infty} \beta_n \sin(nm-1)\theta; \tag{6.1}$$

$$0 \leq \theta \leq 2\pi,$$

where

$$\beta_1 = b, \; \beta_n = b^n \sum_{k=1}^{\infty} \mu_n^{(k)} b^{2k},; \quad n \geq 2 \tag{6.2}$$

and the $\mu_n^{(k)}$ are determined via the schemes of (5.16) as (5.21)-(5.23) (or Table 1) were obtained. The area of the patch is

$$S = \pi\{1 - \sum_{n=1}^{\infty} (nm-1)\beta_n^2\}$$

while the "major-radius" is

$$r_{max} = x(0) = 1 + \sum_{n=1}^{\infty} \beta_n.$$

The "minor-radius", on the other hand, is evaluated as follows:

$$r_{min}^2 = [x(\tfrac{\pi}{m})]^2 + [y(\tfrac{\pi}{m})]^2$$

$$= \{\cos\tfrac{\pi}{m} + \sum_{n=1}^{\infty} \beta_n \cos(nm-1)\tfrac{\pi}{m}\}^2 + \{\sin\tfrac{\pi}{m} - \sum_{n=1}^{\infty} \beta_n \sin(nm-1)\tfrac{\pi}{m}\}^2,$$

and therefore

$$r_{min} = 1 + \sum_{n=1}^{\infty} (-1)^n \beta_n.$$

The angular velocity Ω may be computed by means of (5.24), where in subsequent quantities A_1, B_1 and C_1, ε is replaced by b. We shall always take $\omega_0 \equiv 1$. The period is then determined by

$$T = \frac{2\pi}{\Omega} .$$

We compute the cases of $m = 3, 4$ and compare them with the numerical results of Deem and Zabusky [5]. The present computation is based on taking the first three terms in the series of (6.2) and, therefore, may serve only as a rough comparison with [5]. A more intensive computation based on taking a larger number of terms in the series of (6.2), by using the scheme of (5.16), will be presented elsewhere. However, even with the present limitation, it is apparent that the critical parameter in the computation is b and not r_{max}/r_{min} as was emphasized in [5]. Of course, when b is very small then $r_{max}/r_{min} \simeq (1+b)/(1-b)$. We use Table 1 and we present our numerical calculations in Table 2 (see also Figures 1 and 2).

The computations in [5] are not normalized as in (6.1). In order that they be commensurable with ours, and, in view of the invariance of the present problem under magnification, we may rescale the quantities r_{min} and r_{max} in [5] by an arbitrary magnification factor of a $(a > 0)$ and, therefore, the area S in [5] will be magnified by a factor of a^2. We choose a so that the areas in [5] will be the same as ours. With such a rescaling our calculations may be compared with those of [5], which for convenience are presented in Table 3. Each of the first two cases of $m = 3$ and $m = 4$ in Table 2 should be compared with the corresponding cases of Table 3. The remaining cases of Table 3 were not considered in [5] as the contour-dynamics algorithm employed there seems to be less effective when $r_{max}/r_{min} \simeq (m-1)/(m-2)$, $m \geq 3$. In these cases, as Figures 1(d) and 2(d) indicate, more than three terms in the series of (6.2) are needed when $b \simeq 1/2(m-1)$, $m \geq 3$.

Table 2

m	b	S	r_{min}	r_{max}	Ω	T
3	0.196818	2.8979560	0.8070	1.2009	0.3141592	20
3	0.234274	2.7954880	0.7741	1.2438	0.3064968	20.5
3	0.250000	2.7465040	0.7613	1.2633	0.3030148	20.73557
3	0.300000	2.5600300	0.7275	1.3362	0.2915219	21.55305
4	0.084926	3.0736120	0.9155	1.0854	0.3695989	17
4	0.147084	2.9371100	0.8576	1.1527	0.3590391	17.5
4	0.166667	2.8778090	0.8417	1.1774	0.3547662	17.71078
4	0.200000	2.7514000	0.8197	1.2292	0.3470325	18.10547

Table 3*

m	S	r_{min}	r_{max}	T	a
3	2.8979560	0.8039422	1.2112424	20	0.7613089
3	2.7954880	0.7716556	1.2591744	20.5	0.7477283
4	3.0736120	0.9055689	1.1000114	17	0.7840423
4	2.9371100	0.8561075	1.1672799	17.5	0.7664366

*The present Table 3 is based on the results of Deem and Zabusky [5], rescaled by the tabulated magnifying factor of a. The actual area S of [5] is always 5, the present S is therefore, $5 \times a^2$. The actual minor and major radii of [5] are then r_{min}/a and r_{max}/a, respectively, with r_{min}, r_{max} and a as given in the present Table 3. The values of the period T are unchanged.

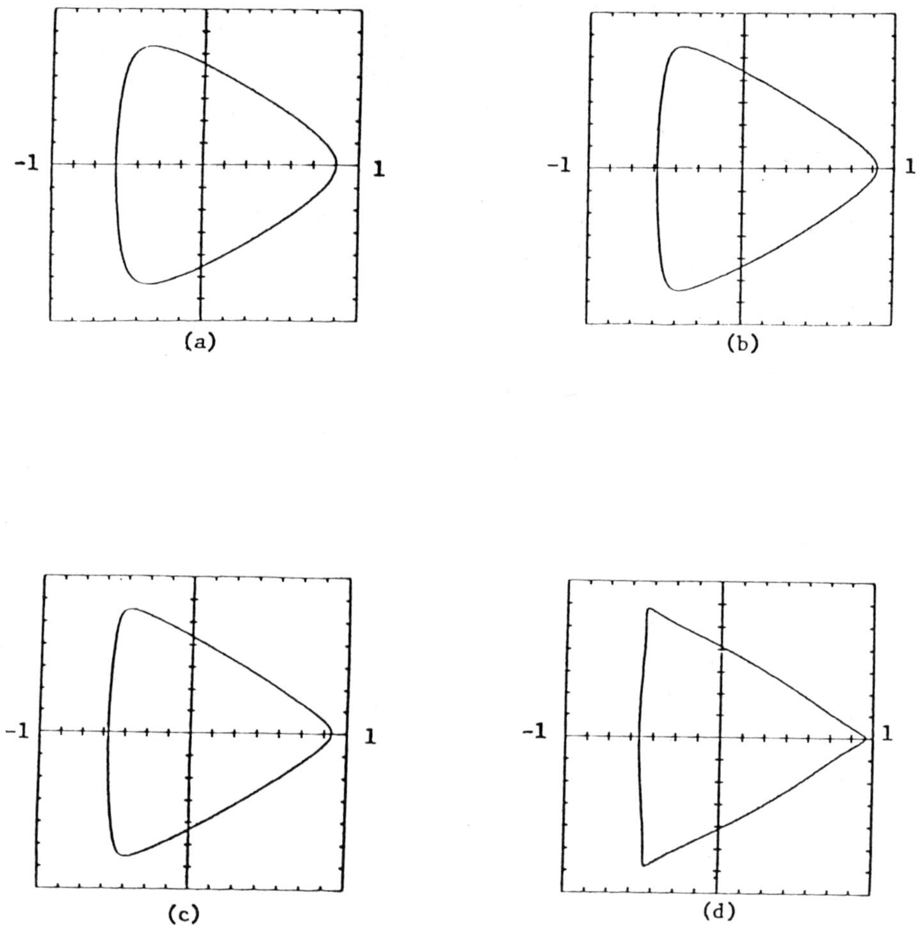

Figure - 1

Rotating m = 3 Kelvin - waves $V_3(b, r_{min}, r_{max}, \Omega)$

(a) $V_3(0.196818, 0.8070, 1.2009, 0.3141592)$

(b) $V_3(0.234274, 0.7741, 1.2438, 0.3064968)$

(c) $V_3(0.250000, 0.7613, 1.2633, 0.3030148)$

(d) $V_3(0.300000, 0.7275, 1.3362, 0.2915219)$.

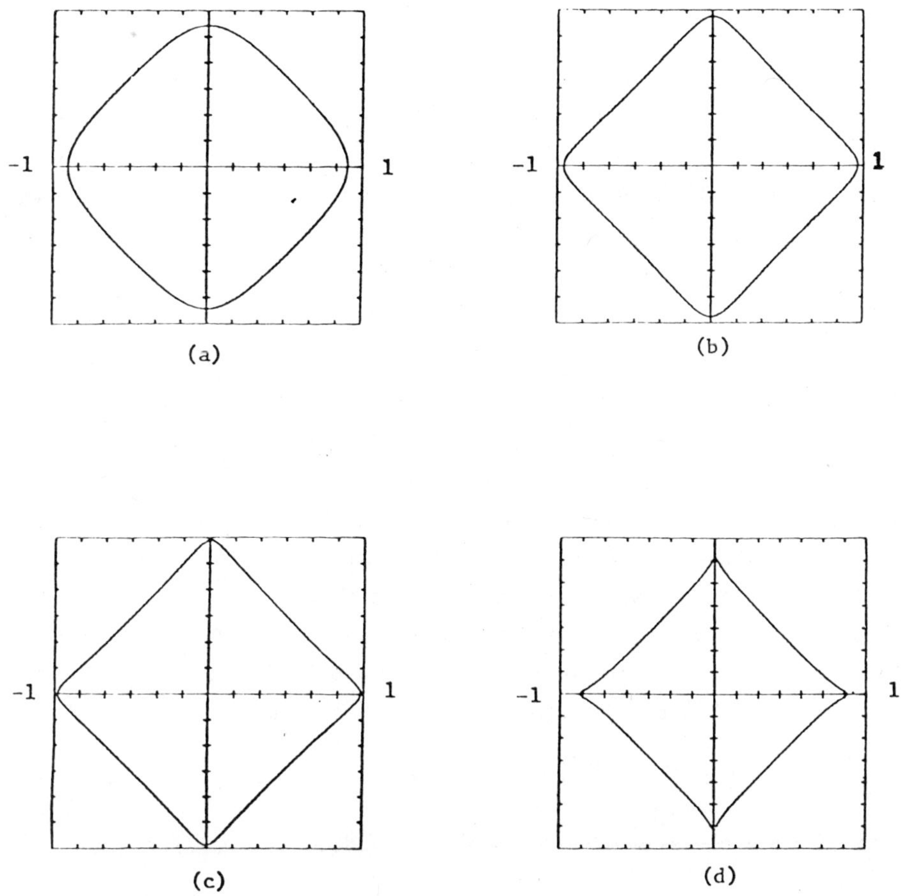

Figure - 2

Rotating m = 4 Kelvin - waves $V_4(b, r_{min}, r_{max}, \Omega)$

(a) $V_4(0.084926, 0.9155, 1.0854, 0.3695989)$

(b) $V_4(0.147084, 0.8576, 1.1527, 0.3590391)$

(c) $V_4(0.166667, 0.8417, 1.1774, 0.3547662)$

(d) $V_4(0.200000, 0.8197, 1.2292, 0.3470325)$.

7. Concluding Remarks

We presented an analytical and "coordinate-free" method for determining the m-waves of Kelvin and their angular velocities Ω. They constitute the solutions of equation (1.3). It is plausible that for $m \geq 3$ these solutions admit a known closed form in terms of some standard transcendental functions. We were not able to identify these proposed functions in the present case but we hope that the methods presented here will shed some light on this question. It is evident that the main difficulty here is connected with the fact that the Grunsky coefficients ε_{nk} are determined by ε in a very complicated way which is exhibited in (5.6) and (5.14'). Nevertheless, it seems to us that the Grunsky coefficients form an essential tool in studying regions of constant vorticity.

Quite recently Su [10] considered a problem similar to ours. His proposed solution is represented in polar coordinates and not in terms of the outer conformal mapping as in the present paper. It is therefore instructive to compare his result with ours. Following Lamb [6, p. 231], Su [10] considers a higher order perturbation solution which corresponds to a distortion of a circular vortex. He states that such a solution may be represented as

$$r(\theta) = a(1 + \delta\, a_1 \cos m\theta + \delta^2\, a_2 \cos 2m\theta + \delta^3 a_3 \cos 3m\theta) + 0(\delta^4) \tag{7.1}$$

where $m \geq 2$ is an integer and δ is a positive number which is sufficiently small. This is computed within the order of δ^3 and he obtains

$$a_2 = \frac{2m-1}{4}\, a_1^{\,2},\quad a_3 = \frac{(3m-1)(m-1)}{8}\, a_1^{\,3} \tag{7.2}$$

and

$$\Omega = \frac{\omega_o}{2}\,\frac{m-1}{m}\,(1 - \frac{m}{2}\, a_1^{\,2}\, \delta^2). \tag{7.3}$$

This, of course, agrees with the previously mentioned linearized result of Lamb, and, as maintained in [10], the above approximate solution can be further improved by adding higher terms in δ.

Specifying our results within $\varepsilon^4 = |b|^4$ our approximate solution is of the form

$$z = a\,\phi(\omega) = a\omega[1 + b\omega^{-m} + 2\,(\tfrac{m}{3})|b|^2\, b^2\, \omega^{-2m}]; \quad \omega = e^{i\theta}, \tag{7.4}$$

and

$$\Omega = \frac{\omega_o}{2}\frac{m-1}{m}\,(1 + 2\,(\tfrac{m}{3})|b|^4)^{-1}(1 - \tfrac{m}{2}\,|b|^2 + 2\,(\tfrac{m}{3})|b|^4). \tag{7.5}$$

This evidently is more precise than (7.1)-(7.3). In fact, as we noted before, (7.4)-(7.5) is a true approximation for the solution, in the sense that it can be indeed improved by adding higher orders of $|b|$. Moreover, for $m = 2$ (7.4)-(7.5) is the exact Kirchoff vortex solution and it reflects the evident fact that the solutions vary drastically with the change of m. These facts are not reflected in (7.1)-(7.3) for there the number m does not play an essential role. Consequently, (7.1) will never converge to the solution. This is indeed the case even if $m = 2$ as we shall see below.

The main problem with (7.1) is that it is not the full Fourier expansion of the solution but rahter only its diagonal "skeleton" (trace). The precise expansion should be

$$f(\theta) = a \sum_{n,k=0}^{\infty} a_{nk}\, \delta^n \cos km\theta; \quad a_{oo} = 1, \tag{7.1'}$$

where (a_{nk}) is an infinite matrix. The exclusion of the sine terms in (7.1') is permitted for we may assume that the m-fold symmetric solution is also symmetric with respect to the x-axis. We shall illustrate these points for the case of $m = 2$.

When $m = 2$, the exact solution is the Kirchoff elliptical solution which may be written as

$$r(\theta) = B(1 - 4\alpha\delta\cos^2\theta)^{-1/2} \tag{7.6}$$

with

$$2\alpha = 1 + B/A, \quad 2\delta = 1 - B/A,$$

where A and B are the major and minor axis, respectively. In this case

$$\Omega = \omega_o \frac{B}{A} \frac{1}{(1+B/A)^2} = \frac{\omega_o}{4} \sum_{n=0}^{\infty} (-1)^n \binom{1/2}{n} \frac{(2n-2)(2n-1)}{(n+1)(n+2)} (4\alpha\delta)^n$$

and thus

$$\Omega = \frac{\omega_o}{4} (1 - \alpha^2 \delta^2 - 4\alpha^3 \delta^3 - 14\alpha^4 \delta^4 - \ldots).$$

This suggests a correction to (7.3) in the term of order δ^3. Further, the expansion of (7.6) in correspondence with (7.1') is

$$r(\theta) = B\{ \sum_{n=0}^{\infty} \frac{1}{n!} \binom{2n}{n} (\tfrac{1}{2})_n (\alpha\delta)^n + 2 \sum_{k=1}^{\infty} [\sum_{n=k}^{\infty} \frac{1}{n!} \binom{2n}{n+k} (\tfrac{1}{2})_n (\alpha\delta)^n] \cos 2k\theta \}.$$

On the other hand the diagonal "skeleton" (trace) of this expansion in accordance with (7.1) is

$$r_o(\theta) = B[1 + 2 \sum_{k=1}^{\infty} \frac{1}{k!} (\tfrac{1}{2})_k (\alpha\delta)^k \cos 2k\theta].$$

This seems to be what Su [10] is considering and, indeed, it does agree with (7.2) when $m = 2$. Clearly, $r_o(\theta)$, which may be also written as

$$r_o(\theta) = B[2\mathrm{Re}(1 - \alpha\delta e^{2i\theta})^{-1/2} - 1],$$

is not an exact solution.

BIBLIOGRAPHY

1. Burbea, J., "Vortex motions and conformal mappings", Nonlinear Evolution Equations and Dynamical Systems, Lecce 1979, Springer-Verlag Lecture Notes in Physics 120 (1980), 276-298.

2. Burbea, J., "Motion of vortex patches", Letters Math. Phys. 6 (1982), (to appear).

3. Burbea, J., "Vortex motions and their stability", Proc. Nonlinear Phenomena in Math. Sci., Arlington 1980, Academic Press, New York, 1982, (to appear).

4. Burbea, J. and Landau, M., "The Kelvin waves in vortex dynamics and their stability", J. Comp. Physics 45 (1982), (to appear).

5. Deem, G. S., and Zabusky, N. J., "Vortex waves, stationary 'V-states', interactions, reccurence and breaking", Phys. Rev. Letters 40 (1978), 854-862.

6. Lamb, H., "Hydrodynamics", Dover Publications, New York, 1945.

7. Miller, A. R., "A machine computation of the Grunsky coefficients of schlicht functions", NRL Report 7225, Naval Research Laboratory, Washington, D.C., 1971.

8. Pommerenke, Ch., "Univalent Functions", Vandenhoeck and Reprecht, Gottingen, 1975.

9. Schur, I1, "On Faber polynomials", Amer. J. Math. 67 (1945), 33-41.

10. Su, C. H., "Motion of fluid with constant vorticity in a singly-connected region", Phys. Fluids 22 (1979), 2032-2033.

ON LIPSCHITZIAN PROXIMITY MAPS*

J. R. Respess and E. W. Cheney

University of Texas at Austin
Austin, Texas
U.S.A.

A proximity map for a subspace G in a normed space X is a map $A : X \to G$ such that

$$\|x - Ax\| = \text{dist}(x, G)$$

for all $x \in X$. Here $\text{dist}(x, G)$ is the infimum of $\|x - g\|$ as g ranges over G. Such an operator A thus produces a best approximation in G for any element in X. A subspace G has a proximity map exactly in the case that G is proximinal; the latter term means that $\inf\|x - g\|$ is attained, for each x, as g ranges over G. Even if G has a proximity map, it need not have a continuous one, and if a proximity map A is continuous, it need not be Lipschitzian. By the latter term we mean that a constant k exists so that for all x and y in X,

$$\|Ax - Ay\| \leq k\|x - y\| . \tag{1}$$

If a subspace has a linear proximity map, then that map is Lipschitzian. This occurs for every closed subspace in Hilbert space and in certain other unusual situations, but in general, linear proximity maps are rare.

In spaces $C(T)$, the situation is rather complicated. Even in

*This work was supported by the U.S. Army Research Office under grant DAAG 29-80-K-0039.

C[0, 1], a satisfactory characterization of subspaces having continuous proximity maps is unknown. Recent progress has been made by Sommer and Nürnberger in [10]. Freud [8] proved in 1958 that the proximity map of C(T) onto a finite-dimensional Haar subspace satisfies a pointwise Lipschitz condition. That means that in Equation (1) above, k is not constant but depends on x. (An n-dimensional subspace in C(T) is called a Haar subspace if each nonzero element of it has at most $n - 1$ zeros in T.) Cline [3] proved that an n-dimensional Haar subspace in C(T) has a Lipschitzian proximity map only if $n = 1$. The converse (presumably well-known) is proved below.

We have discovered the surprising fact that there exists a rich supply of subspaces in C(T) having Lipschitzian proximity maps. These can be constructed with finite or infinite dimension. Our construction in Theorem 7 below produces what may be termed Wulbert Spline Spaces, since these spaces first appeared in the important paper [12].

THEOREM 1. Let H be a one-dimensional Haar subspace in C(T). Then the proximity map of C(T) onto H satisfies the Lipschitz condition $\|Ax - Ay\| \leq k\|x - y\|$, where $2k^{-1} = \inf_{t \in T} |h(t)|$ and h is a generator of norm 1 for H.

Proof. It is known [2, for example] that A is locally Lipschitzian:

$$\|Ax - Ay\| \leq \lambda(x)\|x - y\|$$

where $\lambda(x) = 2/\gamma(x)$, $E(x) = \{t : |x(t) - (Ax)(t)| = \|x - Ax\|\}$, and

$$\gamma(x) = \min_{\substack{g \in H \\ \|g\|=1}} \max_{t \in E(x)} g(t) \operatorname{sgn}[x(t) - (Ax)(t)].$$

If the minimum in this equation is attained by g_0 and the maximum by t_0 then, because $\gamma(x) > 0$, we have

$$\gamma(x) = g_0(t_0) \operatorname{sgn}[x(t_0) - (Ax)(t_0)] = |g_0(t_0)| = |h(t_0)| \geq 2k^{-1}$$

Then $\lambda(x) = 2/\gamma(x) \leq k$. □

It is proved in [3] that a finite-dimensional Haar subspace of dimension greater than 1 in $C(T)$ cannot have a Lipschitzian proximity map. This result can be improved upon slightly as follows.

DEFINITION. Let H be an n-dimensional subspace in $C(T)$. Let $t_0 \in T$. We say that H has the Haar property at t_0 if each neighborhood of t_0 contains n points $t_1, \ldots, t_n$ such that every proper subset of $\{\hat{t}_0|H, \ldots, \hat{t}_n|H\}$ is linearly independent. Here $\hat{t}$ is the point-evaluation functional corresponding to $t : \hat{t}(f) = f(t)$, and $|H$ denotes restriction to H.

THEOREM 2. Let H be an n-dimensional subspace of $C(T)$ having the Haar property at some point. If H has a Lipschitzian proximity map, then $n = 1$.

Proof. Suppose that $n \geq 2$ and that H has the Haar property at t_0. Let A be a proximity map from $C(T)$ onto H. Let $0 < \varepsilon < 1$. We will find functions f and g (depending on ε) such that $\|A(f + g) - Af\| = \frac{1}{\varepsilon}\|g\|$.

Select $h \in H$ so that $\|h\| = 1$ and $h(t_0) = 0$. This is possible because $n \geq 2$. Put $N = \{t : |h(t)| < \varepsilon\}$. Select $t_1, \ldots, t_n \in N$ so that each proper subset of $\{\hat{t}_0|H, \ldots, \hat{t}_n|H\}$ is independent. Since $\dim H = n$, there exist reals $\theta_0, \ldots, \theta_n$ such that $\Sigma\theta_i\hat{t}_i|H = 0$ and $\Sigma|\theta_i| = 1$. Put $\Psi = \Sigma\theta_i\hat{t}_i$. Note that all θ_i are nonzero, that $\|\Psi\| = 1$, and that $\Psi \in H^{\perp}$.

Define $f \in C(T)$ by putting first $f(t_i) = \operatorname{sgn}\theta_i$. Then let $f(t) = h(t) - 1 - \varepsilon$ when $h(t) \geq \varepsilon$, and let $f(t) = h(t) + 1 + \varepsilon$ when $h(t) \leq -\varepsilon$. Finally, extend f by the Tietze Theorem. It is easily verified that $\|f\| = 1$. Hence, $1 \geq \operatorname{dist}(f, H) \geq \Psi(f) = \Sigma\theta_i f(t_i) = \Sigma|\theta_i| = 1$. This shows that 0 is a best approximation to f in H.

In fact, 0 is the only best approximation to f in H. In order to prove this, let $h' \in H\backslash 0$. Then $\|f - h'\| \geq (\operatorname{sgn}\theta_i)(f - h')(t_i) = 1 - (\operatorname{sgn}\theta_i)h'(t_i) > 1$ for some i. In the equation $\Sigma\theta_i h'(t_i) = 0$, no θ_i is 0, and not all $h'(t_i)$ are zero. So at least one summand is negative.

Now define $g \in C(T)$ by letting $g(t) = h(t)$ when $|h(t)| < \varepsilon$, $g(t) = \varepsilon$ when $h(t) \geq \varepsilon$, and $g(t) = -\varepsilon$ when $h(t) < -\varepsilon$. By the arguments used above, h is the unique best approximation to $f + g$. Thus,

$$\|A(f + g) - Af\| = \|h - 0\| = \|h\| = 1 = \frac{1}{\varepsilon}\|g\| . \quad \square$$

Remarks. The preceding proof is essentially that given by Cline. It is easy to give examples of non-Haar subspaces which have the Haar property at some point. For example, let

$$h_i(t) = \begin{cases} t^i & 0 \le t \le 1 \\ 0 & -1 \le t \le 0 \end{cases} .$$

Then the subspace spanned by $\{h_0, \ldots, h_n\}$ has the Haar property at each point of [0, 1]. This subspace is of importance in the theory of spline functions. If T is a space of $n + 1$ points, then each n-dimensional Haar subspace in $C(T)$ has a linear (hence Lipschitzian) proximity map. Cline has shown that if T is finite, each Haar subspace has a Lipschitzian proximity map.

THEOREM 3. If $n \ge 3$, the linear space of continuous piecewise linear functions with n prescribed knots $(t_1 < \ldots < t_n)$ does not have the Haar property at any point. However, it is a weakly Chebyshev subspace of $C[t_1, t_n]$.

<u>Proof</u>. Let G denote the subspace in question. Suppose that G contains an element g such that $g(s_i)g(s_{i-1}) < 0$ $(1 \le i \le n)$ for some ordered set of points $s_0 < s_1 < \ldots < s_n$ in $[t_1, t_n]$. Then g changes sign in at least n points of the interval (t_1, t_n). It follows that some interval $[t_i, t_{i+1}]$ $(1 \le i \le n - 1)$ must contain <u>two</u> points where g changes sign. This is not possible because g is a polynomial of degree 1 on $[t_i, t_{i+1}]$. Hence G is a weakly-Chebyshev subspace.

Suppose that G has the Haar property at some point s_0. Let N be a connected neighborhood of s_0 such that $N\setminus\{s_0\}$ contains no knot. By the definition of the Haar property at s_0, there exist points $s_1 \ldots s_n \in N$ (different from s_0) such that each proper subset of $\{\hat{s}_0|G, \ldots, \hat{s}_n|G\}$ is linearly independent. If s_0 is not a knot, then N is contained in one of the intervals (t_i, t_{i+1}). Since the elements of G are linear on each such interval, $\dim(G|N) = 2$. Since $n \ge 3$, each proper subset of $\{\hat{s}_0|G, \ldots, \hat{s}_n|G\}$ contains at least three elements and must be linearly dependent.

If s_0 is a knot, say $s_0 = t_i$, then N is contained in (t_{i-1}, t_{i+1}). Since $n \ge 3$, one of the two following sets must contain at least three elements

$$[t_i, t_{i+1}] \cap \{s_0, \ldots, s_n\}$$

$$[t_{i-1}, t_i] \cap \{s_0, \ldots, s_n\} .$$

Hence as in the previous case, the corresponding three point functionals must be dependent over G. □

THEOREM 4. For each $n \geq 4$ there is an n-dimensional subspace in $C[0, 1]$ which has the Haar property at 0 but at no other point.

Proof. Put $s_j = 1/j$ and $t_j = \frac{1}{2}(s_j + s_{j+1})$ for $j = 1, 2, \ldots$. Let $S = \{t_1, t_2, \ldots\} \cup \{s_1, s_2, \ldots\}$. Define $h_0 = 1$, and for $i = 1, 2, \ldots, n - 1$, let h_i be a continuous function such that

(1) $h_i(t_j) = t_j^i$

(2) $h_i(s_j) = 0 = h_i(0)$

(3) h_i is linear on the subintervals created by points of S.

Define H as the linear span of $\{h_0, \ldots, h_{n-1}\}$.

To prove that H has the Haar property at 0, let U be a neighborhood of 0, and select m so that $t_m, t_{m+1}, \ldots$ all belong to U. Using $t_0 = 0$, we observe that the matrix with rows $[h_0(t_i), h_1(t_i), \ldots, h_{n-1}(t_i)]$ $(i = 0, m, m + 1, \ldots)$ is a Vandermonde matrix. Hence, each proper subset of $\{\hat{t}_0|H, \hat{t}_m|H, \ldots, t_{m+n-1}|H\}$ is linearly independent.

To prove that H does not have the Haar property at ξ when $\xi \neq 0$, consider first the case $\xi \in S$. Let U be an open interval containing ξ but no other point of S. On U, each h_i is a piecewise linear function having one joint (at ξ). If $\xi = t_j$, then $h_i|U$ is symmetric with respect to t_j. If $\xi = s_j$, then each element h_i $(i > 0)$ vanishes at ξ. Using these facts, it is easy to see that $\{h_i|U\}_{i=0}^n$ generates a space of dimension at most 3.

If $\xi \notin S$, let U be an open interval containing no points of S. On U, each h_i is linear, and $\{h_i|U\}_0^n$ generates a space of dimension at most 2. □

THEOREM 5. The 2-dimensional subspace G in $C[-1, 1]$ spanned by

$g_1(t) = 1$ and $g_2(t) = t^2$ is not a Haar subspace, but it has the Haar property at each point.

Proof. Since $g_1 - g_2$ has $+1$ and -1 as zeros, G is not a Haar subspace. Let t_0 be any point of $[-1, 1]$, and let N be any neighborhood of t_0. Select t_1 and t_2 in N, distinct from t_0, and such that $t_1^2 \neq t_2^2$. Then each proper subset of $\{\hat{t}_0|G, \hat{t}_1|G_1, \hat{t}_2|G\}$ is linearly independent. For example,

$$\begin{vmatrix} g_1(t_1) & g_2(t_1) \\ g_1(t_2) & g_2(t_2) \end{vmatrix} = \begin{vmatrix} 1 & t_1^2 \\ 1 & t_2^2 \end{vmatrix} = t_2^2 - t_1^2 \neq 0. \quad \square$$

THEOREM 6. For each $n = 1, 2, 3, \ldots$ there exists a compact Hausdorff space T and an n-dimensional subspace G in $C(T)$ which has a Lipschitzian proximity map.

Proof. Let T consist of n disjoint closed sets, and let G be the space of functions which are constant on the components of T. Given $x \in C(T)$, define $g \in G$ by stipulating that on any component of T the constant value of g should be the average of the maximum and minimum values of x on that component. Then g is a best approximation to x. (Generally, x will have many best approximations.) The map $x \mapsto g$ is a Lipschitzian proximity map. $\square$

The next theorem provides a large category of infinite-dimensional proximinal subspaces in a space $C(T)$. These are closely related to spline functions, but are much freer in their structure. We deal with an arbitrary normal space T, not necessarily compact. Then $C(T)$ is the space of bounded real continuous functions on T. Let $\{T_i : i \in I\}$ be an upper-semicontinuous decomposition of T into closed sets. Thus, each T_i is closed, $T_i \cap T_j = \emptyset$ when $i \neq j$, $T = \cup T_i$, and if $d(t) = T_i$ when $t \in T_i$, then there corresponds to any neighborhood U of $d(t_0)$ a neighborhood V of t_0 such that $\cup\{d(t) : t \in V\} \subset U$.

Now we suppose that in each space $C(T_i)$ a proximinal linear subspace H_i containing the constants has been prescribed. The "Wulbert spline space" in this situation is then

$$H = \{x \in C(T) : x|T_i \in H_i \quad \text{for all} \quad i\}.$$

The elements of H are then piecewise H_i-functions, so fitted together as to achieve global continuity.

One final assumption is made concerning the decomposition of T. Calling T_j "plural" if it contains 2 or more elements, we assume that each plural set has a neighborhood which intersects no other plural set. In brief, there are no plural limit sets.

THEOREM 7. Under the above circustances, the Wulbert splines form a proximal subspace in C(T).

<u>Proof</u>. If $x \in C(T)$ and $h \in H$, then

$$\|x - h\| \geq \sup_j \|x|T_j - h|T_j\| \geq \sup_j \text{dist}(x|T_j, H_j) = \rho(x)$$

where $\rho(x)$ is defined by the last equation. This calculation establishes that $\text{dist}(x, H) \geq \rho(x)$, and the remainder of the proof will show that for each x there is an $h \in H$ for which $\|x - h\| = \rho(x)$.

It is convenient to describe h by means of certain operators as follows:

1. $A_j : C(T_j) \to H_j$ proximity map
2. $R_j : \ell_\infty(T) \to \ell_\infty(T_j)$ restriction map
3. $E : C(T^*) \to C(T)$ Tietze extension map (here T^* is the closure of the union of the plural sets)
4. $R : \ell_\infty(T) \to \ell_\infty(T^*)$ restriction map
5. $B : C(T) \to \ell_\infty(T)$ defined by $R_jB = A_jR_j$, all $j \in J$
6. $A : C(T) \to H$ defined by $A = I + E(RB - R)$.

Observe that if Bx is continuous, then it is a best approximation to x in H since $\|x - Bx\| = \rho(x)$. However, Bx need not be continuous, and the more elaborate mapping A is required to produce a best approximation.

Let $z = Bx$. We begin by showing that Rz is continuous. Let $s \in T^*$. In the first case, suppose s belongs to a plural set T_j. By one of our hypotheses, there is a neighborhood U of T_j which intersects no other plural sets. Let N be any neighborhood of z(s). Since $R_jz = A_jR_jx \in C(T_j)$, there is a neighborhood V of s in T such that $z(T_j \cap V) \subset N$. If $t \in V \cap U \cap T^*$, then t is the limit of a net of points belonging to plural sets. We may assume that such a net lies in U. By the properties of U, these plural sets can be none other than T_j. Since T_j is closed, $t \in T_j$. Hence $z(t) = (R_jz)(t) \in N$.

In the second case, suppose that $\{s\} = T_j$ for some j. Given $\varepsilon > 0$, let U be a neighborhood of s on which the variation of the function x is less then ε. By the upper-semicontinuity of the decomposition, there is a neighborhood V of T_j such that $d(t) \subset U$ whenever $t \in V$. Let $t \in T^* \cap V \cap U$, and let $d(t) = T_k$. Then $T_k \subset U$, and the variation of $R_k x$ is less then ε. A constant function can then approximate x on T_k with an error at most $\varepsilon/2$. It follows that $\|R_k z - R_k x\| \le \varepsilon/2$ because H_k contains the constants and $R_k z$ is a best approximation to $R_k x$ from H_k. Hence

$$|z(s) - z(t)| \le |x(s) - x(t)| + |x(t) - z(t)| \le 3\varepsilon/2.$$

This concludes the proof that $z|T^*$ is continuous.

Now consider $h = x + E(Rz - Rx)$. Certainly h is continuous. If T_j is any plural set, then $R_j h = R_j x + R_j z - R_j x = R_j z \in H_j$. If T_j is not a plural set, then $R_j h \in H_j$ since $1 \in H_j$. Thus $h \in H$. Finally, since the Tietze extension does not increase the norm, we have

$$\|h - x\| = \|E(Rz - Rx)\| = \|Rz - Rx\| \le \|z - x\| = \rho(x). \quad \square$$

THEOREM 8. In the situation of the preceding theorem, if T is metrizable, and if each H_j has a Lipschitzian proximity map with a common Lipschitz constant, then H has a Lipschitzian proximity map.

Proof. The construction in the proof of Theorem 7 is used here. The extension map $E : C(T^*) \to C(T)$ can now be chosen to be linear and of norm 1, by the Borsuk-Dugundji Theorem [see 11, p. 365 or 6]. Now we have, for $x, y \in C(T)$:

$$\|Bx - By\| = \sup_j \|R_j Bx - R_j By\| = \sup_j \|A_j R_j x - A_j R_j y\|$$

$$\le \sup_j \lambda \|R_j x - R_j y\| = \lambda \|x - y\| . \quad \square$$

Here λ is the common Lipschitz constant for the maps A_j. It is now clear that $\|Ax - Ay\| \le (1 + \lambda) \|x - y\|$.

In 1933, von Neumann [9] described an algorithm for producing best approximations from a vector sum of subspaces. This algorithm, known as the "alternating-direction" method, goes as follows. Suppose that U_1 and U_2 are subspaces of a normed space X, and suppose that proximity maps $M_i : X \to U_i$ are available. Starting with an element x_0 whose best approximation in $U_1 + U_2$ is sought, we form a sequence $x_0, x_1, \ldots,$ by the formulas

$$x_{2n} = x_{2n-1} - M_1 x_{2n-1} \qquad x_{2n+1} = x_{2n} - M_2 x_{2n} .$$

When certain hypotheses are fulfilled, the sequence $x_0 - x_n$ converges to a best approximation of x_0 in $U_1 + U_2$.

In the space $C(S \times T)$, if $U_1 = C(S)$ and $U_2 = C(T)$, the algorithm is effective. The papers of Diliberto and Straus [5] and Aumann [1] give the first proof of this fact. See [4, 7] for related work. We will prove here that the algorithm converges in a number of other cases in the space $C(S \times T)$. The main technique for constructing the proximity maps is described first.

LEMMA. If M is a continuous map of $C(T)$ into $C(T)$, then the map $\overline{M}$ defined on $C(S \times T)$ by the equation $(\overline{M}f)(s, t) = (Mf_s)(t)$ is a continuous map of $C(S \times T)$ into $C(S \times T)$. If M is a proximity map, then so is $\overline{M}$.

Proof. Define $\Phi : C(S \times T) \to C(S, C(T))$ by the equation

$$(\Phi f)(s) = f_s \qquad s \in S, \qquad f \in C(S \times T).$$

By Grothendieck's Theorem [11, p. 357], Φ is an isometry. The definition of $\overline{M}$ given above is equivalent to

$$\overline{M}f = \Phi^{-1}(M \circ (\Phi f)).$$

Hence $\overline{M}$ maps $C(S \times T)$ continuously into $C(S \times T)$.

If M is a proximity map, then

$$\|x - Mx\| \le \|x - My\| \qquad (\text{all } x, y)$$

and conversely. Consequently,

$$\|f - \overline{M}f\| = \sup_s \sup_t |f(s, t) - (\overline{M}f)(s, t)|$$

$$= \sup_s \|f_s - Mf_s\| \le \sup_s \|f_s - MF_s\|$$

$$= \|f - \overline{M}F\| \quad (\text{all } f, F). \quad \square$$

In the following, we have two compact spaces, S and T, each having upper-semicontinuous decompositions. Further structures are defined as in the previous paragraphs. We summarize the notation in the table which follows.

	Space S	Space T
decomposition	$\{S_i : i \in I\}$	$\{T_j : j \in J\}$
approximating subspaces	$G_i \subset C(S_i)$	$H_j \subset C(T_j)$
Wulbert spline space	G	H
Proximity maps	$A_i^S : C(S_i) \twoheadrightarrow G_i$	$A_j^T : C(T_j) \twoheadrightarrow H_j$
Restriction maps	$R_i^S : \ell_\infty(S) \to \ell_\infty(S_i)$	$R_j^T : \ell_\infty(T) \to \ell_\infty(T_j)$
closure of union of plural sets	S^*	T^*
Borsuk-Dugundji maps	$E^S : C(S^*) \to C(S)$	$E^T : C(T^*) \to C(T)$
Restriction maps	$R^S : \ell_\infty(S) \to \ell_\infty(S^*)$	$R^T : \ell_\infty(T) \to \ell_\infty(T^*)$
Proximity maps	$B^S : C(S) \to \ell_\infty(S)$ $R_i^S B^S = A_i^S R_i^S$ all i	$B^T : C(T) \to \ell_\infty(T)$ $R_j^T B^T = A_j^T R_j^T$ all j
Proximity maps	$A^S : C(S) \to G$ $A^S = I + E^S(R^S B^S - R^S)$	$A^T : C(T) \to H$ $A^T = I + E^T(R^T B^T - R^T)$

Now the proximity maps A^S and A^T can be extended as in the lemma to proximity maps $\overline{A}^S$ and $\overline{A}^T$ which act on $C(S \times T)$. Their ranges are respectively $G \otimes_\lambda C(T)$ and $C(S) \otimes_\lambda H$. Here the symbol $\otimes_\lambda$ denotes the completion of the tensor-product space with respect to Schatten's λ-norm. It is the norm which makes the equation

$$C(S, E) = C(S) \otimes_\lambda E$$

valid for arbitrary Banach spaces E.

The alternating direction algorithm is now given by

$$F_0 \in C(S \times T), \; F_{2n} = F_{2n-1} - \overline{A}^S F_{2n-1}, \; F_{2n+1} = F_{2n} - \overline{A}^T F_{2n}.$$

Within the space $C(S_i \times T_j)$ there is also an alternating direction algorithm given by

$$f_0 \in C(S_i \times T_i),\ f_{2n} = f_{2n-1} - \bar{A}_j^T f_{2n-1},\ f_{2n+1} = f_{2n} - \bar{A}_i^S f_{2n}.$$

THEOREM 9. If S_i and T_j are plural sets in the decompositions and if f in the previous equation is $F|(S_i \times T_j)$ then $F_n|(S_i \times T_j) = f_n$.

Proof. The asserted equation is true for $n = 0$, by the definitions. Suppose that the equation is true for $n = 2k$. We verify now that the equation is valid for $n = 2k + 1$. Let $(s, t) \in S_i \times T_j$. Then

$$\begin{aligned} F_{2k+1}(s, t) &= F_{2k}(s, t) - (\bar{A}^S F_{2k})(s, t) \\ &= F_{2k}^t(s) - (A^S F_{2k}^t)(s) \\ &= -[E^S(R^S B^S - R^S) F_{2k}^t](s) \\ &= (-A_i^S R_i^S F_{2k}^t + R_i^S F_{2k}^t)(s) \\ &= (-A_i^S f_{2k}^t + f_{2k}^t)(s) = f_{2k+1}(s, t). \end{aligned}$$

A similar proof applies if the asserted equation is assumed true for $n = 2k + 1$. □

THEOREM 10. In the situation of the preceding theorem, suppose further that the decomposition sets S_i and T_j are all open, closed, and plural. Suppose G_i and H_j are the 1-dimensional spaces of constants. Then the alternating direction method produces best approximations in

$$Z = G \otimes_\lambda C(T) + C(S) \otimes_\lambda H.$$

Proof. Fix (i, j) and let $f_0 = F_0|(S_i \times T_j)$, with $F_0 \in C(S \times T)$. By the preceding theorem

$$F_n|(S_i \times T_j) = f_n.$$

By the theorem of Aumann [1], the sequence f_n converges uniformly to a function which may be denoted by $F|(S_i \times T_j)$. We know that $(F_0 - F)|(S_i \times T_j)$ is a best approximation of f by elements of $C(S_i) + C(T_j)$.

The function F is continuous because $F|(S_i \times T_j)$ is continuous and $S_i \times T_j$ is open.

Next we observe that

$$(F_0 - F)|(S_i \times T_j) \in C(S_i) + C(T_j).$$

This is true because the same assertion is true for F_n in place of F, and F_n converges uniformly on $S_i \times T_j$.

In order to verify that $F_0 - F$ is a best approximation to F_0 in Z, we note that

$$\operatorname{dist}(F_0, Z) \geq \sup_{ij} \operatorname{dist}(F_0|(S_i \times T_j), C(S_i) + C(T_j)).$$

But

$$\begin{aligned} \|F_0 - (F_0 - F)\| = \|F\| &= \sup_{ij} \|F|(S_i \times T_j)\| \\ &= \sup_{ij} \|[F_0 - (F_0 - F)]|(S_i \times T_j)\| \\ &= \sup_{ij} \operatorname{dist}(F_0|(S_i \times T_j), C(S_i) + C(T_j)). \quad \square \end{aligned}$$

REFERENCES

1. G. Aumann, "Uber approximative Nomographie, II", Bayer. Akad. Wiss. Math. Nat. Kl. S.B., 1959, 103-109. MR 22#6968.
2. E. W. Cheney, Introduction to Approximation Theory, McGraw-Hill Book Co., New York, 1966.
3. A. K. Cline, "Lipschitz conditions on uniform approximation operators", J. Approximation Theory 8(1973), 160-172. Also 14 (1975), 245-250.
4. F. Deutsch, "The alternating method of von Neumann", pp. 83-96 in Multivariate Approximation Theory, ed. by W. Schempp and K. Zeller, ISNM vol. 51, Birkhauser Verlag, Basel, 1979.
5. S. P. Diliberto and E. G. Straus, "On the approximation of a function of several variables by the sum of functions of fewer variables", Pacific J. Math. 1 (1951), 195-210. MR 13#334.
6. J. Dugundji, "An extension of Tietze's theorem", Pacific J. Math. 1 (1951), 353-367. MR13#373.
7. N. Dyn, "A straightforward generalization of Diliberto and Straus' algorithm does not work", J. Approximation Theory, 30 (1980), 247-250.

8. G. Freud, "Eine Ungleichung für Tschebysheffsche Approximations-polynome", Acta Scientiarum Math. (Szeged) 19 (1958), 162-164.

9. J. von Neumann, "Functional operators", vol. II, Annals of Math. Studies 21 (1950), Princeton University Press.

10. G. Nürnberger and M. Sommer, "Weak Chebyshev subspaces and continuous selections for the metric projection", Trans. Amer. Math. Soc., to appear.

11. Z. Semadeni, Banach Spaces of Continuous Functions, Polish Scientific Publishers, Warsaw, 1971.

12. D. E. Wulbert, "Some complemented function spaces in C(S)", Pacific J. Math. 24 (1968), 589-602. MR 36#6915.

ON ULTIMATELY NONEXPANSIVE SEMIGROUPS

M. Edelstein

Dalhousie University
Halifax, Nova Scotia
Canada

M. T. Kiang

St. Mary's University
Halifax, Nova Scotia
Canada

Introduction

Let G be a family of selfmappings of a metric space (X, d) and suppose that G is a semigroup under composition. For any $x \in X$, $G(x) = \{g(x) : g \in G\}$ is called the orbit of x under G and a point $z \in X$ such that $G(z) = \{z\}$ is called a fixed point of G. Fixed point properties of semigroups having various contractivity properties have been investigated by several writers (cf. [1], [3], [4], [5], [6]).

Here we discuss semigroups of mappings related to those of [5] but much more general; (e.g., members of the semigroups need not be lipschitzian).

DEFINITION. [2] A semigroup G of continuous selfmappings of a metric space (X, d) is called ultimately nonexpansive if for every pair of points $u, v \in X$ and for every $\alpha > 0$, there is an $f \in G$ such that for all $g \in G$,

$$d(fg(u), fg(v)) \leq (1 + \alpha)d(u, v) \tag{1}$$

It is called asymptotically nonexpansive (cf [3]) if (1) above is satisfied with $\alpha = 0$.

Our main results are as follows:

THEOREM I. Let X be a reflexive Banach space and G an ultimately nonexpansive commutative semigroup of self-mappings of X. If each orbit under G is precompact, then G has a fixed point.

THEOREM II. Let X be a finite dimensional Banach space and G a commutative ultimately nonexpansive semigroup of self-mappings of X. If $G(y)$ is bounded for some $y \in X$, then G has a fixed point.

THEOREM III. Let G be an ultimately nonexpansive commutative semigroup of self-mappings of a reflexive locally uniformly convex Banach space X and suppose that $G(x)$ is precompact for some $x \in X^G$. Then the Chebyshev center of $G(x)$ is a singleton c and $G(c) = \{c\}$ i.e. a fixed point under G.

The proofs of the above theorems are given in [2]. In the following, we state some results used in our proofs.

DEFINITION. The G-closure of X, where (X, d) is a metric space, consists of all points x in X such that for some $z \in X$, $\varepsilon > 0$, and any $f \in G$, there is a $g \in G$ such that $d(fg(z), x) < \varepsilon$. The G-closure of X is denoted by X^G.

The following are results related to X^G and an orbit $G(x)$ when $x \in X^G$.

PROPOSITION 1. Let G and (X, d) be as in proposition 1. Then,
(a) if $x \in X^G$, then the restriction $G|G(x)$ of G to $G(x)$ is a semigroup of isometries
(b) if $\mathrm{cl}G(x)$ is compact, then $G|\mathrm{cl}G(x)$ is a semigroup of isometries (of $\mathrm{cl}G(x)$ onto itself).

PROPOSITION 2. Let X and G be as in proposition 1, then
(a) if $u \in \mathrm{cl}\, G(x)$ with $x \in X^G$ and $v \in X$, $v \neq u$, there is a neighbourhood U of u such that for any positive integer n, there exists an $f \in G$ with the property that for all $g \in G$ and $u' \in U$,

$d(fg(u'), fg(v)) \leq (1 + \frac{1}{n})\, d(u', v)$

(b) if cl $G(x)$ is compact with $x \in X^G$ and $v \neq$ cl $G(x)$, then for any positive integer n, there is a $f \in G$ such that for all $g \in G$ and $u \in$ cl $G(x)$, $d(fg(u), fg(v)) \leq (1 + \frac{1}{n})\, d(u, v)$.

The following are results related to the centers of orbits $G(x)$.

DEFINITION. The Chebyshev center C_A of $G(x)$ with respect to A is the subset of A consisting of all points u such that the radius $R(u)$ of the smallest closed ball about u containing $G(x)$ is minimal, i.e. if

$$R(u) = \sup\{\|u - g(x)\| : g \in G\}$$

$$R_A = \inf\{R(u) : u \in A\}$$

then

$$C_A = \{u \in A : R(u) = R_A\}.$$

LEMMA 1. If $G(x)$ is precompact with $G(x) \neq x$ and $x \in X^G$, then cl $G(x)$ and C_A are disjoint for any closed convex subset A of X.

LEMMA 2. Let C_A be the Chebyshev center of a precompact orbit $G(x)$ where $x \in X^G$, with respect to a closed convex set $A \subset X$ and suppose that $c \in C_A$. If $G(c)$ is a precompact subset then there is a $\bar{c} \in$ cl $G(c)$ such that $G(\bar{c}) \subset C_A$.

LEMMA 3. Let X be a strictly convex Banach space and suppose that for some $x \in X$, $G(x)$ is precompact. Then the Chebyshev center C of $G(x)$ with respect to X consists of one point at most.

REFERENCES

1. M. Edelstein, "Fixed point theorems in uniformly convex Banach spaces", Proc. Amer. Math. Soc. 44 (1974), 369-374.

2. M. Edelstein and M. T. Kiang, "On ultimately nonexpansive semigroups", to appear.

3. R. D. Holmes and P. P. Narayanaswami, "On Asymptotically-nonexpansive semigroups of mappings", Canad. Math. Bull. 13 (1970), 209-214.

4. M. T. Kiang, "Fixed point theorems for certain classes of semigroups of mappings", Trans. Amer. Math. Soc. 189 (1974), 63-76.

5. M. T. Kiang, "A fixed point theorem for eventually nonexpansive semigroups of mappings", J. Math. Analy. and Appl. 56 (1976), 567-569.

6. T. C. Lim, "A fixed point theorem for families of nonexpansive mappings", Pac. J. Math. 53 (1974), 487-493.

NONLINEAR ANALYSIS IN DIFFERENT KINDS OF FRÉCHET SPACES*

E. Dubinsky

Clarkson College of Technology
Potsdam, New York
U.S.A.

Introduction

The scope of this lecture is somewhat more restricted than the title might indicate. The non-linear analysis concerns the inverse function theorem and the method to be considered is Newton's successive approximation scheme with the Nash-Moser modification. By the phrase, "different kinds of Fréchet spaces", I am referring to a classification theory that has been in the process of development by several authors over the past few years.

The inverse function theorem has been well studied in Banach spaces (for example, see [10], p. 20). The fundamental results is that _if a smooth function has a derivative which is an isomorphism at some point, then it is a diffeomorphism in some neighborhood of that point_. This result is false in Fréchet spaces and counterexamples are now known to exist in many different concrete spaces (see J. Eells [4], S. Łojasiewicz and E. Zehnder [7]). Nevertheless, using the Nash-Moser method it is possible to prove certain versions of the inverse function theorem valid in certain spaces for functions satisfying certain conditions.

Most of the research in this area always concerns what is, up to isomorphism, the same space - $C^\infty[-1, 1]$ (see below for definitions). The original goal of the investigations reported here was to see what would happen if a different space was used. It turns out that the restriction

*The research reported here was partially supported by the U.S. National Science Foundation.

to $C^\infty[-1, 1]$ is no accident - it is forced by the estimates used in the Nash-Moser method. On the other hand, as I will show, there are variations of these estimates which can be used to obtain inverse function theorems valid in other spaces. Unexpectedly, the exact form of the theorem seems to depend quite heavily on the particular Fréchet space involved and this relates to some interesting questions about Fréchet spaces. Perhaps the main point that I have to make is that there is a somewhat surprising connection between non-linear analysis and the classification theory mentioned above. This report describes some first applications and I hope that future research will lead to a more thorough exploitation of this connection.

The Classification Theory

It is impossible to give, in this space, a detailed description of the classification or structure theory for Fréchet spaces which has been developed over the last decade. For references one can consult works of D. Vogt, M. J. Wagner and the author [3], [11], [12].

Recall that a Fréchet space E is a complete, metrisable locally convex space. It's topology is defined by a sequence of seminorms $(\|\cdot\|_k)_{k\geq 0}$ and it is usually assumed that $\|x\|_k \leq \|x\|_{k+1}$ $(x \in E,\ k \geq 0)$. An important class of these spaces are the <u>nuclear Fréchet spaces</u> (see [9] for definitions and elementary properties).

Roughly speaking, the classification theory is an attempt to determine exactly which spaces can be subspaces, quotients or complemented subspaces of a given nuclear Fréchet space E. This is done in terms of inequalities on the seminorms. There are some spaces E for which the classification is complete.

As a replacement for the details of the classification theory, we give a list of some important nuclear Fréchet spaces, mention how the classification theory connects them, and give one result with some consequences that will be used later.

First, a list of some nuclear Frechet spaces.

1. $C^\infty(\mathbb{R})$ - the infinitely differentiable real (or complex) valued functions on $\mathbb{R}$. The seminorms are given by

$$\|x\|_k = \sup_{\substack{|t|\le k \\ p\le k}} |x^{(p)}(t)| \qquad (x \in C^\infty(\mathbb{R}),\ k \ge 0).$$

2. $C^\infty[-1, 1]$ - the infinitely differentiable functions on $[-1, 1]$ (with right and left derivatives at $-1, 1$ respectively). The seminorms are given by

$$\|x\|_k = \sup_{\substack{|t|\le 1 \\ p\le k}} |x^{(p)}(t)| \qquad (x \in C^\infty[-1, 1],\ k \ge 0).$$

3. $H(\mathbb{C})$ - the entire functions in one complex variable. The seminorms are given by

$$\|x\|_k = \sup_{|t|\le k} |x(t)| \qquad (x \in H(\mathbb{C}),\ k \ge 0).$$

4. $H_\alpha(\mathbb{C})$ - entire functions with gaps off α. Here $\alpha = (\alpha_n)$ is a fixed, strictly increasing sequence of non-negative integers and $H_\alpha(\mathbb{C})$ consists of those $x \in H(\mathbb{C})$ whose Taylor's series coefficients $\frac{x^{(n)}(0)}{n!}$ vanish except at the indices α_n $(n = 0, 1, 2, \ldots)$. The seminorms are the restrictions of the seminorms on $H(\mathbb{C})$.

5. $H(\mathbb{D})$ - The complex functions analytic on the interior of the unit disk. The seminorms are given by

$$\|x\|_k = \sup_{|t|\le \frac{k}{k+1}} |x(t)| \qquad (x \in H(\mathbb{D}),\ k \ge 0).$$

6. $H_\alpha(\mathbb{D})$ - Defined as a subspace of $H(\mathbb{D})$ analogously to $H_\alpha(\mathbb{C})$.

The inequalties of the classification theory group together the spaces $C^\infty[-1, 1]$, $H(\mathbb{C})$ and $H_\alpha(\mathbb{C})$ in a class disjoint from a class which includes $H(\mathbb{D})$ and $H_\alpha(\mathbb{D})$. The space $C^\infty(\mathbb{R})$ is different from all the others because in this space, $\|x\|_k = 0$ does not imply $x = 0$ whereas the implication clearly holds in the other five spaces. (In this latter case we say that each $\|\cdot\|_k$ is a norm.) An example of the inequalties used is given in the following theorem of D. Vogt [11] which will be useful for us.

THEOREM 1. A nuclear Fréchet space E is isomorphic to a subspace of $C^\infty[-1, 1]$ if and only if every (equivalently, one) system of seminorms $\|\cdot\|_k)$ defining the topology of E is such that each $\|\cdot\|_k$ (k sufficiently large) is a norm and

$$\exists\, k_0 \geq 0 \ni \forall\, k \geq 0 \;\exists\, p \geq 0 \text{ and } C > 0 \ni \|x\|_k^2 \leq C\,\|x\|_{k_0}\,\|x\|_{k+p}$$

$(x \in E)$ (DN).

COROLLARY 1. The spaces $H(\mathbb{C})$ and $H_\alpha(\mathbb{C})$ are isomorpnic to subspaces of $C^\infty[-1, 1]$. The spaces $C^\infty(\mathbb{R})$, $H(\mathbb{D})$ and $H_\alpha(\mathbb{D})$ are not.

Proof. It is clear that $C^\infty(\mathbb{R})$ does not satisfy (DN) because the seminorms are not norms. For the others, one can compute directly, not with the semiforms I gave above but with the ones obtained from representing a function x by its sequence (ξ_n) of Taylor coefficients. They are:

$$H(\mathbb{C}) : \|x\|_k = \sup_n |\xi_n| e^{kn}$$

$$H_\alpha(\mathbb{C}) : \|x\|_k = \sup_n |\xi_{\alpha_n}| e^{k\alpha_n}$$

$$H(\mathbb{D}) : \|x\|_k = \sup_n |\xi_n|\, e^{-\frac{n}{k+1}}$$

$$H_\alpha(\mathbb{D}) : \|x\|_k = \sup_n |\xi_{\alpha_n}|\, e^{-\frac{\alpha_n}{k+1}} .$$

The fact that these seminorms define the same topologies as the ones given is a simple consequence of standard inequalities in complex analysis. Checking (DN) is then straightforward. □

Nash-Moser Inverse Function Theorems

Our basic problem is to solve the equation $f(x) = y$ for x, given y. Here $f : U \to F$ is a function from an open neighborhood U of 0 in a Fréchet space E and $f(0) = 0$. Newton's method sets up a recursion with $x_0 = 0$ and

$$x_{n+1} = x_n - (f'(x_n))^{-1}(y - f(x_n)). \qquad (1)$$

One then proves that (x_n) converges to x and it is easy to see that $f(x) = y$ (assuming sufficient smoothness on f). In the case of normed spaces the key step is to define the function ϕ by $\phi(x) = x - (f'(x))^{-1}$ $(y - f(x))$ and then to prove an inequality of the form

$$\|\phi(u) - \phi(v)\| \leq C\|u - v\| . \qquad (2)$$

To try and do this in a Fréchet space it is first necessary to specify the theory of differentiation. Actually, the Gateaux derivative suffices so we assume that $f'(x)$ $(x \in U)$ is a linear transformation from E to F satisfying, for all $x \in U$, $u \in E$

$$f'(x)(u) = \lim_{t \to 0} \frac{1}{t}(f(x + tu) - f(x)).$$

Then, assuming $f'(x)$ is invertible for all $x \in U$ we can write (1). The main difficulty is that in deriving (2) the best we can do is

$$\|\phi(u) - \phi(v)\|_k \leq C_k\|u - v\|_{\sigma(k)} \qquad (k \geq 0) \qquad (3)$$

where $\sigma(k)$ is, in general, somewhat larger than k. The fact that we have different norms on the two sides of the inequality makes the usual analysis impossible. This phenomenon, which seems endemic to the use of Fréchet spaces, occurs when ϕ is a differentiable operator so we often refer to the function σ as the <u>loss of derivatives</u>.

The Nash-Moser method of dealing with this difficulty is to introduce a one-parameter family $S_t : E \to E$ $(t \geq 1)$ of continuous linear operators such that, for $k \leq j$, $\exists\, C_{kj} > 0 \ni$ for $x \in E$ and $t > 0$, we have

$$\|S_t x\|_j \leq C_{kj} t^{j-k}\|x\|_k , \quad \|x - S_t x\|_k \leq C_{kj} t^{k-j} \|x\|_j . \qquad (4)$$

The recursion (1) is then modified to

$$x_{n+1} = x_n - S_{t_n} (f'(x_n))^{-1}(y - f(x_n)) \qquad (5)$$

for an appropriate choice of (t_n). Then, the first relation in (4) will permit us to obtain (3) with $\sigma(k) = k$. The second relation in (4) guarantees that the error introduced by changing (1) to (5) will vanish in the limit. This is the idea, but the calculations can be formidable as will be seen.

In this situation we will refer to a family of smoothing operators (S_t) which satisfies (4). There is a vast literature giving various versions of inverse function theorems valid for Fréchet spaces E which admit such a family. One particularly clear discussion which contains some additional information is that of Łojasiewicz and Zehnder [7]. We present the result in somewhat less than its full strength.

THEOREM 2. Let E be a Fréchet space which admits smoothing operators satisfying (4). Let U be a neighborhood of 0 in E and $f : U \to E$ a continuous (Gateaux) differentiable function vanishing at 0 for which there exist $d > 0$ and $\lambda \in (1, 2)$ such that for all $k > 0$, we have constants $C_k > 0$ with

$$\|f(x)\|_k \leq C_k \|x\|_{k+d} \qquad (x \in U)$$

$$\|f'(x)v\|_k \leq C_k(\|x\|_{k+d}\|v\|_0 + \|v\|_{k+d}) \qquad (x \in U,\ v \in E)$$

$$\|f(x+v) - f(x) - f'(x)v\|_k \leq C_k(\|x\|_{k+d}\|v\|_0^2 + \|v\|_{k+d}\|v\|_0)$$

$$(x,\ x+v \in U)$$

and moreover that $f'(x)$ has a right inverse $L(x)$ $(x \in U)$ such that

$$\|L(x)y\|_k \leq C_k(\|x\|_{\lambda k+d}\|y\|_d + \|y\|_{\lambda k+d}).$$

Then E has a neighborhood of 0 $V \subset U$ such that $f(W)$ is a neighborhood of 0 for all neighborhoods of 0 $W \subset V$.

Incidentally, in [7] the authors give an example showing that the theorem is false if $\lambda = 2$. Thus, in the most important relation of Theorem 2, the loss of derivatives for the function L, the fastest growth which can be handled is linear with rate $\lambda < 2$.

The first new piece of information in this investigation is that, as a result of the classification theory, Theorem 2 can only be applied to a relatively small collection of nuclear Fréchet spaces.

THEOREM 3. If E is a nuclear Fréchet space which admits a family of smoothing operators satisfying (4), then E is isomorphic to a subspace of $C^\infty[-1, 1]$.

Proof. The calculation is standard. For fixed k apply the first relation in (4) for $k, k-1$ and the second for $k+1, k$. With the triangle inequality this gives

$$\|x\|_k \le C_k(t\|x\|_{k-1} + \frac{1}{t}\|x\|_{k+1}) \qquad (x \in E,\ t \ge 1).$$

Using this with $t = \sqrt{\dfrac{\|x\|_{k+1}}{\|x\|_{k-1}}}$ the relation (DN) follows so the conclusion is a sequence of Theorem 1. □

COROLLARY 2. The spaces $C^\infty[-1, 1]$, $H(\mathbb{C})$, $H_\alpha(\mathbb{C})$ admit smoothing operators satisfying (4) while the spaces $H(\mathbb{D})$, $H_\alpha(\mathbb{D})$ do not.

Proof. The positive result is standard for $C^\infty[-1, 1]$ using either convolution or truncation of Fourier series [10]. For $H(\mathbb{C})$, $H_\alpha(\mathbb{C})$ truncation of the Taylor series expansion along with the second set of seminorms given above yields the smoothing operators. The negative result for $H(\mathbb{D})$ and $H_\alpha(\mathbb{D})$ is immediate from Corollary 1 and Theorem 2. □

Actually more is true. It was pointed out to me by D. Vogt that any nuclear Fréchet space admitting smoothing operators satisfying (4) is a complemented subspace of $C^\infty[-1, 1]$. Moreover, if it is known that the space has a basis then the converse is true and one may conclude that the nuclear Fréchet spaces with bases that admit smoothing operators satisfying (4) are exactly the infinite type power series spaces (see [9] for definitions).

Now I want to consider the question of inverse function theorems which are valid for spaces like $H(\mathbb{D})$, $H_\alpha(\mathbb{D})$. I still use smoothing operators but obviously (4) must be replaced. In view of the above discussion and considering power series spaces of infinite and finite type, it was not hard to find a candidate for the replacement of (4). It is the following: for $k \le j$, $\exists\, C_{kj} > 0 \ni$ for $x \in E$ and $t > 0$ we have

$$\|S_t x\|_j \le C_{kj} t^{(\frac{1}{k+1} - \frac{1}{j+1})} \|x\|_k, \quad \|x - S_t x\|_k \le C_{kj} t^{(\frac{1}{j+1} - \frac{1}{k+1})} \|x\|_j. \qquad (4')$$

One can obtain an inverse function theorem with these estimates and I give it now in full detail. In our analysis we assume that the norms $\|\cdot\|_k$, k a non-negative integer have been augmented to $\|\cdot\|_p$, a non-negative real number.

THEOREM 2'. Let E, F be Fréchet spaces with topologies determined by the systems of norms $(\|\cdot\|_p)_{p\geq 0}$, $(|\cdot|_p)_{p\geq 0}$ respectively and assume that E admits smoothing operators satisfying (4'). Write $U_p = \{x \in E : \|x\|_p < 1\}$ $(p \geq 0)$ and let $f : U_0 \to F$ be a continuous (Gateaux) differentiable function vanishing at 0 for which there exist non-negative constants d_0, d_1 and strictly increasing unbounded functions α, λ_0, λ_1 satisfying

$$0 \leq \lambda_0(0) \leq d_0,\ 0 \leq \lambda_1(0) \leq d_1,\ 0 < d_0 < \tfrac{1}{2},\ \alpha(d_1) \leq d_0 \tag{6}$$

and such that we have $\eta \in (0, 1)$, $C > 0$ such that for all $p \geq 0 \ \exists\ C_p > 0$ such that

$$|f(x)|_p \leq C_p\|x\|_{\alpha(p)} \qquad (x \in \eta U_{d_0},\ p \geq 0) \tag{7}$$

$$|f'(x)v|_{d_1} \leq C\|v\|_{d_0} \qquad (x \in \eta U_{d_0},\ v \in E) \tag{8}$$

$$|f(x + v) - f(x) - f'(x)v|_{d_1} \leq C\|v\|^2_{d_0} \qquad (x,\ x + v \in \eta U_{d_0}) \tag{9}$$

and moreover that $f'(x)$ has a right inverse $L(x)$ $(x \in \eta U_0)$ such that

$$\|L(x)w\|_p \leq C_p(\|x\|_{\lambda_0(p)}|w|_{d_1} + |w|_{\lambda_1(p)}) \quad (x \in \eta U_{d_0},\ w \in F,\ p \geq p_0). \tag{10}$$

Then for each neighborhood of 0 $U \subset \eta\bar{U}_{d_0}$, $f(U)$ is a neighborhood of 0 in F.

Proof. First we will prove that $f(\eta\bar{U}_{d_0})$ is a neighborhood of 0 and then observe that this gives us a right inverse g of f from some neighborhood of 0 in F to $\eta\bar{U}_{d_0}$. Then we will show that g is con- continuous at 0.

Without loss of generality we may assume the constants are all larger than 1, increasing with their indices and satisfy

$$C_{pq}C_p(C_{d_1} + C_q) < C_q^3 \qquad (0 \le p \le q).$$

We select the positive numbers τ, μ, ε, p_0, the function β and the positive numbers s, M, K, δ as follows.

We choose τ to satisfy $1 < \tau < \dfrac{\frac{2}{d_0+1}}{2 - \frac{1}{d_0+1}}$. This is possible by (6).

We choose μ, ε to satisfy

$$\frac{2}{2-\tau}\left(1 - \frac{1}{d_0+1}\right) < \mu < \mu + \varepsilon < \frac{1}{(d_0+1)\tau} \quad .$$

This is possible by the choice of τ which, incidentally, implies $\tau < 2$. We choose p_0 and define β for $p \ge p_0$ by

$$\beta(p) = \min\{p, \lambda_0^{-1}(p), \lambda_1^{-1}(p), (\alpha\lambda_1)^{-1}(p)\}.$$

That is, p_0 is chosen so large that $\beta(p_0) \ge 0$. Clearly β is unbounded. We choose s so that with $\lambda(s) = \max\{\lambda_0(s), \alpha\lambda_1(s)\}$ we have

$$\frac{1}{(s+1)\tau} - \frac{1}{\tau(\tau-1)}\left(\frac{1}{\beta\lambda(s)+1} - \frac{1}{\lambda(s)+1}\right) \le \frac{1}{(d_0+1)\tau} - \mu - \varepsilon.$$

We choose $M \ni$

$$M > \max\{2^{\mu}, \max_{j \ge 0} \{6CC_{d_0 s}C_s(C_{d_1} + C_{\alpha\lambda_1(s)})(3C_{\lambda(s)}^{3j}2^{-\varepsilon\tau^{j+1}}\}\}.$$

We choose

$$K = \sum_{\ell=0}^{\infty} 2^{\tau^{\ell}(1 - \frac{1}{d_0+1} - \mu)}$$

and finally,

$$\delta < \min\{1, (8CC_{0d_0}^2 C_0^2 M)^{-1}, (C_{0d_0}C_0(n + 1)MK)^{-1}n\} .$$

Now we set $\gamma = \max\{\lambda_0(s), \lambda_1(s), \alpha\lambda_1(s), d_1\}$ and $V_\gamma = \{y \in F : |y|_\gamma \leq 1\}$. We will show that $f(\eta\bar{U}_{d_0}) \supset \delta V_\gamma$. In fact, given $y \in \delta V_\gamma$, we will obtain x as the limit of a sequence (x_j) in ηU_{d_0} given by $x_0 = 0$ and

$$x_{j+1} = x_j - S_{t_j} L(x_j)(y - f(x_j)) \qquad (j = 0, 1, 2, \ldots)$$

where $t_j = 2^{\tau^j}$.

The main argument is to use induction on j to establish the following three relations:

(a) $\|x_j\|_{d_0} < \eta \qquad (j \geq 0)$

(b) $\|x_j\|_p \leq (2C_p)^{3j} \left(\prod_{\ell=0}^{j-1} t_\ell\right)^{\frac{1}{\beta(p)+1} - \frac{1}{p+1}} |y|_p \qquad (j \geq 1,\ p \geq p_0)$

(c) $|y - f(x_j)|_{d_1} \leq M(t_j)^{-\mu}|y|_\gamma. \qquad (j \geq 0)$

<u>j = 0</u>

(a) is clear, (b) is vacuous and for (c) we have

$$|y|_{d_1} \leq |y|_\gamma \leq M2^{-\mu}|y|_\gamma .$$

INDUCTION STEP. Assume (a), (b), (c) hold for j. We prove them for $j + 1$. For (a) we calculate, using (a), (4'), (10), (6), (a) and (c),

$$\|x_{j+1}\|_{d_0} \leq \sum_{\ell=0}^{j} \|x_{\ell+1} - x_\ell\|_{d_0} = \sum_{\ell=0}^{j} \|S_{t_\ell} L(x_\ell)(y - f(x_\ell))\|_{d_0}$$

$$\leq C_{0d_0} \sum_{\ell=0}^{j} (t_\ell)^{(1 - \frac{1}{d_0+1})} \|L(x_\ell)(y - f(x_\ell)\|_0$$

$$\leq C_{0d_0} C_0 \sum_{\ell=0}^{j} (t_\ell)^{(1 - \frac{1}{d_0+1})} (\|x_\ell\|_{\lambda_0(0)} |y - f(x_\ell)|_{d_1} + |y - f(x_\ell)|_{\lambda_1(0)})$$

$$\leq C_{0d_0} C_0(\eta + 1) \sum_{\ell=0}^{j} (t_\ell)^{(1 - \frac{1}{d_0+1})} |y - f(x_\ell)|_{d_1}$$

$$\leq C_{0d_0} C_0(\eta + 1)M\delta \sum_{\ell=0}^{j} (t_\ell)^{1 - \frac{1}{d_0+1} - \mu} < \eta.$$

Next we prove (b) for $j = 1$. We have, using (4') and (10),

$$\|x_1\|_p = \|S_{t_0} L(0)\|_p \leq C_{\beta(p),p}(t_0)^{\frac{1}{\beta(p)+1} - \frac{1}{p+1}} \|L(0)y\|_{\beta(p)}$$

$$\leq C_{\beta(p),p} C_{\beta(p)}(t_0)^{\frac{1}{\beta(p)+1} - \frac{1}{p+1}} |y|_{\lambda_1 \beta(p)} \leq 8C_p^3(t_0)^{\frac{1}{\beta(p)+1} - \frac{1}{p+1}} |y|_p$$

as desired.

Now, before proving (b) for $j + 1$ we estimate, using (4'), (10), (7), (6) and (a),

$$\|x_{j+1} - x_j\|_p = \|S_{t_j} L(x_j)(y - f(x_j))\|_p \leq C_{\beta(p),p}(t_j)^{\frac{1}{\beta(p)+1} - \frac{1}{p+1}}$$
$$\|L(x_j)(y - f(x_j))\|_{\beta(p)}$$

$$\leq C_{\beta(p),p} C_{\beta(p)}(t_j)^{\frac{1}{\beta(p)+1} - \frac{1}{p+1}} (\|x_j\|_{\lambda_0 \beta(p)} |y - f(x_j)|_{d_1}$$
$$+ |y - f(x_j)|_{\lambda_1 \beta(p)})$$

$$\leq C_{\beta(p),p} C_{\beta(p)}(C_{d_1} + C_{\alpha^{-1}(p)})(t_j)^{\frac{1}{\beta(p)+1} - \frac{1}{p+1}} (\|x_j\|_p (|y|_{d_1}$$
$$+ \|x_j\|_{\alpha(d_1)})$$

$$+ |y|_p + \|x_j\|_p) \leq 3\tilde{C}(t_j)^{\frac{1}{\beta(p)+1} - \frac{1}{p+1}} (\|x_j\|_p + |y|_p).$$

Hence, to establish (b) for $j + 1$ $(j \geq 1)$ we can calculate, assuming $p \geq p_0$ and with (b),

$$\|x_{j+1}\|_p \leq \|x_{j+1} - x_j\|_p + \|x_j\|_p \leq 3\tilde{C}(t_j)^{\frac{1}{\beta(p)+1} - \frac{1}{p+1}}$$

$$(\|x_j\|_p + |y|_p) + \|x_j\|_p$$

$$\leq 2(3\tilde{C} + 1)(2C_p)^{3j}(\prod_{\ell=0}^{j} t_\ell)^{\frac{1}{\beta(p)+1} - \frac{1}{p+1}}|y|_p$$

which gives (b) in view of our condition on the constants.

Finally we establish (c). We will use (b) in our calculations so it would be necessary to make a separate calculation for $j + 1 = 1$. Instead of repeating similar arithmetic, we will include this case by replacing the use of (b) for $j = 0$ by the convention that the symbol $\prod_{\ell=0}^{j-1}$ means 1 when $j = 0$.

Now we write

$$y - f(x_{j+1}) = f'(x_j)(1 - S_{t_j})L(x_j)(y - f(x_j)) - R_j$$

where

$$R_j = f(x_{j+1}) - f(x_j) - f'(x_j)(x_{j+1} - x_j).$$

Then we begin the proof of (c) by calculating, using (8), (4'), (10), (7), (6), (a) and (b),

$$|f'(x_j)(1 - S_{t_j})L(x_j)(y - f(x_j))|_{d_1} \leq C\|(1 - S_{t_j})L(x_j)(y - f(x_j))\|_{d_0}$$

$$\leq CC_{d_0 s}(t_j)^{\frac{1}{s+1} - \frac{1}{d_0+1}}\|L(x_j)(y - f(x_j))\|_s$$

$$\leq CC_{d_0 s}C_s(t_j)^{\frac{1}{s+1} - \frac{1}{d_0+1}}(\|x_j\|_{\lambda_0(s)}|y - f(x_j)|_{d_1} + |y - f(x_j)|_{\lambda_1(s)})$$

$$\leq CC_{d_0 s}C_s(C_{d_1} + C_{\lambda_1(s)})(t_j)^{\frac{1}{s+1} - \frac{1}{d_0+1}}(\|x_j\|_{\lambda_0(s)}(|y|_{d_1} + \|x_j\|_{\alpha(d_1)})$$

$$+ |y|_{\lambda_1(s)} + \|x_j\|_{\alpha\lambda_1(s)})$$

$$\leq 3\bar{C}(t_j)^{\frac{1}{s+1} - \frac{1}{d_0+1}} (\|x_j\|_{\lambda(s)} + |y|_{\lambda_1(s)})$$

$$\leq 3\bar{C}((2C_{\lambda(s)})^{3j} + 1)(t_j)^{\frac{1}{s+1} - \frac{1}{d_0+1}} (\prod_{\ell=0}^{j-1} t_\ell)^{\frac{1}{\beta\lambda(s)+1} - \frac{1}{\lambda(s)+1}} |y|_\gamma .$$

(Note here that if $j = 0$, then in view of our convention, using (8), (4') and (10) we obtain this last step directly.)

$$\leq \bar{\bar{C}} 2^{\tau^j(\frac{1}{s+1} - \frac{1}{d_0+1}) + \frac{\tau^j-1}{\tau-1}(\frac{1}{\beta\lambda(s)+1} - \frac{1}{\lambda(s)+1})} |y|_\gamma$$

$$\leq \bar{\bar{C}}(t_{j+1})^{\frac{1}{(s+1)\tau} - \frac{1}{(d_0+1)\tau} + \frac{1}{\tau(\tau-1)}(\frac{1}{\beta\lambda(s)+1} - \frac{1}{\lambda(s)+1})} |y|_\gamma$$

$$\leq \bar{\bar{C}}(t_{j+1})^{-\mu-\varepsilon}|y|_\gamma \leq \frac{1}{2}M(t_{j+1})^{-\mu}|y|_\gamma .$$

And for the other half we calculate, using (9), (4'), (10), (6), (a) and (c),

$$|R_j|_{d_1} \leq C\|x_{j+1} - x_j\|_{d_0}^2 = C\|S_{t_j} L(x_j)(y - f(x_j))\|_{d_0}^2$$

$$\leq CC_{0d_0}^2 (t_j)^{2(1 - \frac{1}{d_0+1})} \|L(x_j)(y - f(x_j))\|_0^2$$

$$\leq CC_{0d_0}^2 C_0^2 (t_j)^{2(1 - \frac{1}{d_0+1})} (\|x_j\|_{d_0} |y - f(x_j)|_{d_1} + |y - f(x_j)|_{d_1})^2$$

$$\leq 4CC_{0d_0}^2 C_0^2 (t_j)^{2(1 - \frac{1}{d_0+1})} |y - f(x_j)|_{d_1}^2$$

$$\leq 4CC_{0d_0}^2 C_0^2 M^2 (t_j)^{2(1 - \frac{1}{d_0+1} - \mu)} |y|_\gamma^2$$

$$\leq 4CC_{0d_0}^2 C_0^2 M^2 \delta(t_j)^{-\mu\tau}|y|_\gamma \leq \tfrac{1}{2}M(t_{j+1})^{-\mu}|y|_\gamma$$

so adding the last two estimates gives (c).

This completes the induction. Before we are able to apply (a), (b) and (c), it is necessary to use (4') to derive an interpolation property for the norms in E. Actually this calculation corresponds to the derivation of condition (DN) in the proof of theorem 3.

Let $0 \leq p \leq q \leq r$ and write $\rho = \frac{1}{q+1} - \frac{1}{r+1}$, $\theta = \frac{1}{p+1} - \frac{1}{q+1}$ so from (4') we have,

$$||x||_q \leq ||(1 - S_t)x||_q + ||S_t x||_q \leq C_{pq}C_{qr}(t^{-\rho}||x||_r + t^\theta ||x||_p)$$

and since this holds for all $t > 0$, we may assume that $x \neq 0$ and take $t = (\frac{\rho||x||_r}{\theta||x||_p})^{\frac{1}{\rho+\theta}}$ so we have,

$$||x||_q \leq C_{pq}C_{qr}(||x||_p)^{\frac{\rho}{\rho+\theta}}(||x||_r)^{\frac{\theta}{\rho+\theta}}\left((\tfrac{\theta}{\rho})^{\frac{\rho}{\rho+\theta}} + (\tfrac{\rho}{\theta})^{\frac{\theta}{\rho+\theta}}\right)$$

so we have constants K_{pqr} $(0 \leq p \leq q \leq r)$ such that for $x \in E$ we have

$$||x||_q \leq K_{pqr}(||x||_p)^{\frac{\rho}{\rho+\theta}}(||x||_r)^{\frac{\theta}{\rho+\theta}} \qquad (\rho = \frac{1}{q+1} - \frac{1}{r+1},\ \theta = \frac{1}{p+1} - \frac{1}{q+1}).$$

We apply this with $p = d_0$, $q \geq d_0$ but otherwise arbitrary and $r \geq \max\{d_0, p_0\}$. Then we have, using (4'), (10), (6), (a) and (c),

$$||x_{j+1} - x_j||_{d_0} = ||S_{t_j}L(x_j)(y - f(x_j))||_{d_0} \leq C_{0d_0}(t_j)^{1 - \frac{1}{d_0+1}}$$
$$||L(x_j)(y - f(x_j))||_0$$

$$\leq 2C_{0d_0}C_0(t_j)^{1 - \frac{1}{d_0+1}}|y - f(x_j)|_{d_1}$$

$$\leq 2C_{0d_0}C_0M(t_j)^{1 - \frac{1}{d_0+1} - \mu}|y|_\gamma$$

and from (b),

$$\|x_{j+1} - x_j\|_r \leq \|x_{j+1}\|_r + \|x_j\|_r \leq 2(2C_r)^{3j+3}\left(\prod_{\ell=0}^{j} t_\ell\right)^{\frac{1}{\beta(r)+1} - \frac{1}{r+1}} |y|_r$$

$$\leq 2(2C_r)^{3j+3}(t_j)^{\frac{\tau}{\tau-1}\left(\frac{1}{\beta(r)+1} - \frac{1}{r+1}\right)} |y|_r$$

wo we have K_{qr} such that

$$\|x_{j+1} - x_j\|_q \leq (K_{qr})^j (t_j)^{\frac{\rho}{\rho+\theta}\left(1 - \frac{1}{d_0+1} - \mu\right) + \frac{\theta}{\rho+\theta}\frac{\tau}{\tau-1}\left(\frac{1}{\beta(r)+1} - \frac{1}{r+1}\right)}.$$

Now observe that with q fixed,

$$\lim_{r\to\infty} \frac{\rho}{\rho+\theta} = \frac{d_0+1}{q+1}, \quad \lim_{r\to\infty} \frac{\theta}{\rho+\theta} = \frac{q-d_0}{q+1}, \quad \lim_{r\to\infty} \left(\frac{1}{\beta(r)+1} - \frac{1}{r+1}\right) = 0.$$

Therefore, since $1 - \frac{1}{d_0+1} < \mu$ we can choose r sufficiently large so that we have $\varepsilon > 0$ such that

$$\|x_{j+1} - x_j\|_q \leq (K_{qr})^j (t_j)^{-\varepsilon} = 2^{j\log_2 K_{qr} - \varepsilon\tau^j} |y|_r.$$

Since $\tau > 1$, it follws that (x_j) is a Cauchy sequence in E so it converses to $x \in E$ and by (a), $\|x\|_{d_0} \leq \eta$. Moreover, it follows from (c) and the continuity of f that $f(x) = y$. Thus we have shown that $f(\eta\bar{U}_{d_0})$ contains the neighborhood δV_γ.

All of this has been done for fixed y which we now vary. For each $j = 0, 1, \ldots$ we define $g_j : \delta V_\gamma \to \eta U_{d_0}$ by

$$g_0(y) = 0, \quad g_{j+1}(y) = g_j(y) - S_{t_j} L(g_j(y))(y - f(g_j)y))) \quad (y \in \delta V_\gamma).$$

Condition (a) tells us that $g_j(\delta V_\gamma) \subset \eta U_{d_0}$. Moreover, the above argument tells us that for each q we have r, K, ε such that

$$\|g_{j+1}(y) - g_j(y)\|_q \leq 2^{j\log_2 K - \varepsilon\tau^j} |y|_r$$

so if we define $g(y) = \lim_j g_j(y)$ then $g : \delta V_\gamma \to \overline{U}_{d_0} \subset U_{d_0}$ and $fg(y) = y$. Moreover, for each q we have r and $\tilde{K}$ such that

$$\|g(y)\|_q \leq \tilde{K}|y|_r \qquad (y \in \delta V_\gamma). \tag{11}$$

This implies that $g(0) = 0$ (which also follows from (b)) and g is continuous at the origin. □

REMARKS.

1. We can obtain uniqueness of the solution to the equation $f(x) = y$ by assuming that $f'(x)$ is an isomorphism for $x \in \eta U_0$ and strengthening (8) and (9) as follows. Let $d_3 > 0$ be such that if $d_2 = \max\{d_1, \lambda_1(d_0)\}$ then

$$|f'(x)v|_{d_2} \leq C\|v\|_{d_3} \qquad (x \in \eta U_{d_0}, v \in E) \tag{8'}$$

$$|f(x + v) - f(x) - f'(x)v|_{d_1} \leq C\|v\|_0\|v\|_{d_0} \qquad (x, x + v \in \eta U_{d_0}) \tag{9'}$$

Then we can show that f is $1 - 1$ on $U = \{x \in E : \|x\|_{\max\{d_3, \lambda_0(d_0)\}} \leq (8C^2C_0C_{d_0}(\eta + 1))^{-1}\}$. In fact, if $f(x_1) = f(x_2)\ (x_1, x_2 \in U)$, set $w = f'(x_1)(x_2 - x_1)$ so $L(x_1)w = x_2 - x_1$ and we can calculate, using (9'), (10), (6), (a)

$$|w|_{d_1} \leq C\|x_2 - x_1\|_0\|x_2 - x_1\|_{d_0} = C\|L(x_1)w\|_0\|L(x_1)w\|_{d_0}$$

$$\leq CC_0C_{d_0}(\|x_1\|_{\lambda_0(0)}|w|_{d_1} + |w|_{\lambda_1(0)})(\|x_1\|_{\lambda_0(d_0)}|w|_{d_1} + |w|_{\lambda_1(d_0)})$$

$$\leq CC_0C_{d_0}(\eta + 1)(\|x_1\|_{\lambda_0(d_0)} + 1)|w|_{d_1}|w|_{d_2}$$

$$\leq C^2C_0C_{d_0}(\eta + 1)(\|x_1\|_{\lambda_0(d_0)} + 1)|w|_{d_1}\|x_2 - x_1\|_{d_3}$$

$$\leq 4C^2C_0C_{d_0}(\eta + 1)(8C^2C_0C_{d_0}(\eta + 1))^{-1}|w|_{d_1} = \tfrac{1}{2}|w|_{d_1}$$

and this implies $w = 0$ so $x_2 = x_1$.

2. In the same context as Remark 1 we can show that the inverse function g is continuous on some neighborhood of 0 provided we assume, in addition that E is Montel (see [5] for definitions and basic properties) as is the case in all examples considered in this paper. In this case let U be as in Remark 1 and $V = (\delta V_\gamma) \cap f(U)$. Suppose $\lim_n y_n = y$ in V. Then from (11) and the fact that (y_n) is bounded, it follows that $(g(y_n))$ is also bounded. Since E is Montel, it suffices to show that any limit point of $(g(y_n))$ is $g(y)$. Suppose $\lim_n g(y_n) = x \in U$. Then

$$f(x) = \lim_n fg(y_n) = \lim_n y_n = y = fg(y)$$

and since f is $1 - 1$ on U, it follows that $x = g(y)$.

3. The spaces $H(\mathbb{D})$ and $H_\alpha(\mathbb{D})$ admit smoothing operators satisfying (4'). In fact, using the seminorms given in the proof of corollary 1, we can for example define S_t in $H(\mathbb{D})$ as follows. If $x \in H(\mathbb{D})$ and (ξ_n) is the sequence of coefficients in the Taylor series expansion of x, then $S_t x$ is the element of $H(\mathbb{D})$ whose Taylor series coefficients agree with ξ_n for $n \leq \log t$ and vanish for $n > \log t$. A simple calculation verfies (4').

An analogous definition of S_t can be used for the space $H_\alpha(\mathbb{D})$.

4. It is interesting to compare the hypotheses of Theorems 2 and 2'. Conditions (8) and (9) seem much weaker than their counterparts in Theorem 2 because in Theorem 2' the restriction applies only to the (rather small) norms $\|\cdot\|_{d_0}$ and $|\cdot|_{d_1}$ while in Theorem 2, the restriction applies to all norms. Perhaps more important is that in Theorem 2', the asymptotic behavior of the growth functions α, λ_0 and λ_1 is completely unrestricted whereas in Theorem 2, they are linear with restrictions on the rate. The only conditions at all on the growth functions in Theorem 2' appear in (6) where there are restrictions on their values for small arguments.

Thus, it seems that Theorem 2' is stronger than Theorem 2. On the other hand, it is not so easy to construct concrete examples that are convincing regarding the relative strength of these two results. In the next section, I will indicate what the difficulties are. At this point, it is not clear whether the hypotheses of Theorem 2' are really weaker than their counterparts in Theorem 2 or whether, because of the nature of spaces that admit smoothing operators satisfying (4'), these conditions are, in fact, of equal strength. Hopefully, further research will clarify this point.

Bilinear Functions

Perhaps the simplest example of a non-linear function on a Fréchet space E is a function $f : E \to E$ of the form, $f(x) = x + B(x, x)$ where $B : E \times E \to E$ is bilinear, symmetric and (jointly) continuous. It is easy to check that $f'(x)$ is given by

$$f'(x)v = v + 2B(x, v).$$

We will consider the case in which $f'(x)$ is an isomorphism for all $x \in E$ and try to determine whether E has a neighborhood of zero U such that $f(U)$ is a neighborhood of 0.

In general, this is not the case. The first example of the pathology is due to Lojasiewicz [6], who considers spaces that are like (in the sense of the classification theory) $C^\infty[-1, 1]$, $H(\mathbb{C})$, $H_\alpha(\mathbb{C})$. The basic idea is simple and it is not hard to construct similar examples in spaces like $H(\mathbf{D})$ and $H_\alpha(\mathbb{D})$. In fact, I believe that most nuclear Fréchet spaces will display this phenomenon.

So we try to find restrictions on B that will lead to examples that are interesting in our context. Here is a candidate which we might call <u>jointly bounded</u>:

$$\forall k \; \exists \; C_k > 0 \ni \|B(x, y)\|_k \leq C_k \|x\|_0 \|y\|_0 \qquad (x, y \in E).$$

In this case, it is not hard to show that the hypotheses of both Theorems 2 and 2' are satisfied. Everything is obvious except the loss of derivatives condition on $L(x)$. (In this case, even the fact that $f'(x)$ is an isomorphism for x in some neighborhood of 0 can be proved.) As an illustration we verify (10). We take $d_0 = \frac{1}{4}$, $d_1 = 0$, the functions λ_0, λ_1, α are the identity and $\eta \in (0, 1)$ is taken less than $\frac{1}{2C_0}$. Then if $v = L(x)w$, we have $w = v + 2B(x, v)$ so $v = -2B(x, v) + w$ and for $x \in \eta U_{d_0} \subset \eta U_0$, we have

$$\|v\|_0 \leq 2C\,\|x\|_0\|v\|_0 + \|w\|_0 \leq 2\eta C_0\|v\|_0 + \|w\|_0 .$$

Hence,

$$\|v\|_0 \leq \frac{1}{1-2\eta C_0}\|w\|_0$$

so we have

$$\|L(x)w\|_k = \|v\|_k \leq 2\|B(x, v)\|_k + \|w\|_k \leq 2C_k\|x\|_0\|v\|_0 + \|w\|_k$$

$$\leq \frac{2C_k}{1-2\eta C_0}\|x\|_0\|w\|_0 + \|w\|_k$$

which gives (10).

Now I want to look for a situation which distinguishes Theorems 2 and 2'. Consider the following somewhat weaker condition on B which we may call <u>separately bounded</u>:

$$\forall k\ \exists\ C_k > 0 \text{ and } \sigma(k) \ni \|B(x, y)\|_k \leq C_k\|x\|_0\|y\|_{\sigma(k)} \qquad (x, y \in E).$$

This condition is not strong enough to guarantee the hypotheses of Theorem 2. In fact, in the example given in [7, p. 173], the bilinear map $A^{-1}B$ is separately bounded but ϕ maps no neighborhood of 0 onto a neighborhood of 0 so the same holds for $A^{-1}\phi$ (which is of the form $x + A^{-1}B(x, x)$) because A is an isomorphism. In this example, $E = H(\mathbb{C})$.

On the other hand, if we assume $\sigma(0) = 0$, then the hypotheses of Theorem 2' hold. The details are easy but, the next result suggests that there may not exist any non-trivial examples. We observe that nothing is changed in the definitions of separately and jointly bounded if the norm $\|\cdot\|_0$ is replaced by $\|\cdot\|_{k_0}$ for some fixed k_0.

THEOREM 4. In the spaces $H(\mathbb{D})$, $H_\alpha(\mathbb{D})$ every symmetric, bilinear, separately bounded function is jointly bounded.

Proof. We give the proof for $H(\mathbb{D})$. (The argument for $H_\alpha(\mathbb{D})$ is similar. In fact, this is a much more general result which holds for any nuclear Fréchet space with a regular basis of type (d_2) [1].)

Let e_n be the monomial of degree n in $H(\mathbb{D})$ so every $x \in H(\mathbb{D})$ has the unique expansion $x = \sum_{n=0}^{\infty} \xi_n e_n$ and using the norms $\|\cdot\|_k$ of Corollary 1 we have

$$\|x\|_k = \sup_n |\xi_n| \, \|e_n\|_k \qquad \|e_n\|_k = e^{-\frac{n}{k+1}} .$$

Let $B : H(\mathbb{D}) \times H(\mathbb{D}) \to H(\mathbb{D})$ be symmetric, bilinear and suppose we have $k_0 \geq 0$ and $C_k > 0$, $\sigma(k)$ $(k \geq 0)$ such that

$$\|B(x, y)\|_k \leq C_k \|x\|_{\sigma(k)} \|y\|_{k_0} \qquad (k \geq 0,\ x,\ y \in H(\mathbb{D})).$$

Clearly, if $k_1 \geq 2k_0 + 1$ we have $\|e_n\|_{k_0} \leq (\|e_n\|_{k_1})^2$ so for any k, n we have

$$\|B(e_n, e_n)\|_k \leq C_k \|e_n\|_{\sigma(k)} \|e_n\|_{k_0} \leq C_k (\|e_n\|_{k_1})^2 .$$

Similarly, if $m < n$, then for any k,

$$\|B(e_m + e_n, e_m + e_n)\|_k \leq C_k \|e_m + e_n\|_{\sigma(k)} \|e_m + e_n\|_{k_0}$$

$$= C_k \|e_m\|_{\sigma(k)} \|e_m\|_{k_0}$$

$$\leq C_k \|e_m\|_{k_1}^2 \leq C_k (\|e_m\|_{k_1}^2 + \|e_n\|_{k_1}^2).$$

By the symmetry this holds for $m > n$ and it is clear (with larger C_k) for $m = n$. Hence, we have for any m, n, k

$$\|B(e_m, e_n)\|_k \leq \tfrac{1}{2}(\|B(e_m + e_n, e_m + e_n)\|_k + \|B(e_m, e_m)\|_k$$

$$+ \|B(e_n, e_n)\|_k)$$

$$\leq C_k (\|e_m\|_{k_1}^2 + \|e_n\|_{k_1}^2).$$

Next, I will show that for each $k \geq 0$ we have,

$$\sup_{m,n} \frac{\|B(e_m,e_n)\|_k}{\|e_m\|_{k_1}\|e_n\|_{k_1}} < \infty .$$

If this were false for some $k \geq 0$, we would have a sequence $(m_\nu, n_\nu) \ni$

$$\lim_\nu \frac{\|B(e_{m_\nu},e_{n_\nu})\|_k}{\|e_{m_\nu}\|_{k_1}\|e_{n_\nu}\|_{k_1}} = \infty.$$

Now suppose we had constants $M > \delta > 0 \ni \delta \leq \dfrac{\|e_{m_\nu}\|_{k_1}}{\|e_{n_\nu}\|_{k_1}} \leq M \;\forall\; \nu$. Then for each ν,

$$\frac{\|B(e_{m_\nu},e_{n_\nu})\|_k}{\|e_{m_\nu}\|_{k_1}\|e_{n_\nu}\|_{k_1}} \leq \frac{C_k(\|e_{m_\nu}\|_{k_1}^2 + \|e_{n_\nu}\|_{k_1}^2)}{\delta\|e_{n_\nu}\|_{k_1}^2} \leq \frac{C_k(M+1)}{\delta} < \infty$$

which is a contradiction. Suppose the upper bound did not exist (the other case is similar). Then by passing to a subsequence we would have,

$$\lim_\nu \frac{\|e_{m_\nu}\|_{k_1}}{\|e_{n_\nu}\|_{k_1}} = \infty.$$

But then,

$$\frac{\|B(e_{m_\nu},e_{n_\nu})\|_k}{\|e_{m_\nu}\|_{k_1}\|e_{n_\nu}\|_{k_1}} \leq C_k \frac{\|e_{m_\nu}\|_{\sigma(k)}\|e_{n_\nu}\|_{k_0}}{\|e_{m_\nu}\|_{k_1}\|e_{n_\nu}\|_{k_1}} \leq C_k \frac{\|e_{n_\nu}\|_{k_1}}{\|e_{m_\nu}\|_{k_1}}$$

and the last quantity goes to 0 as ν goes to ∞ which is a contradiction so the claim is proved.

Finally, we choose $k_2 > k_1$ so that $\sum_n \frac{\|e_n\|_{k_1}}{\|e_n\|_{k_2}} = C < \infty$ and writing $x = \sum_n \xi_n e_n$, $y = \sum_n \eta_n e_n$ we can compute for any k,

$$\|B(x, y)\|_k \leq \sum_{m,n} |\xi_m| \, |\eta_n| \, \|B(e_m, e_n)\|_k \leq C_k' \sum_{m,n} |\xi_m| \, |\eta_n|$$

$$\|e_m\|_{k_1} \, \|e_n\|_{k_1}$$

$$\leq CC_k' \|x\|_{k_2} \, \|y\|_{k_2}. \qquad \square$$

At this point we do not have a concrete example of a space E and a function f such that the loss of derivatives in (10) is non-linear but Theorem 2' still applies. We consider Theorem 4 to be an interesting result about Fréchet spaces obtained in an unsuccessful attempt to find such an example. In the next section we will see another property of some Fréchet spaces which seems to be related to this difficulty.

Tame Spaces

In trying to find examples of functions in which the loss of derivatives function is more rapidly increasing, it does not help to replace the expression $f(x) = x + B(x, x)$ by $f(x) = Ax + B(x, x)$ where A is a fixed continuous linear operator. The reason is that the requirement that $f'(0)$ be an isomorphism forces A to be an isomorphism and then $A^{-1}f$ has the original form. We could however consider a function $f(x) = A(x) + B(x, x)$ where $A(x)$ is a non-linear function whose derivative is an isomorphism. This leads us to a consideration of continuous linear operators $T : E \to E$ which satisfy a relation of the form

$$\|Tx\|_k \leq C_k \|x\|_{\sigma(k)} \qquad (x \in E, \; k \geq 0)$$

where σ is some function which presumably grows rapidly. It is a little surprising that there are spaces in which this cannot happen. Specifically, we say that a Fréchet space E is _tame_ if there is a sequence of seminorms $(\|\cdot\|_k)_{k=0}^{\infty}$ defining the topology and a function σ such that for every continuous linear operator $T : E \to E$ there exists a sequence of positive constants (C_k) and k_0 such that

$$\|Tx\|_k \leq C_k \|x\|_{\sigma(k)} \qquad (x \in E,\ k \geq k_0).$$

We collect the known facts about tame spaces in the following result.

THEOREM 5. The spaces $C^{\infty}[-1, 1]$ and $H(\mathbb{C})$ are not tame. The spaces $H(\mathbb{D})$ and $H_{\alpha}(\mathbb{D})$ are tame. If $\lim_n \frac{\alpha_{n+1}}{\alpha_n} = \infty$, then $H_{\alpha}(\mathbb{C})$ is tame.

Proof. The statement about $H(\mathbb{D})$ and $H_{\alpha}(\mathbb{D})$ is contained in the arguments of [8]. The statement about $H_{\alpha}(\mathbb{C})$ follows from the results of [2] and a simple analysis of diagonal operators. We give the proof that $H(\mathbb{C})$ is not tame. The argument for $C^{\infty}[-1, 1]$ is similar.

Again we let e_n be the monomial of degree n in $H(\mathbb{C})$ so every $x \in H(\mathbb{C})$ has the unique expansion $x = \sum_{n=0}^{\infty} \xi_n e_n$ and using the norms $\|\cdot\|_k$ of Corollary 1 we have,

$$\|x\|_k = \sup_n |\xi_n| \ \|e_n\|_k, \quad \|(e_n\|_k = e^{kn}.$$

Let σ be any function. We will construct a continuous linear operator $T : E \to E$ such that $\sup_{x \in E} \frac{\|T_x\|_k}{\|x\|_{\sigma(k)}} = \infty$ for all k. There is no loss of generality if we assume that $\sigma(k + 2) - \sigma(k + 1) \geq 2(\sigma(k + 1) - \sigma(k))$, and that $\sigma(2) > \sigma(1)$.

For each n, k let q_n^k be the smallest positive integer such that

$$q_n^k \geq (\sigma(k + 1) - \sigma(k))n.$$

Clearly $q_n^{k+1} \geq q_n^k$ and

$$q_n^k \leq 2(q_n^k - 1) < 2(\sigma(k + 1) - \sigma(k))n \leq (\sigma(k + 2) - \sigma(k + 1))n.$$

Set $t_{nq_n^1} = e^{\sigma(2)n - q_n^1}$ and for $k = 1, 2, \ldots$

$$t_{nq_n^{k+1}} = t_{nq_n^k} e^{(k+1)(q_n^k - q_n^{k+1})} .$$

It follows from [3, p. 36] that if we define $Te_n = \sum_{j=1}^{\infty} t_{nq_n^j} e_{q_n^j}$, then

$$\|Te_n\|_k = t_{nq_n^k} e^{kq_n^k} .$$

We will complete the proof by establishing, for all k, the two relations

$$\sup_n \frac{\|Te_n\|_k}{\|e_n\|_{\sigma(k+1)}} < \infty, \qquad \sup_n \frac{\|Te_n\|_k}{\|e_n\|_{\sigma(k)}} = \infty.$$

The first guarantees that T extends to a continuous linear operator from E to itself and the second shows that "σ does not suffice". We use induction on k.

For $k = 1$ we calculate

$$\|Te_n\|_1 \quad t_{nq_n^1} e^{q_n^1} = e^{\sigma(2)n}, \; \|e_n\|_{\sigma(1)} = e^{\sigma(1)n}, \; \|e_n\|_{\sigma(2)} = e^{\sigma(2)n}$$

which gives both statements. Now assuming that they hold for k, we calculate

$$\sup_n \frac{\|Te_n\|_{k+1}}{\|e_n\|_{\sigma(k+2)}} = \sup_n \frac{\|Te_n\|_{k+1}}{\|Te_n\|_k} \frac{\|Te_n\|_k}{\|e_n\|_{\sigma(k+1}} \frac{\|e_n\|_{\sigma(k+1)}}{\|e_n\|_{\sigma(k+2)}}$$

$$= \sup_n e^{q_n^k - (\sigma(k+2) - \sigma(k+1))n} \frac{\|Te_n\|_k}{\|e_n\|_{\sigma(k+1)}} < \infty$$

and,

$$\sup_n \frac{\|Te_n\|_{k+1}}{\|e_n\|_{\sigma(k+1)}} = \sup_n \frac{\|Te_n\|_{k+1}}{\|Te_n\|_k} \frac{\|Te_n\|_k}{\|e_n\|_{\sigma(k)}} \frac{\|e_n\|_{\sigma(k)}}{\|e_n\|_{\sigma(k+1)}}$$

$$= \sup_n e^{q_n^k - (\sigma(k+1) - \sigma(k))n} \frac{\|Te_n\|_k}{\|e_n\|_{\sigma(k)}} = \infty. \quad \square$$

There is more to be said about tameness - especially in relation to problems concerning complemented subspaces - but this will be developed elsewhere.

EPILOGUE

My main goal in this lecture has been to point out connections between the classification theory and certain questions concerning inverse functions. In conclusion, I would like to point out one imperfection that may lead to a refinement of the classification theory. Theorems 2 and 2' separate the spaces $C^\infty[-1, 1]$, $H(\mathbb{C})$ and $H_\alpha(\mathbb{C})$ from the spaces $H(\mathbb{D})$ and $H_\alpha(\mathbb{D})$. This corresponds nicely to the classification theory (spaces of type (d_1) and (d_2) [1]). The same division holds for the property described in Theorem 4 which is false for $C^\infty[-1, 1]$, $H(\mathbb{C})$ and $H_\alpha(\mathbb{C})$. On the other hand, in Theorem 5 the spaces $H(\mathbb{D})$, $H_\alpha(\mathbb{D})$ and $H_\alpha(\mathbb{C})$ $(\lim_n \frac{\alpha_{n+1}}{\alpha_n} = \infty)$ are grouped together and separated from $C^\infty[-1, 1]$, $H(\mathbb{C})$. This suggests that in the classification theory, we will have to incorporate the behavior of these gaps, as well as the norm inequalities.

My final statement is that there are many things to be investigated regarding inverse function theorems in Fréchet spaces. For example, a more detailed analysis of concrete examples such as polynomials of degree more than 2, entire functions, partial differential equations. Also, there are other Fréchet spaces to be considered and it is also possible that duals of Fréchet spaces will behave better than Fréchet spaces.

REFERENCES

1. M. M. Dragilev, "On regular bases in nuclear spaces", Amer. Math. Soc. Transl. (2) 93 (1970), 61-82.

2. M. M. Dragilev and V. P. Kondakov, "A class of nuclear spaces" Matematicheskie Zametki 8, 2 (1970), 169-179 (Russian).

3. Ed Dubinsky, "The structure of nuclear Fréchet spaces", LNIM 720, Springer-Verlag (1979).

4. J. Eells, Jr., "A setting for global analysis", Bull. Amer. Math. Soc., 72 (1966), 751-807.

5. G. Köthe, "Topologische lineare Räume", Springer-Verlag (1960).

6. S. Łojasiewicz, Jr., "An example of a continuous injective polynomial map with nowhere dense range whose differential at each point is an isomorphism", Bull. Acad. Polon. Sci. 24, 12 (1976), 1109-1111.

7. S. Łojasiewicz, Jr. and E. Zehnder, "An inverse function theorem in Fréchet-spaces", J. Funct. Anal. 33 (1979), 165-174.

8. B. S. Mitiagin, "Equivalence of bases in Hilbert scales", Studia Math. 37 (1971), 111-137, (Russian).

9. A. Pietsch, "Nukleare lokalkonvexe Räume", Berlin (1965).

10. J. T. Schwartz, "Non-linear functional analysis (1963-1964)", Courant Institute, NYU (1965).

11. D. Vogt, "Subspaces and quotient spaces of (s), in: Functional Analysis, Surveys and Recent Results", (Proc. Conf. Paderborn 1976), North Holland (1977), 167-187.

12. D. Vogt and M. J. Wagner, "Charakterisierung der Unterräume und Quotientenräume der nukleare stabilen potenzreihen Räume Von Unendlichen Typ", Studia Math. (to appear).

A REVIEW OF SOME RECENT RESULTS CONCERNING FACTORIZATION OF MATRIX AND OPERATOR VALUED FUNCTIONS

P. Lancaster

University of Calgary
Calgary, Alberta
Canada

1. PROLOGUE

Before beginning with the main topic of this review, consider an easy factorization problem for a complex-valued function. This will serve as a prototype and help to motivate our subsequent ventures into more complicated situations.

PROPOSITION 1 (after Fejér, 1916 [8]). *Let* $\varphi(z) = \sum_{k=-n}^{n} c_k z^k$, *with* $c_{-k} = \bar{c}_k \in \mathbb{C}$ $(k = 0,1,\ldots,n)$, *and* $c_n \neq 0$. *Suppose also that* $\varphi(z) \geq 0$ *on the unit circle* U. *Then there is a polynomial* q *of degree* n *such that* $\varphi(z) = q(z)\bar{q}(1/z)$.

Note that the conditions on the coefficients c_k ensure that φ is real on U. Also, the factorization yields $\varphi(z) = q(z)\overline{q(z)} = |q(z)|^2$ on U. A proof is quite easily obtained by first considering the case $\varphi(z) > 0$ on U, and using the symmetry of the zeros of φ with respect to U to settle this case. Then the case $\varphi(z) \geq 0$ on U is settled by continuity. In fact, the polynomial q can be chosen so that all of its zeros are outside or on U.

By using a transformation $\zeta = i(1-z)(1+z)^{-1}$ the unit circle is mapped onto the real line and a corresponding factorization result is obtained; one which yields even more readily to a direct argument using symmetries of the zeros.

PROPOSITION 2. *Let $p(\zeta)$ be a polynomial of degree $2n$ with real coefficients and suppose that $p(\zeta) \geq 0$ on the real line. Then there is a polynomial m of degree n for which $p(\zeta) = m(\zeta)\bar{m}(\zeta)$.*

Note that polynomial m is not real, in general, and that when ζ is real $p(\zeta) = |m(\zeta)|^2$. In this case m could be chosen so that all of its zeros are in the closed upper half-plane.

Later in the twentieth century deeper results were obtained concerning the factorization of analytic functions with respect to the unit circle. Important contributions were made by Szegö, and by Wiener, for example. However, the direction that we take is not in generalizing the dependence of φ (or p) on z (or ζ), so much as changing the nature of the coefficients of φ and p to operators or matrices. Thus, our main emphasis will be on operator and matrix-valued functions which are polynomials in a complex independent variable. Many generalizations of the results reviewed here to analytic, or rational, operator functions are also available and some references to such results are made as appropriate.

The main body of the paper is written in two parts. Part A includes those results known to be true in a general Hilbert space setting. (Some results may be more appropriately formulated in Banach spaces, but we present results uniformly in the Hilbert space setting for convenience.) In Part B the context is that of a finite dimensional space. Most of the results here either do not extend, or are not known to extend, to spaces of infinite dimension.

The theory described in this paper has its roots firmly planted in several different areas of application. However, for the purpose of this review we shall leave aside such motivation and try to indicate only the main structural ideas which have been developed in recent years.

PART A: MAINLY ON OPERATOR POLYNOMIALS

2. Preliminaries

Let X be a complex Hilbert space and $L(X)$ denote the algebra of all bounded linear operators on X. Given $A_0, A_1, \ldots, A_\ell \in L(X)$ with $A_\ell \neq 0$ define the operator polynomial $L\colon \mathbb{C} \to L(X)$ by

$$L(\lambda) = \sum_{j=0}^{\ell} A_j \lambda^j . \tag{1}$$

When $A_\ell = I$, L is said to be *monic*. The *adjoint* operator polynomial L^* is defined by $L^*(\lambda) = \sum_{j=0}^{\ell} A_j^* \lambda^j$, and L is said to be *selfadjoint* if $L^* = L$, i.e. if $A_j^* = A_j$ for $j = 0,1,\dots,\ell$. Note that $L(\lambda) = I\lambda - A_0$ is a monic operator polynomial which plays a vital role in the spectral analysis of A_0.

Looking ahead, the kind of question arising as a generalization of Proposition 2 might be: If $L^* = L$ and, for all $\lambda \in \mathbb{R}$, $(L(\lambda)x,x) \geq 0$ for all $x \in X$, is there an operator polynomial M for which $L(\lambda) = M^*(\lambda)M(\lambda)$? We return to this question later but note here that points of the *spectrum* of L play the role of the zeros of p in the proof of Proposition 2. The spectrum of L is defined by $\sigma(L) = \{\lambda \in \mathbb{C} | L(\lambda)^{-1} \notin L(X)\}$.

The first idea that we need is that of a linearization for an operator polynomial. This is defined in terms of equivalence of operator polynomials. For this, we say that operator polynomials $A(\lambda)$, $B(\lambda)$ are equivalent if there exist operator polynomials $E(\lambda)$, $F(\lambda)$ for which

$$B(\lambda) = F(\lambda)A(\lambda)E(\lambda)$$

and $E(\lambda)^{-1}$, $F(\lambda)^{-1}$ are also operator polynomials. Then a linearization of L is linear function $P-\lambda Q$ with $P,Q \in L(X^\ell)$ (X^ℓ denotes the direct sum of ℓ copies of X) which is equivalent to an extension of L of the form $L(\lambda) \oplus I$, and acting in $L(X^\ell)$ of course. Thus, for operator polynomials $E(\lambda)$, $F(\lambda)$ whose inverses are also operator polynomials:

$$L(\lambda) \oplus I = E(\lambda)(P-\lambda Q)F(\lambda) .$$

When this is the case it is clear that the spectra of $L(\lambda)$ and $P-\lambda Q$ coincide, and the analysis of this and other spectral properties for L can be attacked via corresponding properties for the linearization. A simple linearization is given by the operator matrices

$$P = \begin{bmatrix} 0 & I & 0 & \cdots & 0 \\ 0 & 0 & I & & \\ \vdots & \vdots & & \ddots & \vdots \\ 0 & 0 & & & I \\ -A_0 & -A_1 & & \cdots & -A_{\ell-1} \end{bmatrix} \qquad Q = \begin{bmatrix} I & 0 & \cdots & & 0 \\ 0 & I & & & \\ \vdots & & \ddots & & \vdots \\ & & & I & 0 \\ 0 & \cdots & & 0 & A_\ell \end{bmatrix} . \tag{2}$$

When A_ℓ is invertible there is a linearization in which $Q = I$. Thus, with P, Q as defined above $Q^{-1}P-\lambda I$ is a linearization for L in this case.

For extensions of the notion of linearization to analytic operator functions we refer to [9] and [18], especially, and also [5] and [26]. Important ideas in the finite dimensional case with A_ℓ not invertible are developed in [7] and part II of [15].

Suppose now L is *monic*. Then $P-\lambda I$ is a linearization for L and P (as defined in (2)) is called the *first companion operator* for L, and we shall rename it: $P = C_L$. We couple C_L with the operator

$$X_0 = [I\ 0\ \ldots\ 0]:\ \mathsf{X}^\ell \to \mathsf{X}$$

and then define a pair of operators X, T to be a *standard pair* for L if there is an invertible $S \in L(\mathsf{X}^\ell)$ for which

$$X = X_0 S, \qquad T = S^{-1}C_L S . \tag{3}$$

It is clear that, if X, T is a standard pair for L, then T has the property that $T-\lambda I$ is a linearization of L. Note also that, in the finite dimensional case, S can be chosen to produce a standard pair $\hat{X}$, J in which J is a matrix in Jordan normal form. It turns out that such a *Jordan pair* contains complete information on the spectrum and right eigenvector structure of L.

Returning to the general case, standard pairs can be characterized in another way:

A pair of operators X: $\mathsf{X}^\ell \to \mathsf{X}$ and $T \in L(\mathsf{X}^\ell)$ is a standard pair for the monic operator polynomial L (of equation (1)) if and only if the operator $Q \in L(\mathsf{X}^\ell)$ defined by

$$Q = \begin{bmatrix} X \\ XT \\ \cdot \\ \cdot \\ \cdot \\ XT^{\ell-1} \end{bmatrix} \overset{\text{def}}{=} \operatorname{col}[XT^j]_{j=0}^{\ell-1} \tag{4}$$

is invertible and also

$$\sum_{j=0}^{\ell} A_j XT^j = 0 . \tag{5}$$

We must extend these ideas a little further: we associate with the standard pair X_0, C_L a third operator $Y_0 : X \to X^\ell$ defined by

$$Y_0 = \begin{bmatrix} 0 \\ \cdot \\ \cdot \\ \cdot \\ 0 \\ I \end{bmatrix} .$$

Then extend the definition (3) as follows: three operators (X,T,Y) form a *standard triple* for L if there is an invertible $S \in L(X^\ell)$ for which

$$X = X_0 S , \qquad T = SC_L S^{-1} , \qquad Y = SY_0 .$$

(In the finite dimensional case we can choose S so that T is in Jordan form and we obtain a *Jordan triple* which summarizes the spectral structure of L including both left and right eigenvectors.)

Now define operator $R \in L(X^\ell)$ by

$$R = [Y \; TY \; \ldots \; T^{\ell-1}Y] \overset{\text{def}}{=} \operatorname{row}[T^j Y]_{j=0}^{\ell-1} \tag{6}$$

and it is found that R is invertible and (the analogue of (5))

$$\sum_{j=0}^{\ell} T^j Y A_j = 0.$$

3. First results

The standard pairs and triples introduced above can be used to describe L itself. These results are known as *representation theorems* and play an important part in the theory. Results for operator polynomials (see [11] for more detail) are as follows:

Let (X,T,Y) be a standard triple for monic operator polynomial L. Then the coefficients of L are given by

$$[A_0 \ A_1 \ \dots \ A_{\ell-1}] = -XT^{\ell}Q^{-1} \tag{7}$$

and by

$$\begin{bmatrix} A_0 \\ A_1 \\ \cdot \\ \cdot \\ \cdot \\ A_{\ell-1} \end{bmatrix} = -R^{-1}T^{\ell}Y \ . \tag{8}$$

Furthermore, the inverse (or resolvent) of $L(\lambda)$ is given by

$$L(\lambda)^{-1} = X(I\lambda - T)^{-1}Y \ , \qquad \lambda \notin \sigma(L) \ . \tag{9}$$

Note that Q and T appearing in (7) and (8) are as defined in (5) and (6). To take advantage of (7) one needs only a standard *pair* (X,T) for L and a similar remark applies to (8). The resolvent form of (9) turns out to be particularly useful.

The significance of the results is well illustrated in the very special case of a finite dimensional space and a monic polynomial of first degree. Then $L(\lambda) = I\lambda - A_0$ and a Jordan triple for L is just (X,J,Y) where J is a Jordan form for A_0, $Y = X^{-1}$ and $A_0 = XJY$.

Representation theorems of the above kind might also be seen as special cases of the "realization theorems" of system theory in which rational dependence on the parameter arises most naturally. For the development of these ideas in the Banach space context see [2], where the connection with the notion of a "node" is also developed (see also [6]).

Returning to the polynomial case it is clear that the invertibility of operators Q and R of (5) and (6) plays an important role. It turns out that interesting new ideas develop if we relax the invertibility condition somewhat. What if the operator Q of (5) is merely *left* invertible, for example?

The heuristics of this situation are revealed if one considers the finite dimensional case in which Q can be interpreted as a rectangular matrix with ln rows and less columns. The fact that there are *less* columns suggests that the pair X, T contains inadequate information to completely determine a monic polynomial. This line of thought can be made precise and it is found that if Q is merely left invertible then there exist *extensions* of X and T which form a standard pair for some monic operator polynomial. For the general result see [11] and for more detailed information in the finite dimensional case see [17] or chapter 6 of [15].

The dual problem is that in which Q has a *right* inverse. In this case it can be shown that X, T have *restrictions* which form a standard pair for some monic operator polynomial. Heuristically, X and T can be thought of as containing redundant information.

4. First results on factorization

The principal factorization problem to be considered is that of writing a monic operator polynomial as a product of two monic operator polynomials of lower degree. Thus, for a given monic operator polynomial L we seek others, say L_1 and L_2, for which $L(\lambda) = L_2(\lambda)L_1(\lambda)$ for all $\lambda \in \mathbb{C}$. In this case L_1 is called a *right divisor* of L.

It turns out that right divisors of L can be characterized by certain invariant subspaces of any (fixed) operator C for which $C-\lambda I$ is a linearization of L. Furthermore this characterization provides a useful machinery for the further analysis of existence theorems, stability properties, and so on. To describe this characterization we first introduce the operators Q_k: $X^{\ell} \to X^k$, $k = 1,2,\ldots,\ell-1$ by writing

$$Q_k = \begin{bmatrix} X \\ XT \\ \cdot \\ \cdot \\ \cdot \\ XT^{k-1} \end{bmatrix} . \tag{10}$$

As before (X,T) is a standard pair for a monic operator polynomial L. The result (see [11] and also [23]) is that, *if there exists a* k *and an invariant subspace* S *of* T *for which the restriction of* Q_k *to* S, $Q_k|_S : S \to X^k$ *is invertible, then* L *has a (monic) right divisor of degree* k.

Conversely, if L *has a monic right divisor of degree* k *then* $Q_k|_S$ *is invertible.*

The subspace appearing in this theorem is called the *supporting subspace* for the right divisor (say L_1) *with respect to* T. In the finite dimensional case, a Jordan pair (X,J) can be used in place of (X,T) in which case the explicit description of the invariant subspaces of the linearization J is a triviality.

A simple finite-dimensional (but negative) example may be helpful at this point. If

$$L(\lambda) = \begin{bmatrix} 1 & 0 \\ 0 & 1 \end{bmatrix} \lambda^2 + \begin{bmatrix} 0 & 0 \\ 1 & 0 \end{bmatrix}$$

a little direct calculation convinces one that there are *no* complex matrices A and B for which

$$L(\lambda) = (I\lambda + A)(I\lambda + B) . \tag{11}$$

In this case it is easily seen that we may take

$$X = \begin{bmatrix} 0 & 0 & 1 & 1 \\ 1 & 1 & 0 & 0 \end{bmatrix}, \qquad J = \begin{bmatrix} 0 & 1 & 0 & 0 \\ 0 & 0 & 1 & 0 \\ 0 & 0 & 0 & 1 \\ 0 & 0 & 0 & 0 \end{bmatrix}$$

and the invariant subspaces of J are generated by the unit vectors as follows.

$$\text{span}\{e_1\},\quad \text{span}\{e_1,e_2\},\quad \text{span}\{e_1,e_2,e_3\},\quad \text{span}\{e_1,e_2,e_3,e_4\}.$$

For the existence of a factorization (11) we would need the restriction of $Q_1 = X$ to a two-dimensional invariant subspace to be invertible. But there is only one such subspace S and obviously $X|_S$ is not invertible.

Returning once more to the Hilbert space version of the theorem we remark that explicit formulae for the divisor L_1 and quotient L_2 can be obtained using the concepts of the theorem. For example, if $L_1(\lambda) = \sum_{j=0}^{k} B_j\lambda^j$, $(B_k = I)$, then

$$[B_0\ B_1\ \ldots\ B_{k-1}] = -X|_S\, T|_S^k (Q_k|_S)^{-1},$$

and this should be compared with the representation for L itself given by (7).

An interesting special case concerns the solution of the equation

$$\sum_{j=0}^{\ell} A_j D^j = 0 \tag{12}$$

for $D \in L(X)$. That this is related to a factorization problem follows immediately from the remainder theorem: *There is a solution D for* (12) *if and only if $I\lambda-D$ is a right divisor of $L(\lambda)$.*

An application of the factorization theorem using the linearization C_L leads at once to the result: *There is a solution $D \in L(X)$ of* (12) *if and only if there is an invariant subspace S of C_L of the form*

$$S = Im\ \text{col}[D^j]_{j=0}^{\ell-1}.$$

This is due to Langer [22], see also [25] and, as one might expect, the subspace S is the supporting subspace for $I\lambda-D$ with respect to C_L.

Another important insight concerning supporting subspaces and their right divisor arises when $L = L_2L_1$ and $\sigma(L_1)$, $\sigma(L_2)$ have no points in common. To describe this we need some more definitions. A point $\lambda \in \mathbb{C}$ is said to be a *regular point* of monic operator polynomial L if $L(\lambda)^{-1} \in L(X)$. Let Γ be a contour of regular points. Then a monic right divisor L_1 of L (say $L = L_2L_1$) is said to be a *Γ-spectral right divisor* of L if $\sigma(L_1)$ is inside Γ and $\sigma(L_2)$ is outside Γ. The following result can then be proved: *Let* (X,T) *be a standard pair for* L, *and* L_1 *be a* Γ-*spectral right divisor having supporting subspace* S *with respect to* T. *If* R_Γ *is the Riesz projector for* L *associated with* Γ:

$$R_\Gamma = \frac{1}{2\pi i}\int_\Gamma (I\lambda - T)^{-1} d\lambda , \tag{13}$$

then $S = Im\ R_\Gamma$.

This result gives a useful characterization of the supporting subspace when a Γ-spectral divisor is known to exist. It is, however, only a necessary condition and there is no converse statement yielding sufficient conditions for existence of a right divisor. Even in the finite dimensional case, one can use a contour Γ to enclose a subset of $\sigma(L)$ so that R_Γ exists and $Im\ R_\Gamma$ has dimension kn (where n is the dimension of X); but there is no guarantee that $Im\ R_\Gamma$ will then be supporting for a divisor of degree k, i.e. Q_k of (10) restricted to $Im\ R_\Gamma$ need not be invertible. The extra conditions needed in order to assert the existence of either a right or a left spectral divisor, or both, are investigated in [10]. For analysis of spectral divisors for non-monic matrix polynomials, see [22] and [3], for example.

More generally, if S is a supporting subspace for a right divisor of an operator polynomial, any projector onto S is called a *supporting projector* and makes a useful tool. This concept was introduced in [1] and was extended in [2] for the analysis of rational operator functions. Note that in the special case of a spectral divisor the Riesz projector of [13] gives an example of a supporting projector; one for which the kernel of R_Γ is also invariant under T.

5. Perturbations of divisors

The concept of supporting subspace for a right divisor is particularly useful and suggestive in the context of perturbation theory. Here the first companion operator C_L for L plays a special role, because the effect of variations in the coefficients of L is immediately apparent in the linearization C_L (recall $C_L = P$ of equations (2)). Consequently, we consider now supporting subspaces for divisors *with respect to* C_L. Introduce a topology on the set of subspaces of X^ℓ, namely, the spherical metric (a close relation of the "gap"), and a natural metric on the set of monic operator polynomial of fixed degree k by writing

$$\beta_k(L_1,L_2) = \sum_{j=0}^{k-1} \|A_j^{(1)} - A_j^{(2)}\| . \tag{14}$$

Then it is possible to investigate, for example, the continuous dependence of the supporting subspace (with respect to C_L) on the coefficients. Several questions of this kind are studied in [12]. For brevity, and to give some of the flavour of the results, we describe just two results on analytic dependence. But first consider a very simple example which serves to suggest that some curious phenomena are to be expected.

Early in section three an example of non-existence of a right divisor is presented. We elaborate on that: the family of monic operator polynomials

$$L(\lambda,\mu) = I\lambda^2 + \begin{bmatrix} 0 & 0 \\ \mu & 0 \end{bmatrix}$$

has *no* right divisor as long as $\mu \neq 0$. However, it trivially has a monic right divisor when $\mu = 0$. This would be described as an *isolated* divisor. (Note that it is certainly not a Γ-spectral divisor for any Γ.)

On the other hand, the family

$$M(\lambda,\mu) = I\lambda^2 + \begin{bmatrix} 0 & 1 \\ \mu & 0 \end{bmatrix}$$

has a monic right divisor for all μ *except* $\mu = 0$.

More generally, consider

$$L(\lambda,\mu) = I\lambda^\ell + \sum_{i=0}^{\ell-1} A_i(\mu)\lambda^i \tag{15}$$

where the coefficients $A_0(\mu),\ldots,A_{\ell-1}(\mu)$ are analytic operator-valued functions defined on a connected domain Ω of $\mathbb{C}$ with values in $L(X)$. Briefly, we say that $L(\lambda,\mu)$ is analytic on Ω. A family $M(\mu)$ of subspaces of X is said to be an *analytic family* in Ω if, for every $\mu_0 \in \Omega$ there is a neighbourhood $U(\mu_0) \subset \Omega$ such that $M(\mu) = A(\mu)M_0$ for all $\mu \in U(\mu_0)$, where $A(\mu)$ is an invertible analytic operator-valued function with values in $L(X)$ and M_0 is a fixed subspace of X (depending only on μ_0).

With these concepts the analyticity of divisors can be coupled with the fact that the corresponding supporting subspaces of $C_L(\mu)$ form an analytic family. More precisely: *for each* $\mu \in \Omega$ *let* $L_1(\lambda,\mu)$ *be a right divisor of* $L(\lambda,\mu)$ *with supporting subspace* $M(\mu)$ *with respect to* $C_2(\mu)$, *and let* $L(\lambda,\mu) = L_2(\lambda,\mu)L_1(\lambda,\mu)$. *Then*

(i) $\{M(\mu): \mu \in \Omega\}$ *is an analytic family in* Ω *implies that* $L_1(\lambda,\mu)$ *and* $L_2(\lambda,\mu)$ *are analytic in* Ω.

(ii) $L_1(\lambda,\mu)$ *analytic in* Ω *implies* $L_2(\lambda,\mu)$ *is analytic in* Ω *and* $\{M(\mu): \mu \in \Omega\}$ *is an analytic family in* Ω.

(iii) *As in case* (ii) *with subscripts* "1" *and* "2" *reversed.*

Recall now the possibility of describing the supporting subspace as the image of a Riesz projector as in (13) (with T now replaced by $C_L(\mu)$) in the case that the spectrum of the divisor is separated from that of L, as in the definition of a Γ-spectral divisor. Because of the continuity properties of the spectrum of $C_L(\mu)$ it is to be anticipated that Γ-spectral divisors will be relatively "stable" under perturbations. This is the intuition behind the next result.

If, for each $\mu \in \Omega$, *there is a separated part of the spectrum of* L, *say* σ, *and a corresponding* Γ_μ*-spectral divisor* $L_1(\lambda,\mu)$ and *for each* $\mu_0 \in \Omega$ *there is a neighbourhood* $U(\mu_0)$ *such that for each* $\mu \in U(\mu_0)$, σ_0 *is inside* Γ_{μ_0}, *then* L_1 *and* L_2 *are analytic in* Ω.

PART B: MAINLY ON MATRIX POLYNOMIALS

6. Preliminary remarks

In part A our main topic concerned operator valued functions, but we occasionally abused the notation to replace operators by matrices. For the rest of this review we will generally adopt the matrix point of view so that the underlying Hilbert space X is the linear space of n-tuples from $\mathbb{C}$ (i.e. $\mathbb{C}^n$) viewed as column vectors together with the usual inner product:

$$(x,y) = \sum_{j=1}^{n} x_j \bar{y}_j \ . \tag{16}$$

Comments on problems defined on infinite-dimensional spaces will now become the exception.

Obviously all the concepts and conclusions described in Sections 2, 3 and 4 have a meaning in the new context when operators are replaced by their matrix representations in a fixed basis for X. Before going on to new topics a couple of remarks are in order on the additional information that can be obtained for factorization problems in this context.

It has been noted that the factorization of a monic matrix polynomial $L = L_2L_1$ is particularly delicate when $\sigma(L_1)$ and $\sigma(L_2)$ have points in common and even when, say, n eigenvalues are separated, the associated invariant subspace (for any linearization T) is not necessarily supporting. However, if all elementary divisors are *linear* (i.e. there may be multiple eigenvalues but no Jordan chains of length two or more) then if L has degree ℓ, it can be factored into a product of linear factors:

$$L(\lambda) = \prod_{j=1}^{\ell}(I\lambda - B_j) \ .$$

It is interesting that such a complete factorization is obtained with a condition on the nature of the root subspaces rather than the multiplicities of the eigenvalues. The result appears to originate in [24]. See also [19] as well as [10] where a proof is given using the techniques described here.

In the matrix case we have, of course, a complete spectral characterization of any matrix polynomial. This can be achieved by introducing a Jordan pair (X,J) as mentioned in Section 2. Thus, J is in Jordan form displaying all the eigenvalues and their partial multiplicities, while the columns of X are precisely the associated eigenvectors and generalized eigenvectors. Furthermore, in the case of a parameter dependent matrix polynomial, as in (15), the possibility arises of constructing an associated family of Jordan pairs (X_μ, J_μ). Fortunately, a comprehensive perturbation theory for matrix polynomials has been developed elsewhere (see [4]) which admits some description of the dependence of X_μ and J_μ on μ. Combining these ideas with those of Section 5 it is possible to obtain results on global analytic perturbations and, in particular, the global extension of divisors in the complex μ-plane. This program is also carried out in [12].

In the case of *rational* matrix valued functions more detail is available in the resolution of the *realization* problem, in particular. This important problem area arises in system theory and the reader is referred to [2] for more detail.

7. Stability

Another line of attack on perturbation problems for divisors depends on the formal notion of stability, and this turns out to be sharp enough to characterize those factorizations which are stable even though the divisor and quotient polynomials have points of spectrum in common.

Recall the metric suggested by (14) for the set M_k of all monic matrix polynomials of fixed degree k. Suppose that $L \in M_\ell$, $L_1 \in M_k$, $L_2 \in M_{\ell-k}$ with $L = L_2L$. This factorization is said to be *stable* if given $\varepsilon > 0$, there is a $\delta > 0$ such that whenever $L' \in M_\ell$ and $\rho_\ell(L,L') < \delta$, then L' admits a factorization $L' = L_2'L_1'$ with $L_1' \in M_k$, $L_2' \in M_{\ell-k}$ and

$$\rho_k(L_1,L_1') < \varepsilon, \quad \rho_{\ell-k}(L_2,L_2') < \varepsilon .$$

This seems a natural definition and, when coupled with the characterization of right divisors by supporting subspaces, it will be seen to be equivalent to a natural definition of stability for invariant subspaces. This is defined as follows: if $A \in L(X)$, an A-invariant subspace S is said to be *stable* if, given $\varepsilon > 0$ there is a $\delta > 0$ such that, whenever $\|B-A\| < \delta$, B has an invariant subspace J such that $\rho(S,J) < \varepsilon$. (Here, ρ denotes the spherical metric on subspaces of X referred to in Section 4.)

With this definition a complete description of all stable invariant subspaces of a matrix A can be found in [2]. *Let* $\sigma(I\lambda-A) = \{\lambda_1,\lambda_2,\ldots,\lambda_r\}$.

Then an A-invariant subspace S is stable if and only if $S = \sum_{k=1}^{r} \dot{+} S_k$ *where each subspace* S_k *is one of:*

(a) $\{0\}$

(b) *a complete root-subspace of* λ_k*, say* E_k*,*

(c) *any A-invariant subspace of* E_k provided *that*

$$\dim[\operatorname{Ker}(I\lambda_k - A)] = 1 \ . \tag{17}$$

For a negative illustration observe that the zero matrix can have *no* non-trivial invariant subspace which is stable. A more constructive example is an arbitrary companion matrix:

$$A = \begin{bmatrix} 0 & 1 & 0 & & 0 \\ 0 & 0 & 1 & & \\ \cdot & \cdot & \cdot & & \\ \cdot & \cdot & & \cdot & \\ \cdot & \cdot & & & \cdot \\ 0 & 0 & \cdots & & 1 \\ -a_0 & -a_1 & \cdots & & -a_{n-1} \end{bmatrix} \tag{18}$$

where $a_0, a_1, \ldots, a_{n-1} \in \mathbb{C}$. Such a matrix is known to be non-derogatory, i.e. each eigenspace satisfies condition (17). It follows that *all* invariant subspaces of a companion matrix are stable.

With this machinery stable factorizations of a monic matrix polynomial can be described as follows (see [2]): *If* L, L_1, L_2 *are monic matrix polynomials and* $L = L_2 L_1$*, then the following are equivalent*

(i) *the factorization is stable,*

(ii) *the supporting subspace with respect to linearization* T *of* L *is a stable invariant subspace of* T*,*

(iii) *if* $\lambda_0 \in \sigma(L_1) \cap \sigma(L_2)$ *then* $\dim[\operatorname{Ker} L(\lambda_0)] = 1$.

Note that if the divisor is Γ-spectral then condition (iii) is trivially satisfied and the factorization is stable. Another simple (and otherwise obvious) conclusion is that all factorizations of a scalar polynomial are stable. This follows from description (ii) since the first companion matrix is like (18), and it also follows trivially from description (iii).

8. Self-adjoint matrix polynomials

Several important applications give rise to the study of monic matrix polynomials with coefficients (in equation (1)) which are hermitian matrices; $A_j^* = A_j$ for $j = 0,1,\ldots,\ell-1$. Then, in the sense of the inner product (16), the polynomial is said to be self-adjoint. This is consistent with the terminology introduced in Section 2, and we write $L^* = L$.

It is easily seen that, in this case, the spectrum of L is symmetric with respect to the real line. However, non-linear elementary divisors, and the associated Jordan chains of generalized eigenvectors can and do occur, and these are phenomena more usually associated with non-self-adjoint problems. But some structure can be preserved when it is recognized that the linearization C_L, while certainly not hermitian, can be described as self-adjoint in an *indefinite* scalar product.

To see this, introduce the $\ell n \times \ell n$ (block Hankel) matrix

$$B = \begin{bmatrix} A_{\ell-1} & \cdots & A_1 & I \\ & & I & 0 \\ \vdots & \cdot^{\cdot^{\cdot}} & & \vdots \\ A_1 & I & & \\ I & 0 & \cdots & 0 \end{bmatrix}. \tag{19}$$

Observe that $B^* = B$ when $L^* = L$ and that B is invertible. This means that B can be used to define an indefinite scalar product $[\cdot,\cdot]$ on $\mathbb{C}^{\ell n}$ by means of

$$[x,y] = (x,By)$$

for all $x,y \in \mathbb{C}^{\ell n}$. Next, recall that the linearization C_L is just P of equation (2) and a simple calculation shows that $BC_L = C_L^*B = (BC_L)^*$. This property can be described as saying that C_L is *self-adjoint* with respect to B. More generally, any matrix M is self-adjoint with respect to B if $[Mx,y] = [x,My]$ for all $x,y \in \mathbb{C}^{\ell n}$, and this is easily seen to be equivalent to the relation $BM = M^*B$; as satisfied by $M = C_L$.

A useful observation is that, in the sense of Section 2, $L(\lambda) \oplus I_{(\ell-1)n}$ is equivalent not only to $I\lambda - C_L$ but also to the *hermitian* pencil $B\lambda - BC_L$. The analysis of the latter would be relatively simple and familiar if B were positive definite, but this can never be the case and, for most problems of interest, it is not a reasonable hypothesis for BC_L either.

In order to proceed further it is necessary to make a diversion and formulate a general result on the simultaneous reduction of C_L and B (or of B and BC_L).

9. Canonical pairs of matrices

Let H be an invertible hermitian $n \times n$ matrix. As above, define an indefinite scalar product on $\mathbb{C}^n$ by writing $[x,y] = (x,Hy)$ and let A be an H-self-adjoint matrix, i.e. one for which $HA = A^*H$. In order to lead up to the desired canonical form we consider some canonical examples.

If $J(\lambda_0,\alpha)$ is a Jordan block of size α and P_α is an $\alpha \times \alpha$ matrix with $p_{ij} = 1$ if $i+j = \alpha+1$ and $p_{ij} = 0$ otherwise, then it is easily seen that, *if* λ_0 *is real*, then $J(\lambda_0,\alpha)$ is P_α-self-adjoint and also $(-P_\alpha)$-self-adjoint.

If $\lambda_0 \neq \bar{\lambda}_0$ then the Jordan form

$$\begin{bmatrix} J(\lambda_0,\alpha) & 0 \\ 0 & J(\bar{\lambda}_0,\alpha) \end{bmatrix} \tag{20}$$

is self-adjoint with respect to the indefinite scalar product defined by

$$\begin{bmatrix} 0 & P_\alpha \\ P_\alpha & 0 \end{bmatrix} . \tag{21}$$

Now since $A^* = HAH^{-1}$, it follows that the eigenvalues of A (with their partial multiplicities) are symmetric relative to the real axis and a Jordan form for A can be built up as a direct sum of Jordan blocks with real eigenvalues together with blocks of type (20). Let $P_{\varepsilon,J}$ be a corresponding direct sum of signed matrices of type $\pm P_\alpha$, one of appropriate size for each real Jordan block in J, and a block of type (21) for each block of type (20) in J. Here, ε denotes the ordered set of signs associated one with each real Jordan block of J (i.e. in assigning $+P$ or $-P$).

Note that $P_{\varepsilon,J}^{*} = P_{\varepsilon,J}$, $P_{\varepsilon,J}^{2} = I$ and $P_{\varepsilon,J}J = J^{*}P_{\varepsilon,J}$, i.e. J is $P_{\varepsilon,J}$ self-adjoint. The result we are to describe says that, in an appropriate sense, $(J,P_{\varepsilon,J})$ is a canonical form for pairs of the form (A,H). A complete statement with proof can be found in [15], although the result has appeared elsewhere in a variety of forms over a period of fifty years or so. The formal statement includes a uniqueness result for ε, when (A,H) is fixed; this uniqueness is only up to permutations of signs on blocks P_{α} corresponding to Jordan blocks having the same size and same real eigenvalue.

Let A be H-self-adjoint and J be a Jordan normal form for A. Then there is a unique ordered set of signs ε and an invertible matrix T such that

$$T^{-1}AT = J \quad \textit{and} \quad T^{*}HT = P_{\varepsilon,J} . \tag{22}$$

Note that A and H are simultaneously reduced; A by similarity and H by congruence.

For the factorization of self-adjoint polynomials we make use of (22) for the construction of invariant subspaces of A which have specified sign (or neutrality) with respect to H. To illustrate how this can be done, let t_j be a column of T which is also an eigenvector of A. Thus $AT = TJ$ will yield $At_j = t_j\lambda_j$ for some eigenvalue λ_j and the span of t_j is an A-invariant subspace, say S. But the second relation in (22) tells us that

$$[t_j,t_j] = t_j^{*}Ht_j = 0, \text{ or } 1, \text{ or } -1.$$

We get 0 if λ_j is non-real or if λ_j is real and t_j leads a Jordan chain of length two or more. We get +1 or -1 if λ_j is real and t_j represents a complete Jordan chain; the choice of sign is determined by the corresponding entry in ε. Thus, depending on the jth diagonal element of $P_{\varepsilon,J}$ it can be asserted that S is H-neutral, or H-positive or H-negative.

Suffice it to say that this simple example can be extended to examine any invariant subspace of A and its "sign" with respect to H. (A subspace S is H-nonnegative, for example, if $[x,x] = (x,Hx) \geq 0$ for all $x \in S$.)

10. Self-adjoint matrix polynomials again

Return now to the self-adjoint polynomials introduced in Section 8. We have seen that the linearization C_L is self-adjoint with respect to B of (19). Consequently the result above applies and we can assert the existence of an invertible $\ell n \times \ell n$ Q such that

$$C_L Q = QJ \quad \text{and} \quad Q^* B Q = P_{\varepsilon,J} . \tag{23}$$

Furthermore, the set of signs ε is (essentially) uniquely determined by the polynomial L. Consequently, it is called the *sign-characteristic* of L.

Before considering the factorization problem itself observe an interesting bi-product. The second relation of (23) implies that (if sig denotes the *signature* of a matrix) sig $P_{\varepsilon,J}$ = sig B. But sig B can be determined directly from (19) and is, in fact, 0 or n according as ℓ is even or odd. Thus, for odd degrees ℓ, the fact that sig $P_{\varepsilon,J} = n$ immediately implies the existence of *real* eigenvalues for L and, furthermore, some constraint on the sign-characteristic. In fact, *if $L^* = L$ and ℓ is odd, then L has at least n elementary divisors associated with real eigenvalues.*

Two special manifestations of this result are (a) every real polynomial of odd degree has at least one real zero, (the case $n = 1$), and (b) every hermitian matrix has only real eigenvalues (the case $\ell = 1$).

As described at the end of Section 9, the relations (23) can now be utilised to investigate those invariant subspaces S of C_L which are of one sign with respect to B, i.e. subspaces S for which

$$C_L S \subset S , \quad \text{and} \quad (x,Bx) \geq 0 \quad \forall x \in S .$$

Clearly, the sign characteristic plays an essential part in these investigations.

We get a clue to the role of the "sign" attached to invariant subspaces of C_L from a special case of "symmetric" factorization. Thus, if $L^* = L$ and there are monic matrix polynomials L_1, L_3 for which $L_3^* = L_3$ and $L = L_1^* L_3 L_1$ the factorization is described as *symmetric*. It is not difficult to see, using (a) the fact that both L_1 and $L_3 L_1$ are right divisors and (b) the relationship between supporting subspaces for divisor L_1 of L and $L_3^*L_1$ of L^*, that the supporting subspace M for L_1 with respect to C_L is B-neutral, i.e.

$$(Bx,y) = 0 \quad \forall x,y \in M .$$

The direction of argument for non-symmetric factorization is then to start with the construction of B-neutral invariant subspaces of C_L and then to extend them to form B-nonnegative invariant subspaces. Indeed, if one completes the process by constructing a *maximal* B-nonnegative invariant subspaces of C_L (maximal with respect to inclusion), it can be shown that such a subspace is necessarily supporting, i.e. it determines a right divisor.

This line of argument is used in [13] (see also [15]) to prove the following theorem due, originally, to Langer [23].
Let L be a self-adjoint monic operator polynomial of degree ℓ.
(a) *If $\ell = 2k$, then L has a factorization $L = L_2L_1$ where L_1 is monic of degree k and the supporting subspace of L_1 with respect to C_L is a maximal B-nonnegative invariant subspace of C_L of dimension kn.*
(b) *If $\ell = 2k+1$, the conclusion is as in case* (a) *except that k is replaced by $k+1$.*

It is perhaps, remarkable, that even in the case $\ell = 2$ this theorem was not known before 1964. It first appeared in [21], see also [20], where generalizations are proved (when $\ell = 2$) in the context of infinite dimensional spaces.

The sign characteristic admits other interesting characterizations and the concept extends readily to the case of analytic matrix-valued functions (see [13]). A particularly interesting description of the sign characteristic can be given in terms of the eigenvalues $\mu_j(\lambda)$ of the self-adjoint matrix $L(\lambda)$ seen as functions of the *real* parameter λ. More precisely, *let $\mu_1(\lambda),\ldots,\mu_n(\lambda)$ be real analytic functions of λ for which*

$$\det[I\mu_j(\lambda) - L(\lambda)] = 0 , \qquad \lambda \in \mathbb{R} ,$$

and $j = 1,2,\ldots,n$. Let $\lambda_1 < \lambda_2 < \ldots < \lambda_r$ be the real eigenvalues of $L(\lambda)$ and for each $i = 1,2,\ldots,r$ write

$$\mu_j(\lambda) = (\lambda-\lambda_i)^{m_{ij}}\nu_{ij}(\lambda)$$

where $\nu_{ij}(\lambda_i) \neq 0$ *and is real. Then the non-zero numbers among* $m_{i1},\ldots,m_{in}$ *are the degrees of the elementary divisors of* $L(\lambda)$ *associated with* λ_i *and the sign of* $\nu_{ij}(\lambda_i)$ *(for* $m_{ij} \neq 0$*) is the sign attached to the elementary divisor* $(\lambda-\lambda_i)^{m_{ij}}$ *of* $L(\lambda)$ *in the sign characteristic of* L.

11. Epilogue

Enough machinery has now been developed to allow us to formulate results which generalize the scalar problems mentioned in the prologue. If $L(\lambda)$ is self-adjoint and we are interested in factorizations of the form $L(\lambda) = M^*(\lambda)M(\lambda)$ then this is a special case of symmetric factorization as described in Section 10, and we can bring to bear all the techniques developed to this point. The self-adjoint polynomial L is said to be *nonnegative* if $(L(\lambda)x,x) \geq 0$ for all real λ (cf. Proposition 2).

The next theorem is proved in [13]; note that the *monic* hypothesis can be relaxed here.

If $L^* = L$ *then the following are equivalent:*

(i) L *is nonnegative and* $\det L(\lambda) \not\equiv 0$.

(ii) L *admits a factorization* $L(\lambda) = M^*(\lambda)M(\lambda)$ *for some matrix polynomial* M *with* $\det M(\lambda) \not\equiv 0$, $\sigma(M)$ *is in the closed upper half-plane, and the degree of* M *does not exceed the integer part of* $\frac{1}{2}(\ell+1)$.

(iii) *The partial multiplicities for all real eigenvalues of* L *are even and* $L(a)$ *is positive definite for some* a.

(iv) *The sign characteristic of* L *consists only of* $+1$*'s, and* $L(a)$ *is positive definite for some real* a *less than all real eigenvalues of* L.

It is not difficult to see that when the above statements hold $\operatorname{sig} L(\lambda) = n$ for all real $\lambda \notin \sigma(L)$. More generally, the hypothesis $\operatorname{sig} L(\lambda) = k \leq n$ for all real $\lambda \notin \sigma(L)$ is of interest. This has been investigated in [14] where references to earlier work of Jakubovic and of Coppel can be found.

REFERENCES

1. H. Bart, I. Gohberg and M.A. Kaashoek, *A new characteristic operator function connected with operator polynomials*, Integral Equations and Operator Theory, **1**, 1-12 (1978).

2. H. Bart, I. Gohberg and M.A. Kaashoek, *Minimal Factorizations of Matrix and Operator Functions*, Birkhauser Verlag, Basel, 1979.

3. H. Bart and G. van der Beek, *Spectral Divisors of Matrix Polynomials*, Proc. R. Ir. Acad. 81A, 99-108 (1981).

4. H. Baumgärtel, *Endlich-dimensionale Analytische Störungstheorie*, Akademie-Verlag, Berlin, 1972.

5. B. den Boer, *Linearization of operator functions on arbitrary open sets*, Integral Equations and Operator Theory, 1, 19-27 (1978).

6. M.S. Brodskii, *Triangular and Jordan representations of linear operators*, Vol. 32, AMS Translation of Mathematical monographs, Providence, Rhode Island, 1971.

7. N. Cohen, *On spectral analysis of matrix polynomials*, M.Sc. Thesis, Weizmann Institute of Science, Israel, 1979.

8. L. Fejér, *Über trigonometrische Polynome*, Jour. f. reine u. ang. Mathematik 146, 53-82 (1916).

9. I. Gohberg, M.A. Kaashoek and D. Lay, *Equivalence, linearization and decomposition of holomorphic operator functions*, J. of Functional Analysis, 28, 102-144 (1978).

10. I. Gohberg, P. Lancaster and L. Rodman, *Spectral analysis of matrix polynomials - I. Canonical forms and divisors*, Lin. Alg. & Applic. 20, 1-44 (1978).

11. I. Gohberg, P. Lancaster and L. Rodman, *Representations and divisibility of operator polynomials*, Canadian J. Math., 30, 1045-1069 (1978).

12. I. Gohberg, P. Lancaster and L. Rodman, *Perturbation theory for divisors of operator polynomials*, SIAM Journal of Math. Anal. 10, 1161-1183 (1979).

13. I. Gohberg, P. Lancaster and L. Rodman, *Spectral analysis of self-adjoint matrix polynomials*, Annals of Mathematics, 112, 34-71 (1980).

14. I. Gohberg, P. Lancaster and L. Rodman, *Factorization of matrix polynomials with constant signature*, Linear & Multilinear Algebra, 1982 (to appear).

15. I. Gohberg, P. Lancaster and L. Rodman, *Matrix Polynomials*, Academic Press, New York, 1982.

16. I. Gohberg, L. Lerer and L. Rodman, *Stable factorization of operator polynomials and spectral divisors simply behaved at infinity, I and II*, J. of Math. Analysis and Appl. 74, 401-431 (1980); 75, 1-40 (1980).

17. I. Gohberg and L. Rodman, *On the spectral structure of monic matrix polynomials and the extension problem*, Linear Algebra Appl. 24, 157-172 (1979).

18. M.A. Kaashoek, C.V.M. van der Mee and L. Rodman, *Analytic operator functions with compact spectrum. I. Spectral nodes, linearization and equivalence*, Integral Equations and Operator Theory 4, 504-547 (1981).

19. P.S. Kasimirsky, *On the factorization of a matrix polynomial*, Ukr. Math. J. 24, 316-327 (1972) (In Russian).

20. M.G. Krein and H. Langer, *On some mathematical principles in the linear theory of damped oscillations of continua, I, II*, Integral Equations and Operator Theory, 1, 364-399, 539-566 (1978) (Translation).

21. H. Langer, *Spektraltheorie linearer Operatoren in J-Räumen und einige Anwendungen auf die Schar* $L(\lambda) = \lambda^2 I + \lambda B + C$. Habilitationsschrift, Tehnischen Universität Dresden, 1964.

22. H. Langer, *Über eine nichtlinearer Eigenwertprobleme*, Acta Sci. Math. (Szeged) 35, 79-93 (1973).

23. H. Langer, *Factorization of operator pencils*, Acta Sci. Math. (Szeged) 38, 83-96 (1976).

24. A.S. Marcus and I.V. Mereutsa, *On some properties of λ-matrices* Matem. Issled., Kishinev 3, 207-213 (1975) (Russian).

25. I.V. Mereutsa, *On the properties of the roots of an operator equation corresponding to an operator pencil*, Mat. Issled. Kishinev 1(27), 96-115 (1973) (In Russian).

26. B. Mityagin, *Linearization of holomorphic operator functions I and II*, Integral Equations and Operator Theory, 1, 114-131, 226-249 (1978).

SMOOTH BUMP FUNCTIONS AND RENORMINGS OF BANACH SPACES: A SURVEY

J. H. M. Whitfield

Lakehead University
Thunder Bay, Ontario
Canada

Throughout this survey X will denote a real Banach space, X^* its continuous dual, $B(X)$ and $S(X)$ its closed unit ball and sphere, respectively. If X is finite dimensional, it is well known that there exists a function $f: X \to \mathbb{R}$ (the reals) which is of class C^k (in fact C^∞) and f has bounded nonempty support, that is, $c\ell\{x \in X: f(x) \neq 0\}$ is nonempty and bounded. We call such a function a C^k-bump function on X. On the other hand, if X is infinite dimensional, the situation is more delicate.

The purpose of this survey is to present some of the results in this setting with particular emphasis on the relationship between smooth renormings and the existence of smooth bump functions. The results are not all given in their greatest generality and will not be accompanied by proofs in most cases.

Differentiable will mean Frechet differentiable as we will restrict our attention to that case. For an introduction to the differential calculus in the infinite dimensional setting see [5] or [12]. Definitions and terminology concerning Banach spaces can be found in [3] or [4].

The easiest, and perhaps only, way of obtaining a C^k-bump function on X is via a C^k-norm. (By a C^k-norm on X we mean one that is of class C^k everywhere on X except at the origin.) Namely, one chooses a C^∞-bump function on $\mathbb{R}$ equal to one on some neighbourhood of zero and composes it with the C^k-norm to obtain such a function on X. Thus, the ability to renorm X with an equivalent C^k-norm is of interest and importance here. In fact, it may be true, but is unknown in general at this time, that if X admits a C^k-bump function, then X can be renormed with an equivalent C^k-norm.

A list of some infinite dimensional Banach spaces which admit (or do not admit) smooth bump functions is given below.

(a) Hilbert space admits C^∞-bump functions.

(b) $c_0(\Gamma)$, for Γ any set of infinite cardinality, admits C^∞-bump functions, since $c_0(\Gamma)$ admits an equivalent C^∞-norm (see [2]).

(c) ℓ^p and L^p, $2 \le p < \infty$ and p an even integer, admit C^∞-bump functions (see [2]).

(d) ℓ^p and L^p, $1 < p < \infty$, p is not an even integer:

(i) admit C^{p-1}-bump functions, if p is an odd integer;

(ii) admit $C^{[p]}$-bump functions, if p is not an integer; ([p] is the greatest integer less than p.)

(iii) do not admit a C^n-bump function, if $n \ge p$. (Once again, see [2]; actually the results there are somewhat more precise than those given here.)

(e) ℓ^1 and $C[0, 1]$ do not admit differentiable bump functions. (Kurzweil [11] shows that these spaces do not admit C^1-bump functions; the result given above is due to Bonic and Frampton. However, Kurzweil's technique for showing the nonexistence of such bump functions, suitably refined, is that used to obtain subsequent similar results, eg. [2, 13, 14].)

(f) Restrepo [16] shows that a separable X admits an equivalent C^1-norm if and only if X^* is separable. Consequently if X^* is separable, then X admits a C^1-bump function.

The converse of the final results in (f) obtains and follows (as do the results in (e)) from the theorems given below (see Corollaries 1 and 2). First some definitions and preliminary results are needed.

DEFINITION 1. A norm on X is said to be rough if there is $\varepsilon > 0$ such that for all $x \in X$, $x \ne 0$, and $\eta > 0$ there are $x_1, x_2 \in X$, $u \in S(X)$ such that $||x_i - x|| < \eta$, $i = 1, 2$, and $(d^+||x_1|| - d^+||x_2||)(u) \ge \varepsilon$, where $d^+||x||(u) = \lim_{t\to 0+} \frac{1}{t}(||x + tu|| - ||x||)$. (Recall that for $x \ne 0$ this limit exists for all $u \in S(X)$.)

Note that a norm is rough if $d^+||\cdot||$ is uniformly discontinuous on $S(X)$, so such a norm cannot be Frechet differentiable. In fact, as is seen below, if X admits an equivalent rough norm, then it cannot be renormed with a differentiable norm. The usual norms on ℓ^1 and $C[0, 1]$ are both rough with $\varepsilon = 2$.

DEFINITION 2. $B(X^*)$ is weak*-dentable if for any given $\varepsilon > 0$ there are $\alpha > 0$ and $x \in S(X)$ such that $\operatorname{diam}(\{x^* \in B(X^*): x^*(x) > 1 - \alpha\}) < \varepsilon$.

LEMMA 1. The following are equivalent.

(i) The norm on X is rough.

(ii) $B(X^*)$ is not weak*-dentable.

(iii) There is $\varepsilon > 0$ such that for each $x \in S(x)$ and $\eta > 0$ there is $v \in S(X)$ such that $||x + tv|| \geq ||x|| + |t|\varepsilon - \eta$ for all t, $|t| \leq 1$.

LEMMA 2. If X admits an equivalent rough norm and $f: X \to \mathbb{R}$, $f(0) = 0$, is differentiable, then $\{x: f(x) \leq ||x||\}$ is unbounded.

The proofs of these lemmas can be found in [10] and [13].

THEOREM 1. If X admits an equivalent rough norm, then X does not admit a differentiable bump function.

Proof. Assume X admits a rough norm and there is a differentiable function $f: X \to \mathbb{R}$, $f(0) = 2$, $f[B(X)] \subseteq [-1, 1]$ and $f \equiv 0$ outside $B(X)$. Let $g(x) = 1 - f(x)$ and use it to contradict Lemma 2.

THEOREM 2. X admits an equivalent rough norm if and only if there is a separable subspace $Y \subseteq X$ with Y^* nonseparable.

The proof of the "only if part" is due to Leach and Whitfield; details are in [18]. The "if part" follows from Lemma 1 and some results of Namioka and Phelps [15]. In the case X is separable and X^* is non-separable (or more generally, if dens $X <$ dens X^*) there is a construction of a rough norm on X given in [13].

COROLLARY 1. If X admits a differentiable bump function, then every separable subspace of X has a separable dual.

This immediate consequence of Theorems 1 and 2 was first proved by Ekeland and Lebourg [6] using much different methods.

The situation for separable spaces is summarized in the following.

COROLLARY 2. If X is separable, then the following are equivalent:

(i) X^* is separable.

(ii) X admits an equivalent C^1-norm.

(iii) X admits a C^1-bump function.

An outstanding problem is the converse of Corollary 1, as is the possibly more difficult problem of whether or not the conclusion of Corollary 1 implies X can be renormed with a C^1-norm. In the case X is reflexive, Trojanski [17] shows that X can be so renormed. This is a consequence of the following deep results.

DEFINITION 3. X is said to be weakly compactly generated (WCG) if there is a weakly compact set $K \subseteq X$ whose closed linear span is X .

The class of WCG spaces includes all separable and reflexive spaces, $c_0(\Gamma)$, $L_1(\mu)$, where μ is σ-finite and many others. (There is a nice discourse on WCG spaces in [4].) In 1968 Amir and Lindenstrauss obtained the fundamental result that if X is WCG then there is a set Γ and a continuous, linear, one-to-one mapping $T: X \to c_0(\Gamma)$. Using this result and an intricate construction of a 'long sequence' of projections, Trojanski [17] proved the following.

THEOREM 3. If X is WCG, then X can be renormed with an equivalent locally uniformly rotund (LUR) norm.

Recall that a norm $||\cdot||$ is LUR if $||x|| = ||x_n|| = 1$ and $||x_n + x|| \to 2$, then $||x_n - x|| \to 0$. Also if X can be renormed such that the dual norm on X^* is LUR, then the norm on X is differentiable.

A consequence of Theorem 3 is:

COROLLARY 3. If X is reflexive, then X can be renormed with a C^1-norm. Consequently, reflexive spaces admit C^1-bump functions.

By refining Trojanski's techniques, John and Zizler [8, 9] obtained the following.

THEOREM 4. (i) If X is WCG then there is an equivalent C^1-norm on X if (and only if) X admits a C^1-bump function.
(ii) If X and X^* are both WCG, then X admits an equivalent C^1-norm and, consequently, a C^1-bump function.

There are examples of spaces X such that X is WCG and X^* is not and spaces such that X^* is WCG and X is not. Also if X^* is WCG, then every separable subspace of X has a separable dual. So the problems stated following Corollary 2 can be specialized to spaces X such that X^* is WCG.

We conclude with the following result that shows that both X and

X^* can admit C^k-bump functions, $k \geq 2$, only for a restricted class of spaces.

THEOREM 5. The following are equivalent.

(i) X is isomorphic to Hilbert space.

(ii) Both X and X^* can be renormed with C^2-norms.

(iii) Both X and X^* admit C^2-bump functions.

The implications (i) $\Rightarrow$ (ii) $\Rightarrow$ (iii) are immediate. (iii) $\Rightarrow$ (i) is proven by Meshkov [14] and a more general result appears in [7].

REFERENCES

1. D. Amir and J. Lindenstrauss, The structure of weakly compact sets in Banach spaces, Ann. of Math. 88 (1968), 35-46.
2. R. Bonic and J. Frampton, Smooth functions on Banach manifolds, J. Math. and Mech. 15 (1966), 877-898.
3. M. M. Day, Normed Linear Spaces, 3rd ed., Springer-Verlag, Berlin and New York, 1973.
4. J. Diestel, Geometry of Banach Spaces - Selected Topics, Lecture Notes in Math., vol. 485, Springer-Verlag, Berlin and New York, 1975.
5. J. Dieudonné, Foundations of Modern Analysis, Academic Press, New York, 1960.
6. I. Ekeland and G. Lebourg, Generic Frechet differentiability and perturbed optimization problems in Banach spaces, Trans. Amer. Math. Soc. 224 (1976), 193-216.
7. M. Fabian, J. H. M. Whitfield and V. Zizler, Norms with locally Lipschitzian derivatives, preprint.
8. K. John and V. Zizler, A renorming of dual spaces, Israel J. Math. 12 (1972), 331-336.
9. K. John and V. Zizler, Smoothness and its equivalents in weakly compactly generated Banach spaces, J. Functional Analysis 15 (1974), 1-11.
10. K. John and V. Zizler, On rough norms on Banach spaces, Comment. Math. Univ. Carolinae 19 (1978), 335-349.
11. J. Kurzweil, On approximation in real Banach spaces, Studia Math. 14 (1954), 213-231.
12. S. Lang, Introduction to Differentiable Manifolds, Interscience, New York, 1962.
13. E. B. Leach and J. H. M. Whitfield, Differentiable functions and rough norms on Banach spaces, Proc. Amer. Math. Soc. 33 (1972), 120-126.
14. V. Z. Meshkov, Smoothness properties in Banach spaces, Studia Math. 63 (1978), 111-123.
15. I. Namioka and R. R. Phelps, Banach spaces which are Asplund spaces, Duke Math. J. 42 (1975), 735-750.

16. G. Restrepo, Differentiable norms in Banach spaces, Bull. Amer. Math. Soc. 70 (1964), 413-414.

17. S. L. Trojanski, On locally uniformly convex and differentiable norms in certain non-separable Banach spaces, Studia Math. 37 (1971), 173-180.

18. J. H. M. Whitfield, Rough norms on Banach spaces, Proceedings, 25th Meeting, Can. Math. Congress, (1972), 595-601.

CERTAIN EXTREME VALUE PROBLEMS IN STATISTICS*

C. S. Wong

University of Windsor
Windsor, Ontario
Canada

1. Introduction

Examples of extreme value problems in statistics include the finding of uniformly most powerful tests, least square estimates, maximum likelihood estimates, Bayes rules, minmax rules, maximum likelihood ratio procedures and optimal designs of experiments. Let f be a real-valued function on a convex subset C of a locally convex topological (Hausdorff) linear space E over the real field $\mathbf{R}$. Each of the above problems may fall into one of the following two categories: f attains its extreme (maximum or minimum) value at some point on the boundary or in the interior of C. The tools used can be differentials, inequalities or combinations of both. For example, the likelihood estimate can often be found by differentials and when f is convex, the notion of monotone operators can be used to justify that the point obtained is indeed a maximum likelihood estimate [13]. If f attains its extreme value on the boundary of C, sometimes the notion of Lagrange multipliers can be introduced so that the method of differentials is still applicable; more often, however, various inequalities are needed to find the extreme value of f. The separation theorem (or equivalently, Hahn-Banach theorem) is one of the most important inequalities. For example, a generalization of the Neyman-Pearson Lemma [10], existence of certain Bayes rules, minimax rules and the values of certain games can be proved by the separation

*This research is partially supported by NSERC Grant A8518.

theorem [7]. (The existence of the value of a game may also be shown by a fixed point theorem (see e.g. [1], [8]). Indeed, it was known more than forty years ago that, with effort, the Hahn-Banach theorem can be rephrased as a fixed point theorem. Many statistical problems, such as the existence of optimal equivariant rules, can be treated as fixed point problems. However, it is often not the case that the fixed point theory, at its present stage, can be applied to solve a given fixed point problem in statistics.) Certain complete class theorems in statistical decision theory are consequences of the separation theorem through versions of Jensen's inequality. To illustrate this, let me state the following result in [12].

PROPOSITION. Let $A \subset R^n$. Then A is convex if and only if for any random vector X of a probability space $(\Omega, \mathcal{A}, P)$ into R^n, $E(X) \in A$ whenever $E(X) \in R^n$ and $P(X \in A) = 1$.

The above result includes a mean-value theorem which is a consequence of the separation theorem. The following version of Jensen's inequality follows immediately from the above mean-value theorem.

COROLLARY. Let A be a convex subset of R^n, f be a convex function of A into R and X be a random vector of a probability space $(\Omega, \mathcal{A}, P)$ into R^n such that $P(X \in A) = 1$, $E(X)$ and $E(f(X))$ are finite. Then

$$f(E(X)) \leq E(f(X)).$$

In the remaining talk, we shall merely demonstrate how certain optimal designs of experiments can be obtained.

2. Optimal Designs Of A Regression Experiment

Let $\{Y(t)\}_{t \in T}$ be a vector-valued stochastic process on $T = [a, b]$, $a < b$, such that each $Y(t)$ is a random vector of a probability space $(\Omega, \mathcal{A}, P)$ into R^n with

$$E(Y(t)) = f(t)\theta, \ \mathrm{Cov}(Y(s), Y(t)) = (R_i(s, t)\delta_{ij}), \ s, t \in T,$$

where $\theta = (\theta_j) \in R^p$, $f = (f_j)$ is a continuous function of T into

$R^{n\times p}$, $f\theta = \sum_{j=1}^{p} \theta_j f_j$, R_i's are known, positive definite and are continuous on $T \times T$. Let $\varepsilon(t) = Y(t) - f(t)\theta$. Then $E(\varepsilon(t)) = 0$ and $R(s, t) = E(\varepsilon(s)\varepsilon(t)')$, $s, t \in T$. For brevity, the model $\{Y(t)\}_{t\in T}$ will be denoted by $m(f)$. (If we allow $a = b$, then $m(f)$ is the usual linear model.) Following Parzen [11], we associate each R_i with a reproducing kernel Hilbert space $H(R_i)$. If we treat R_i as a linear operator on $L^2([a, b])$, then $H(R_i)$ is the range of $R_i^{1/2}$, with the following inner product $(f, g)_{R_i}$:

$$(f, g)_{R_i} = (f, R_i^{-1}(g)), \quad f, g \in H(R_i),$$

where (f, g) is the usual inner product for $L^2([a, b])$. Let $H(R)$ be the product Hilbert space of all $H(R_i)$. Thus, the inner product $(\ ,\ )_R$ and the norm $\|\ \|_R$ for $H(R)$ are defined by

$$(f, g)_R = \sum_{j=1}^{n} (f_j, g_j)_{R_j}, \quad \|f\|_R = \left(\sum_{j=1}^{n} \|f_j\|^2_{R_j}\right)^{1/2}, \quad f, g \in H(R).$$

We assume that for each $\omega \in \Omega$, $Y(\cdot)(\omega) \in H(R)$ and $f \in D$, where $D = \{(f_j) : \|f_j\|_R \leq \ell\}$ and $\ell > 0$. Let $\omega \in \Omega$, $z \in R^p$. Then there exists a unique $\hat{\theta}(f)(\omega) \in R^p$ such that

$$\|Y(\cdot)(\omega) - f\hat{\theta}(f)(\omega)\|^2_R = \min_{\theta\in R^p} \|Y(\cdot)(\omega) - f\theta\|^2_R .$$

The least square estimator $\hat{\theta}(f)$ of the unknown parameter θ is often referred to as the Gauss-Markov estimator of θ and can also be defined as follows.

$$\hat{\theta}(\omega) = M(f)^{-1}((Y(\cdot)(\omega), f_j)_R), \quad M(f) = ((f_j, f_k)_R).$$

An estimator $W = (W_j)$ of θ is said to be a continuous linear estimator of θ if there exist continuous linear functionals L_j on $H(R)$ such that

$$W_j(\omega) = L_j(Y(\cdot)(\omega), \omega \in \Omega, j = 1, 2, \ldots, p.$$

$\hat{\theta}$ is BLUE in the sense that among all continuous linear unbiased estimators W of θ, $\sum_W - \sum_{\hat{\theta}(f)}$ is nonnegative definitive, where $\sum_W$ is dispersion matrix of W. We shall equip $M \equiv \{m(f) : f \in D\}$ with a certain practical total order $\leq$. For $f \in D$, $m(f)$ is said to be $\leq$-optimal if $m(f) \geq m(g)$, $g \in D$. Maxe A will denote the largest eigenvalue of a symmetric matrix A over R. For $f, g \in D$, we define that

$$m(f)\ E\ m(g) \quad \text{if} \quad \text{maxe}\textstyle\sum(f) \geq \text{maxe}\sum(g);$$

$$m(f)\ D\ m(g) \quad \text{if} \quad |\textstyle\sum(f)| \geq |\sum(g)|;$$

$$m(f)\ A\ m(g) \quad \text{if} \quad \text{tr}\textstyle\sum(f) \geq \text{tr}\sum(g),$$

where $\sum(f)$ is the dispersion matrix of $\hat{\theta}(f)$, and is equal to $M(f)^{-1}$. Let $\{\lambda(f)_j\}_{j=1}^p$ be the spectrum of $M(f)$ and ϕ be a function of $(0, p\ell^2]$ into R. We define for $f, g \in D$,

$$m(f) \leq_\phi m(g) \quad \text{if} \quad \sum_{j=1}^p \phi(\lambda(f)_j) \geq \sum_{j=1}^p \phi(\lambda(g)_j).$$

THEOREM 1. Let $f \in D$ and let ϕ be a continuous strictly decreasing function of $(0, p\ell^2]$ into R such that $\lim_{x \to 0} \phi(x) = \infty$ and ϕ' is a one-to-one function of $(0, p\ell^2)$ into R. Then the following conditions are equivalent.

(i) $f_1, f_2, \ldots, f_p$ are orthogonal in $H(R)$ with norm ℓ.

(ii) f is $\leq_\phi$-optimal.

Proof. We shall prove the case where $p \geq 2$. Consider the function H on $K = \{\lambda \in R^p : \lambda_j > 0, \sum_{j=1}^p \lambda_j \leq p\ell^2\}$ defined by

$$H(\lambda) = \sum_{j=1}^p \phi(\lambda_j), \quad \lambda \in K.$$

Since ϕ is a continuous function of $(0, p\ell^2]$ into $\mathbb{R}$ such that $\lim_{x\to 0} \phi(x) = \infty$, there exists $\lambda_0 = (\lambda_{oj})$ in K such that

$$H(\lambda_0) = \min H(K).$$

Since ϕ is strictly decreasing, $\sum_{j=1}^{p} \lambda_{oj} = p\ell^2$. Consider

$$F(\lambda) = \sum_{j=1}^{p-1} \phi(\lambda_j) + \phi(p\ell^2 - \sum_{j=1}^{p-1} \lambda_j),\ \lambda \in S = \{\mu \in \mathbf{R}^{p-1} : \mu_j > 0,$$

$$\sum_{j=1}^{p-1} \eta_j < p\ell^2\}.$$

Then

$$F((\lambda_{oj})_{j=1}^{p-1}) = \min F(S)$$

and

$$\frac{\partial F(\lambda)}{\partial \lambda_k} = \phi'(\lambda_k) - \phi'(p\ell^2 - \sum_{j=1}^{p-1} \lambda_j),\ k = 1, 2, \ldots, p - 1.$$

Thus, all $\frac{\partial F(\lambda)}{\partial \lambda_k} = 0$ at $\lambda = (\lambda_{oj})_{j=1}^{p-1}$. Since ϕ' is one-to-one on $(0, p\ell^2)$,

$$\lambda_{ok} = p\ell^2 - \sum_{j=1}^{p-1} \lambda_{oj},\ k = 1, 2, \ldots, p - 1.$$

So all $\lambda_{ok} = \ell^2$, $k = 1, 2, \ldots, p$. Now note that there exist $f_1, f_2, \ldots, f_p$ such that (i) holds. For this f, each $\lambda(f)_j = \ell^2 = \lambda_{oj}$. The desired result follows routinely.

COROLLARY. Let $f \in D$. Then (i) in Theorem 1 is equivalent to each of the following.

(iii) f is A-optimal.

(iv) f is D-optimal.

THEOREM 2. Let $f \in D$. Then (i) in Theorem 1 is equivalent to

(v) f is E-optimal.

The above results correct and generalize the results in [3] (see also [4]). Our proofs are much simpler than those in [3]. D is constructed from the covariance function R; it is abstract and may not be practical. A different approach to the above problem is given in [5] and for A-optimality, solved in [2]. However, $\hat{\theta}$ in [5] may not be a BLUE. Thus, it is desirable to find an optimal experiment for the above $\hat{\theta}$ with D replaced by something like $\{(f_{ij}) : a_j \leq \|f_{ij}\|_\infty \leq b_j\}$, $a_j < b_j$. This problem is unsolved (even for the usual linear regression model [6], [9]) and is expected to be very difficult.

REFERENCES

1. F. E. Browder, "The fixed point theory of multivalued mappings in topological vector spaces", Math. Ann. 177 (1968), 283-301.

2. N. N. Chan and C. S. Wong, "Existence of an A-optimal model for a regression experiment", J. Math. Anal. and Appli. 77 (1980), 403-415.

3. D. S. Cheng, "Design of optimal control for a regression problem", Ann. Statist. 7 (1979), 1078-1085.

4. D. S. Cheng and C. S. Wong, Correction to "Design of optimal control for a regression problem", Ann. Statist. 8 (1980), 1403.

5. D. S. Cheng and C. S. Wong, "A general approach to optimal control of a regression experiment", J. Multivariate Anal. 11 (1981), 85-101.

6. C. S. Cheng, "Optimality of some weighing and 2^n fractional factorial designs", Ann. Statist. 8 (1980), 436-446.

7. T. S. Ferguson, "Mathematical Statistics", Academic Press, NY, (1967).

8. C. W. Ha, "Minimax and fixed point theorems", Math. Ann. 248 (1980), 73-77.

9. A. Hedayat and W. D. Wallis, "Hadamard matrices and their applications", Ann. Statist. 6 (1978), 1184-1238.

10. E. L. Lehmann, "Testing statistical hypotheses", John Wiley & Sons, Inc., New York, (1959).

11. E. Parzen, "An approach to time series analysis", Ann. Math. Statist. 32 (1961), 951-989.

12. C. S. Wong, "Mathematical Statistics", 3 vols., Tamkang Chair 24 (1980), Tamkang University.

13. C. S. Wong and K. S. Wong, "Minima and maxima in multivariate analysis", Canad. J. Statist. 8 (1980), 103-113.

FIXED POINT THEOREMS AND RELATED RESULTS

SOME GEOMETRIC PROPERTIES RELATED TO NORMAL STRUCTURE

D. van Dulst

University of Amsterdam
Amsterdam
The Netherlands

It is our aim here to call attention to some recent work on the geometry of Banach spaces and its relation to fixed point theory. We first discuss the work of Bernal ([2]) and then, in the second part, describe some fixed point results involving a property which is an outgrowth of recent work of Huff ([6]). Let us first agree on the following terminology. We say that a [dual] Banach space X has the [dual] fixed point property (FPP) [(FPP*)] if for every weakly [weak*] compact convex set $K \subset X$ and for every nonexpansive $T : K \to K$ there exists an $x \in K$ such that $Tx = x$. X is said to have normal structure if every convex bounded $K \subset X$ contains a point x satisfying $\sup\{\|x - y\| : y \in K\} < \operatorname{diam} K$.

1. In his thesis ([2]) Bernal studies k-dimensional convexity moduli, continuing the work of Sullivan ([9]) and others ([5]). Let us recall the definition of uniform convexity. For a space X to be uniformly convex it is required that for every $\varepsilon > 0$ there exists a $\delta > 0$ such that the midpoint $\frac{x + y}{2}$ of any segment $[x, y]$ in the unit ball B_X having length $\|x - y\| \geq \varepsilon$ satisfies $\|\frac{x + y}{2}\| \leq 1 - \delta$. Now let $k \in \mathbb{N}$ be arbitrary. Let us replace the two point set $\{x, y\} \subset B_X$ with any set $\{x_1, \ldots, x_{k+1}\}$ of $k + 1$ points in B_X and let us assume, instead of $\|x - y\| \geq \varepsilon$, that the "volume" $A(x_1, \ldots, x_{k+1})$ of the set $\{\sum_{j=1}^{k} \gamma_j (x_j - x_{k+1}) : 0 \leq \gamma_j \leq 1, j = 1, \ldots, k\}$ is $\geq \varepsilon$ ($\varepsilon > 0$ arbitrary). If it follows from this assumption that the centroid $\frac{x_1 + \ldots + x_{k+1}}{k + 1}$ has

norm $\leq 1 - \delta$ for some $\delta > 0$ depending only on ε, then we say that X is k-uniformly convex (k-UC). It is not immediately clear how the volume $A(x_1, \ldots, x_{k+1})$ should be defined in an arbitrary Banach space (cf. [9]). For our purposes it suffices to mention that in Hilbert space $A(x_1, \ldots, x_{k+1})$ equals, and in the general case is at least as large as the number

$$\text{dist}(x_{k+1},[x_k,\ldots,x_1])\cdot\text{dist}(x_k,[x_{k-1},\ldots,x_1])\ldots\text{dist}(x_3,[x_2,x_1])\cdot$$

$$\|x_2-x_1\|,$$

where $[x_i, \ldots, x_1]$ denotes the affine span of $x_1, \ldots, x_i$ $(i = 1, \ldots, k)$. The modulus δ_k of k-uniform convexity is now defined for every $\varepsilon > 0$ as $\delta_k(\varepsilon) := \inf\{1 - \|\frac{x_1+\ldots+x_{k+1}}{k+1}\| : x_1, \ldots, x_{k+1} \in B_X, A(x_1,\ldots, x_{k+1}) \geq \varepsilon\}$. In terms of the modulus δ_k the definition of k - UC now reads: $\delta_k(\varepsilon) > 0$ for every $\varepsilon > 0$. Note that uniformly convex = 1 - UC. It is known ([9]) that k - UC implies $(k + 1)$ - UC for every $k \in N$ and that the converse is false.

One of the main results of Bernal is, loosely stated, that if for some $k \in \mathbb{N}$ the modulus of k-uniform convexity satisfies $\delta_k(\varepsilon) > 0$ for some $0 < \varepsilon < 1$, then X has normal structure.

THEOREM 1 ([2]). A Banach space X has normal structure if there exist $0 < \varepsilon < 1$, $\delta > 0$ and $k \in \mathbb{N}$ such that for any set $\{x_1, \ldots, x_{k+1}\}$ in the unit ball B_X of X,

$$\text{dist}(x_{k+1},[x_k,\ldots,x_1])\ldots\text{dist}(x_3,[x_2,x_1])\cdot\|x_2 - x_1\| \geq \varepsilon$$

implies

$$\|\frac{x_1+\ldots x_{k+1}}{k+1}\| \leq 1 - \delta.$$

The condition $\varepsilon < 1$ is crucial here. Observe that, for $x_1, \ldots x_{k+1} \in B_X$, the quantity

$$\text{dist}(x_{k+1}, [x_k, \ldots x_1]) \ldots \text{dist}(x_3, [x_1]) \cdot \|x_2 - x_1\|$$

is bounded above by 2^k. As we shall presently see, the value 2^k can be attained in any Banach space, after suitable renorming. Therefore, the following property suggests itself for consideration.

(*) There exist $k \in \mathbf{N}$, $0 < \varepsilon < 2^k$ and $\delta > 0$ such that for any set $\{x_1, \ldots, x_{k+1}\} \subset B_X$.

$$\text{dist}(x_{k+1}, [x_k, \ldots, x_1]) \ldots \text{dist}(x_3, [x_2, x_1]) \cdot \|x_2 - x_1\| \geq \varepsilon$$

implies $\left\|\dfrac{x_1 + \ldots x_{k+1}}{k+1}\right\| \leq 1 - \delta$.

It is a remarkable fact, established by Bernal ([2]) that in Hilbert space we have $\lim_{k \to \infty} \delta_k(\varepsilon) = 1$ for every $\varepsilon > 0$. Using this, he proves the following.

THEOREM 2 ([2]). Let $(X, \|\cdot\|)$ be a Hilbert space and let $|||\cdot|||$ be an equivalent norm on X satisfying

$$A\, |||\cdot||| \leq \|\cdot\| \leq C\, |||\cdot||| \text{ with } \frac{A}{C} > \frac{1}{2}.$$

Then $(X, |||\cdot|||)$ satisfies (*).

In this connection let us recall the renormings of Hilbert space considered by Karlovitz ([7]). For $\beta > 1$ he defines an equivalent norm $\|\cdot\|_\beta$ on l^2 by

$$\|\cdot\|_\beta := \max(\beta\|\cdot\|_\infty, \|\cdot\|_2),$$

where

$$\|x\|_\infty := \sup_{i \in \mathbf{N}} |x_i| \text{ and } \|x\|_2 := \left(\sum_{i=1}^{\infty} x_i^2\right)^{\frac{1}{2}}.$$

Clearly $\frac{1}{\beta}\|\cdot\|_\beta \leq \|\cdot\|_2 \leq \|\cdot\|_\beta$, so that Theorem 2 yields that $(l^2, \|\cdot\|_\beta)$ satisfies (*) for $1 \leq \beta < 2$. On the other hand, Baillon and Schöneberg

([1]) have shown that $X_\beta(=(l^2, \|\cdot\|_\beta))$ has the (FPP) for $1 \le \beta < 2$. This last result would follow from Bernal's work if the following question would have an affirmative answer.

QUESTION 1 ([2]). Does (*) imply the (FPP)? Note that for $\sqrt{2} \le \beta < 2$, X_β does not have normal structure ([1]), so that (*) is strictly weaker than the condition appearing in Theorem 1.

A final observation should be made here. It can be shown ([2]), using a technique of James, that (*) implies superreflexivity. (Therefore also the spaces to which Theorem 1 applies are all superreflexive.) The converse, however, is false. This follows from the following renorming theorem.

THEOREM 3 ([4]). Any Banach space $(X, \|\cdot\|)$ (so in particular Hilbert space) admits an equivalent norm $|||\cdot|||$ for which there exists a sequence $\{x_n\}$ with the following properties.

(i) $|||x_n||| = 1$ for all $n \in \mathbb{N}$

(ii) $|||\frac{x_1+\ldots+x_n}{n}||| = 1$ for all $n \in \mathbb{N}$

(iii) $|||x_i - z||| = 2$ for all $i \in \mathbb{N}$ and every $z \in [x_1,\ldots,x_{i-1},x_{i+1},\ldots]$.

In the proof of this theorem two sequences $\{x_n\} \subset X$ and $\{x_n^*\} \subset X^*$ and an equivalent norm $|||\cdot|||$ on X are constructed so that

$$|||x_n||| = |||x_n^*||| = 1 \quad \text{for all } n \in \mathbb{N}$$

and

$$x_m^*(x_n) = \begin{cases} 1 & \text{if } m \neq n \\ -1 & \text{if } m = n \end{cases}.$$

(ii) and (iii) easily follow from this.

It is clear that $(X, |||\cdot|||)$ does not satisfy (*). Also, of course, it fails to have normal structure. It is even possible, by a slight modification in the proof of Theorem 3 to construct on any Banach space an equivalent norm $|||\cdot|||$ so that $(X, |||\cdot|||)$ fails to have asymptotic normal structure (cf. [1]) and does not satisfy (*). Specializing this construction to l^2 yields the norm

$$|||x||| = \max(\tfrac{1}{3}\|x\|_2, \sup_{n=2,3,\ldots} |x_1 + x_n + x_{n+1}|).$$

Note that, by Theorem 2, the Banach-Mazur distance of $(l^2, |||\cdot|||)$ and $(l^2, \|\cdot\|_2)$ is ≥ 2. All this suggests

QUESTION 2. Does l^2 with the above norm have the (FPP)?

2. We now pass to another weakening of uniform convexity that is related to fixed point theory. There is some analogy with Bernal's work but whereas the conditions discussed so far (such as (*) and the stronger one appearing in Theorem 1) are finite dimensional, the property (P) below establishes a link between the weak and the norm topology, and is therefore essentially infinite-dimensional.

We say that the norm of a [dual] Banach space is a Kadec-Klee (KK) norm [(KK*)-norm] provided that on the unit sphere sequences converge in norm whenever they converge weakly [weak*]. Clearly every uniformly convex norm is (KK). In a recent paper ([6]) Huff introduces the intermediate notions of uniformly Kadec-Klee (UKK) and nearly uniformly convex (NUC) and proves the following implications

(UC) $\Rightarrow$ (NUC) $\Rightarrow$ (UKK) $\Rightarrow$ (KK).

Also, for each pair of these properties he gives an example of a space having the weaker property but failing the stronger one for every equivalent norm. We now propose a further property of similar type, denoted (P), and discuss its connection to fixed points. Before defining it, let us record the following equivalent formulation of (KK) (cf. [6]).

$$\left.\begin{array}{l} \|x_n\| \leq 1 \ (n=1,2\ldots) \\ x_n \xrightarrow{w} x \\ \operatorname{sep}(x_n) > 0 \end{array}\right\} \Rightarrow \|x\| < 1,$$

where $\operatorname{sep}(x_n)$ denotes $\inf\{\|x_n - x_m\| : n \neq m\}$. The definition of (UKK) now easily suggests itself: A norm is (UKK) if for every $\varepsilon > 0$ there exists a $\delta < 1$ such that

$$\left.\begin{array}{l}\|x_n\| \leq 1 \ (n=1,2\ldots) \\ x_n \overset{w}{\to} x \\ \operatorname{sep}(x_n) \geq \varepsilon\end{array}\right\} \Rightarrow \|x\| \leq \delta$$

In the definition of (P) below we do not require the above implication to hold for <u>all</u> $\varepsilon > 0$ (and suitable δ) but only for <u>some</u> $0 < \varepsilon < 1$. (Note the similarity with the requirement in Theorem 1.)

DEFINITION. The [dual] norm of a [dual] Banach space is said to have the <u>property (P)</u> [(P*)] if there exists an $\varepsilon < 1$ and a $\delta < 1$ such that

$$\left.\begin{array}{l}\|x_n\| \leq 1 \ (n=1,2\ldots) \\ x_n \overset{w}{\to} x \ [x_n \overset{w^*}{\to} x] \\ \operatorname{sep}(x_n) \geq \varepsilon\end{array}\right\} \Rightarrow \|x\| \leq \delta$$

Clearly (UKK) $\Rightarrow$ (P), but it can be shown that neither (KK) $\Rightarrow$ (P) nor (P) $\Rightarrow$ (KK) is true. The property (P) derives its interest from

THEOREM 4 ([3]). If a Banach space X satifies (P), then X has normal structure, and consequently X has the (FPP).

The proof is easy. More difficult to answer, and also more interesting, is the question of whether (P*) implies (FPP*). We can show that this implication is true for dual Banach spaces with boundedly complete bases satisfying a strong monotonicity condition. Let us recall that a basis $\{x_n\}$ of X is called boundedly complete if, for any sequence $\{\alpha_n\}$ of scalars, $\sum_{n=1}^{\infty} \alpha_n x_n$ converges whenever $\{\sum_{n=1}^{k} \alpha_n x_n\}_{k=1}^{\infty}$ is bounded. It is well-known that every X with a boundedly complete monotone basis $\{x_n\}$ is canonically isometric to the dual of the subspace of X^* spanned by the coefficient functionals $\{x_n^*\}$ of the basis $\{x_n\}$. Whenever we consider X as a dual space and deal with its w*-topology, we shall have this identification in mind. The strong monotonicity condition mentioned above is that for all $n, p \in \mathbb{N}$ we have

$$(**)\quad \left\| \sum_{i=1}^{n} \alpha_i x_i + \sum_{i=n+p}^{\infty} \alpha_i x_i \right\| \leq \left\| \sum_{i=1}^{\infty} \alpha_i x_i \right\| \text{ for every } \sum_{i=1}^{\infty} \alpha_i x_i \in X.$$

We are now ready to state

THEOREM 5 ([3]). Let X be a Banach space with a boundedly complete basis satisfying (**). If X satisfies (P^*), then X has (FPP*).

An essential tool in the proof is the fact that, under the stated hypotheses, for any bounded sequence $\{y_n\}$ in X the function

$$z \to r(z) := \limsup_{n\to\infty} \|z - y_n\| \quad (z \in X) \quad \text{is } w^*\text{-l.s.c.}$$

This allows a dualization of a lemma of Karlovitz ([7]). In its dual form it says that if T is a nonexpansive map from a w*-compact convex set C into itself and if C contains no proper T-invariant w*-compact convex subset, then $\lim_{n\to\infty} \|x - y_n\| = \operatorname{diam} C$ for all $x \in C$, whenever $\{y_n\} \subset C$ is a sequence satisfying $\|Ty_n - y_n\| \to 0$. For details of proof we refer to [3].

A standard example to which Theorem 5 applies is l^1. Also, if one replaces the usual norm $\|\cdot\|_1$ by

$$|||x||| = \max(\|x\|_1, (1+\alpha)\|x\|_\infty),$$

then for sufficiently small $\alpha > 0$ Theorem 5 still applies. There is some interest in this last example since (as opposed to the case of l^1 with its standard norm) the w^* Opial condition fails here, so that no other proof of (FPP*) is readily available.

Finally, let us examine the example given in [8] by Lim in the light of Theorem 5. Lim defines an equivalent (dual) norm on l^1 by

$$|||x||| = \max(\|x^+\|_1, \|x^-\|_1),$$

when x^+ and x^- denote the positive and the negative part of x, respectively. The standard basis is boundedly complete and also satisfies (**), for the norm $|||\cdot|||$. But (P*) fails, as it must, since Lim has shown that $(l^1, |||\cdot|||)$ does not have (FPP*). However, it is not difficult to show that for every $\varepsilon > 1$ there exists a $\delta < 1$ such that $|||x_n||| \leq 1$, $x_n \overset{w^*}{\rightarrow} x$, $\text{sep}(x_n) \geq \varepsilon$ imply $|||x||| \leq \delta$. This shows that the condition $\varepsilon < 1$ in (P^*) cannot be relaxed. In some sense therefore, Theorem 5, at least as far as (P^*) is concerned, is the sharpest possible result.

We do not know of any result linking (KK) to (FPP).

REFERENCES

1. J. B. Baillon and R. Schöneberg, "Asymptotic normal structure and fixed points of nonexpansive mappings", Proc. Amer. Math. Soc. 81 (1981), 257-264.
2. J. Bernal, "Behavior of k-dimensional convexity moduli", Thesis, Catholic University of America, 1980.
3. D. van Dulst, "Nonexpansive mappings in Banach spaces with norms of type (KK)", preprint.
4. D. van Dulst and A. J. Pach, "A renorming of Banach spaces", Rev. Roumaine Math. Pures Appl. (to appear).
5. R. S. Geremia, "K-uniformly convex Banach spaces", Thesis, Georgetown University (1979).
6. R. Huff, "Banach spaces which are nearly uniformly convex", Rocky Mountain J. of Math. 10 (1980), 743-749.
7. L. A. Karlovitz, "Existence of fixed points of nonexpansive mappings in a space without normal structure", Pacific J. Math. 66 (1976), 153-159.
8. T. C. Lim, "Asymptotic centers and nonexpansive mappings in conjugate Banach spaces", Pacific J. Math. 90 (1980), 135-143.
9. F. Sullivan, "A generalization of uniformly rotund Banach spaces", Can. J. Math. 31 (1979), 628-636.

COMMON FIXED POINTS OF MAPPINGS AND SET-VALUED MAPPINGS

B. Fisher

Leicester University
Leicester
United Kingdom

Abstract

Let (X, d) be a complete metric space and let $B(X)$ be the set of all non-empty subsets of X. The function $\delta(A, B)$ with A and B in $B(X)$ is defined by

$$\delta(A, B) = \sup\{d(a, b) : a \in A, b \in B\}.$$

A sequence $\{A_n\}$ of non-empty subsets of X is said to <u>converge</u> to the subset A of X if:

(i) each point a in A is the limit of a convergent sequence $\{a_n\}$, where a_n is in A_n for $n = 1, 2, \ldots$,

(ii) for arbitrary $\epsilon > 0$ there exists an integer N such that $A_n \subset A_\epsilon$ for $n > N$, where A_ϵ is the union of all spheres with centres in A and radius ϵ.

A mapping F of (X, d) into $B(X)$ is said to be <u>continuous</u> if for each point x in X and each sequence $\{x_n\}$ in X converging to x, the sequence of sets $\{Fx_n\}$ in $B(X)$ converges to Fx in $B(X)$.

The following theorems are given.

THEOREM 1. Let F be a continuous mapping of a complete metric space (X, d) into $B(X)$ and let I be a mapping of X into itself satisfying the inequality

$$\delta(Fx, Fy) \leq c.\max\{d(Ix, Iy), \delta(Ix, Fx), \delta(IY, Fy), \delta(Ix, Fy), \delta(Iy, Fy)\}$$

for all x, y in X, where $0 \leq c < 1$. If F and I commute and the range of I contains the range of F, then F and I have a unique common fixed point z and further $Fz = \{z\}$.

THEOREM 2. Let F be a continuous mapping of a compact metric space (X, d) into $B(X)$ and let I be a continuous mapping of X into itself satisfying the inequality

$$\delta(Fx, Fy) < \max\{d(Ix, Iy), \delta(Ix, Fx), \delta(Iy, Fy), \delta(Ix, Fy), \delta(Iy, Fx)\}$$

for all x, y in X for which the right-hand side of the inequality is positive. If F and I commute and the range of I contains the range of F, then F and I have a unique common fixed point z and further $Fz = \{z\}$.

Details will appear in Rostocker Mathematisches Kolloquium.

FIXED POINT PRINCIPLES FOR CONES OF A BANACH SPACE*

G. Fournier

Universite de Sherbrooke
Sherbrooke, Quebec
Canada

In [6] and [7], Krasnosel'skii proved several fundamental fixed point principles for operators leaving invariant a cone in a Banach space. In [12], Nussbaum extended one of the results, the theorem about compression and expansion of a cone, to k-set contraction maps, $k < 1$. Other versions for completely continuous maps were given by Fournier-Peitgen [4] and G. Fournier [2], and for k-set contractions, $k < 1$, in Fournier-Peitgen [5].

The purpose of this paper is to generalize the more comprehensive results of [2] to non completely continuous maps.

1. Measure Of Noncompactness

The notion of measure of noncompactness is due to Kuratowski [8, 9].

Let (Y, d) be a metric space. For $A \subset Y$, we define the *measure of noncompactness* $\gamma(A)$ *of* A to be

$$\gamma(A) = \inf \{r > 0 \mid \text{there exists a covering of } A \text{ by subsets of diameter at most } r\}.$$

Notice that $\gamma(Y) < \infty$ if and only if Y is bounded. Let $f : X \to Y$

*This research has been supported by grants from the NSERC and by a Bourse d'action concertée du ministère de l'éducation du Québec.

be a continuous map between two metric spaces, then the *measure of noncompactness* $\gamma(f)$ *of* f is

$$\gamma(f) = \inf\{k \mid \gamma(f(A)) \leq k\gamma(A) \quad \text{for all } A \subset X\}.$$

If $\gamma(f) < 1$ we say that f is a k-*set contraction*, $k < 1$.

We have the following properties:

(1.1) $0 \leq \gamma(A) \leq \delta(A)$ where $\delta(A)$ is the diameter of A.

(1.2) $\gamma(A \cup B) = \max\{\gamma(A), \gamma(B)\}$.

(1.3) $\gamma(N_\varepsilon(A)) \leq 2\varepsilon\gamma(A)$ where $N_\varepsilon(A) = \{x \in X \mid \exists\, a \in A\ d(x, a) < \varepsilon\}$.

(1.4) $\gamma(g \circ f) \leq \gamma(g)\,\gamma(f)$.

(1.5) $\gamma(A) = 0$ if and only if A is relatively compact.

Furthermore, if Y is a linear normed space, then (cf. [1])

(1.6) $\gamma(A + B) \leq \gamma(A) + \gamma(B)$.

(1.7) $\gamma(\text{co } A) = \gamma(A)$ where co A is the closed convex hull of A.

In order to state one final property, we need the following definition.

DEFINITION 1.8. (See Nussbaum [11]). Let $f : X \to X$ be a continuous map. A compact nonempty subset $M \subset X$ is a *compact attractor for* f if and only if $f(M) \subset M$ and for any neighbourhood U of M and any compact subset $K \subset X$ there exists an integer $n_{U,K}$ such that $f^m(K) \subset U$ for all $m \geq n_{U,K}$.

(1.9) (See [11]). Let $f : X \to X$ be such that $\lim_{n\to\infty} \gamma(f^n(X)) = 0$ and X is complete and bounded, then f has a compact attractor.

2. Strongly Compacting Maps

In this paper, we shall make an essential use of the fixed point index by R. D. Nussbaum [11] with further properties given by G. Fournier-R. Fournier [3]. In this paragraph, we define the class of functions for which they are all true.

As in Nussbaum [12], we shall write $X \in \mathcal{F}$ if X is a closed subset of a Banach space from which it inherits its metric and if X has a locally finite covering by closed convex subsets of X. If in addition the covering is a finite covering, we shall write $X \in F_o$.

Note that if $X \in F$, it follows that X is an ANR.

DEFINITION 2.1. (See [3]). Let $X \in F$ and $f : U \to X$ be continuous such that $\mathrm{Fix}(f) = \{x \in U \mid f(x) = x\}$ is compact where U is open in X. Then f is *strongly compacting* if for any compact subset M of U with $\mathrm{Fix}(f) \subset f(M) \subset M$, there exist a decreasing sequence $K_n \in F_o$ of subsets of X, an open subset W of U and a sequence of strictly positive numbers $\{\varepsilon_n\}$ such that

(2.1.1) $M \subset W \subset K_1$

(2.1.2) $f(W \cap N_{\varepsilon_n}(K_n)) \subset K_{n+1}$

(2.1.3) $\lim_{n\to\infty} \gamma(K_n) = 0.$

The following is due to G. Fournier-R. Fournier [3].

PROPOSITION 2.2. *Let* $f : U \to X$ *be continuous, where* $X \in F$. *Assume that* $\mathrm{Fix}(f)$ *is compact, if* $\gamma(f) < 1$ *then* f *is strongly compacting.*

Proof. See [3, proposition 2.2.3]. □

DEFINITION 2.3. Let $g : U \to X$ be a continuously Fréchet differentiable map, where X is a convex subset of a Banach space. Then g is *homogeneously eventually condensing* if there exist $0 < k < 1$ and $m \in N$ such that for any $\{x_0, x_1, \ldots x_m\} \subset U$ we have that

$$\gamma(Dg(x_0) \circ \ldots \circ Dg(x_m)) < k.$$

By a *wedge* in a Banach space, we shall mean a nonempty closed convex set K such that $\gamma K \subset K$ for all $\gamma \geq 0$.

PROPOSITION 2.4. *Let* $f, g : U \to X$ *be two maps such that* f *is compact and* g *is homogeneously eventually condensing, where* X *is a wedge in a Banach space. Then* $f + g$ *is a strongly compacting map, provided that* $\mathrm{Fix}(f + g)$ *is compact.*

Proof. See [3, proposition 2.2.9]. □

Let us give some examples of homogeneously eventually condensing maps $f : U \to X$.

(2.5) If f is linear eventually condensing (cf. Nussbaum [10]).

(2.6) There is a finite sequence $\{E_i\}_{i=1,\ldots,n}$ of subspaces and $k, L > 0$ such that

(a) E_1 is the whole space

(b) $Df(x)(E_i) \subset E_{i+1} \subset E_i$ for all $i = 1, \ldots, n - 1$ and all $x \in U$

(c) $\gamma(Df(x)|_{E_n}) < k < 1$ for all $x \in U$

(d) $\gamma(Df(x)) \leq L$ for all $x \in U$.

(2.7) Let $X = \{f : [0, 1] \to R \mid f$ is continuous positive increasing with $f(0) = 0\}$ with the sup norm. Define $L_t : X \to X$ by

$$L_t f(x) = \begin{cases} 0 & \text{if } 0 \leq x \leq t \\ f(x - t) & \text{if } t \leq x \leq 1 \end{cases}$$

and $L_{t,n} : X \to X$ by $L_{t,n}f(x) = (L_t f(x))^n$. Then L_t is linear and $L_t(fg) = L_t(f)\, L_t(g)$ and

$$DL_{t,n}(f)(g)(x) = n(L_t f(x))^{n-1}\, (L_t g)(x);$$

and so if $m = [t^{-1}]$ and $f_1, \ldots, f_m, g \in X$, we have that $L_t^m(g)(x) = 0$ for all x and

$$DL_{t,n}(f_1) \circ \ldots \circ DL_{t,n}(f_m)(g)(x)$$

$$= n^m (L_t f_1(x))^{n-1} (L_t^2 f_2(x))^{n-1} \ldots (L_t^m f_m(x))^{n-1} L_t^m(g) = 0$$

thus, $\gamma(DL_{t,n}(f_1) \circ \ldots \circ DL_{t,n}(f_m)) = 0 < 1$. That is $L_{t,n}$ is homogeneously eventually condensing on X.

3. Fixed Point Index

In this paragraph we state the properties of Nussbaum's fixed point index [13] for the class of strongly compacting maps. But first we shall need some definitions.

Let H denote the singular homology functor with rational coefficients and f_* denote $H(f)$, where $f : X \to X$ is a continuous map. The map f is a *Lefschetz map* if and only if

$$H_n(X) = Nf_{*n} = \cup \{\ker f^i_{*n} \mid i \in N\}$$

for all but a finite number of n for which $H_n(X)/Nf_{*n} = \tilde{H}_n(X)$ is finite dimensional; in that case $\Lambda(f : X \to X) = \Lambda(f) = \sum_q (-1)^q \operatorname{tr}(\tilde{f}_{*q})$, where $\tilde{f}_{*q} : \tilde{H}_n(X) \to \tilde{H}_n(X)$ is induced by f_{*q}.

A toplogical space X is *acyclic* if $\dim H_0(X) = 1$ and for all $q > 0$, $H_q(X) = 0$. Notice that any contractible space is acyclic (X is contractible if 1_X is homotopic to a constant map); thus any convex subset of a Banach space is acyclic. Furthermore if f is homotopic to g, then $f_* = g_*$ and $\Lambda(f) = \Lambda(g)$.

We now give some properties of Nussbaum's fixed point index.

(3.1) Let $f : U \to X$ be a strongly compacting map, then the number ind (X, f, U) is defined ([13]) and called the fixed point index of the map $f : U \to X$.

This index satisfies a number of properties among which are the following:

(3.1.1) Additivity. Assume that $U = U_1 \cup U_2$ and that $\operatorname{Fix}(f) \cap U_1 \cap U_2 = \phi$, then ind $(X, f, U) =$ ind $(X, f, U_1) +$ ind (X, f, U_2).

(3.1.2) Excision. Assume that $\operatorname{Fix}(f) \subset U_1 \subset U$, then ind $(X, f, U) =$ ind (X, f, U_1).

(3.1.3) Fixed Point. Assume that $\operatorname{Fix}(f) = \phi$, then ind $(X, f, U) = 0$.

(3.1.4) Normalization. Assume that $f(U) \subset U$ and that $f : U \to U$ has a compact attractor, then ind $(X, f, U) = \Lambda(f : U \to U)$.

(3.1.5) Homotopy. (See G. Fournier-R. Fournier [3]). Let $H : U \to X \times [0, 1]$ be a strongly compacting map of the form $H(x, t) = (h(x, t), t)$ then ind $(X, h_0, U_0) =$ ind (X, h_1, U_1) where $h_t(x) = h(x, t)$ and $U_t = \{x \in X \mid (x, t) \in U\}$.

4. The Main Results

Two maps $f, g : U \to X$ are said to be *homotopic relative to* $A \subset U$ if there exists $h : U \times [0, 1] \to X$ continuous such that $h(x, 0) = f(x)$ and $h(x, 1) = g(x)$ for all $x \in U$ and $h(x, t) = f(x) = g(x)$ for all $x \in A$.

The following lemma, although stated with more conditions on f is proved in ([2] lemma 3.7).

LEMMA 4.1. *Let* $f : X \to X$ *be continuous where* X *is a complete* ANR. *Let* A *be a closed acyclic subset of* X *such that there exists a continuous map* $r : A \to \partial A$ *for which* $i \circ r$ *is homotopic to* 1_A, *relative to* A, *where* $i : \partial A \to A$ *is the inclusion map. Furthermore, assume that the following conditions are satisfied:*

(4.1.1) *for any* $a \in A$ *there exists an integer* $n(a) \geq 1$ *such that* $f^{n(a)}(a) \in \mathring{A}$

(4.1.2) *there exists an integer* $n \geq 1$ *such that* $f^n(\partial A) \subset A$

(4.1.3) *for any* $a \in \partial A$ *and for any integer* $1 \leq i \leq n$ *(where* n *is the same as in condition 2), there exist an integer* m *and* V *a neighbourhood of* a *such that* $f^m(V \cap \partial A) \subset A$ *and* $m \equiv i \pmod n$. *Then* $g(W) \subset W$ *and* $\Lambda(g : W \to W) = 1$ *where* $g = f \circ \rho$ *and*

$$W = \cup \{f^{-i}(\mathring{A}) \mid i \geq 1\} = \cup \{g^{-i}(\mathring{A}) \mid i \geq 1\}$$

and $\rho : X \to X$ *is defined by* $\rho|_A = i \circ r$ *and* $\rho|_{X\setminus\mathring{A}} = 1_{X\setminus\mathring{A}}$.

Proof. See [2, lemma 3.7]. □

In order to apply the proofs of [2] one needs the following lemma.

LEMMA 4.2. *Assume the conditions of* (4.1) *and that* $\gamma(\delta A) < \infty$, *that* f *is bounded on bounded subsets and that for some* m *we have that* $\gamma(g^m) < 1$, *then* $g : W \to W$ *has a compact attractor.*

<u>Proof.</u> Denote $V = \bigcup_{i=1}^{n} g^i(\partial A)$, we have that $g(V) \subset V \subset \bigcup_{i=1}^{n} f^i(\partial A)$

which is bounded: in fact, if $x \in \partial A$ then $g^{i+1}(x) = f^{i+1}(x)$ provided that $f^j(x) \notin \overset{o}{A}$ for all $j < i$; hence,

$$g^{i+1}(A) = g^{i+1}(\partial A) \subset \bigcup_{j=1}^{i} g^j(\overset{o}{A}) \cup f^{i+1}(\partial A)$$

and so $g^{n+1}(\partial A) \subset g\left(\bigcup_{j=1}^{n-1} g^j(A) \cup f^n(\partial A)\right) \subset \bigcup_{j=1}^{n} g^j(A)$ since $f^n(\partial A) \subset A$.

Notice that $\overline{V} \subset \overline{\bigcup_{i=1}^{n} f^i(\partial A)} \subset W$ since $f^i(\partial A) \subset f^{i-n}(A) \subset W$ and $f^{i-n}(A)$ is closed for all $i = 1, \ldots, n$. Since

$$\lim_{i\to\infty} \gamma(g^{im}(\overline{V})) = \lim_{i\to\infty} (\gamma(g^m))^i \, \gamma(\overline{V}) = 0,$$

by (1.9), we have that $g^m : \overline{V} \to \overline{V}$ has a compact attractor L.

Denote $M = \bigcup_{i=0}^{m-1} g^i(L)$, we claim that M is a compact attractor for $g : W \to W$: in fact $g(M) \subset M$, since $g^m(L) \subset L$, and if K is compact in W, then there exists n_K such that

$$g^{n_K}(K) \subset \bigcup_{i=1}^{\infty} g^i(\overset{o}{A}) = \bigcup_{i=1}^{\infty} g^i(\partial A) \subset V \subset \overline{V};$$

let Y be a neighbourhood of M, then $Y' = \bigcap_{i=0}^{m} g^{-i}(Y) \supset M \supset L$ and so there exists m_K such that $g^{mi+n_K}(K) \subset Y'$ for all $i \geq m_K$; that is $g^i(K) \subset Y$ for all $i \geq mm_K + n_K$. □

Let us call a wedge P, a *special wedge* if there exists $y \in P\backslash\{0\}$ such that $\|x + y\| \geq \|x\|$ for all $x \in P$. For any $R > 0$ denote

$$B_R = \{x \in P \mid \|x\| \leq R\} \text{ and } S_R = \{x \in P \mid \|x\| = R\}.$$

The following lemma is due to Fournier-Peitgen [5].

LEMMA 4.3. *If* P *is a special wedge in a linear normed space, then for any* $\varepsilon > 0$, *there exists a retraction* $r : B_R \to S_R$ *such that* $\gamma(r) \leq 1 + \varepsilon$.

Proof. See [5, lemma 1.5.1]. □

The next theorem is similar to one of Nussbaum [12] and one of Fournier-Peitgen [5].

THEOREM 4.4. *Let* P *be a special wedge in a Banach space, let* $f : P \to P$ *be continuous such that* $\gamma(f) < 1$. *Assume furthermore that conditions* (4.1.1) *to* (4.1.3) *are satisfied for* $A = B_{R_1}$ *and* $A = P\backslash \overset{\circ}{B}_{R_2}$, *then if* $R_1 > R_2$ *we have that* $\text{ind}\,(P, f, \overset{\circ}{B}_{R_1}\backslash B_{R_2}) = 1$ *and if* $R_2 > R_1$ *we have that* $\text{ind}\,(P, f, \overset{\circ}{B}_{R_2}\backslash B_{R_1}) = -1$, *and in both cases* f *has a non trivial fixed point.*

Proof. In both cases, there exists a retraction $r : A \to \partial A$ satisfying the conditions of (4.1) and such that $\gamma(r) < (\gamma(f))^{-1}$ hence $\Lambda(g : W \to W) = 1$.

In both cases ∂A is bounded $\gamma(\partial A) < \infty$ and $\gamma(f) < \infty$ implies that f is bounded on bounded subsets. By lemma 4.2 and (3.1.4), since $\gamma(g) \leq \gamma(f)\ \gamma(\rho) \leq \gamma(f)\ \gamma(r) < 1$, we have $\Lambda(g : W \to W) = \text{ind}\,(P, g, W)$. By (3.1.2), since $W = \cup \{g^{-i}(\overset{\circ}{A}) \mid i = 1, \ldots\}$, we get $1 = \text{ind}\,(P, g, W) = \text{ind}\,(P, g, \overset{\circ}{A})$.

Now, in the case $A = B_{R_1}$, consider the homotopy $h(x, t) = tf(x) + (1 - t)g(x)$ and let $H(x, t) = (h(x, t), t)$ and $D \subset P \times [0, 1]$; denote by D_1 the projection of D onto P, we have that $D \subset D_1 \times [0, 1]$ and $\gamma(D) = \gamma(D_1)$; thus

$$\gamma(H(D)) \leq \gamma(H(D_1 \times I)) \leq \gamma(\text{co}\ \{(f(D_1) \times \{1\}) \cap (g(D_1) \times \{0\})\})$$

$$\leq \max\ \{\gamma(f(D_1)),\ \gamma(g(D_1))\}$$

$\leq \max \{\gamma(f)\ \gamma(D_1),\ \gamma(f)\ \gamma(r)\ \gamma(D_1)\}$

and so $\gamma(H) < 1$, and H is strongly compacting and

$$\text{ind}\ (P,\ g,\ \overset{\circ}{B}_{R_1}) = \text{ind}\ (P,\ f,\ \overset{\circ}{B}_{R_1})$$

since $\text{Fix}(H)$ is closed in B_{R_1} and contained in $\overset{\circ}{B}_{R_1}$, furthermore $\text{Fix}(H) \subset \bigcap_{i=1}^{\infty} \overline{H^i(\overset{\circ}{B}_{R_1})} \cap B_{R_1}$ which is compact since $\gamma(H^i(\overset{\circ}{B}_{R_1})) \leq (\gamma(H))^i R_1 \to 0$ if $i \to \infty$.

Finally, in the case $A = P \backslash B_{R_2}$, since $\gamma(g) < 1$ and the set of fixed points is closed in $\bigcap_{i=1}^{\infty} g^i(P) = \bigcap_{i=1}^{\infty} \overline{g^i(B_{R_2})}$ which is a compact attractor for $g : P \to P$, we get

$$\text{ind}\ (P,g,P\backslash B_{R_2}) + \text{ind}\ (P,g,\overset{\circ}{B}_{R_2}) = \text{ind}\ (P,g,P) = \Lambda(g : P \to P) = 1$$

by (3.1.2) and (3.1.4). Thus $\text{ind}\ (P,\ f,\ \overset{\circ}{B}_{R_2}) = \text{ind}\ (P,\ g,\ \overset{\circ}{B}_{R_2}) = 0$ since $f|_{B_{R_2}} = g|_{B_{R_2}}$ and so we get the conclusion. □

For our next and last main result, we need the following notions.

A *cone* is a wedge P such that if both $y \in P$ and $-y \in P$, then it follows that $y = 0$.

A map $f : P \to P$ is said to be *bounded on bounded subsets* if for any R the set $f(B_R)$ is bounded. It is said to be *eventually bounded* if for some ℓ, $f^{\ell}(P)$ is bounded.

LEMMA 4.5. *Assume that* $f(W) \subset W = \cup \{f^{-i}(\overset{\circ}{A}) \mid i = 0, \ldots\}$, *that* $A \subset W$, *that for some* k, *the set* $\cup \{f^{k+i}(A) \mid i = 0, \ldots\}$ *is bounded and that for some* m, $\gamma(f^m) < 1$. *Then* $f : W \to W$ *has a compact attractor.*

Proof. Consider $D = \cup \{f^{k+i}(A) \mid i = 0, \ldots\} \subset W$; D is bounded and $f(D) \subset D$, thus $f(\overline{D}) \subset \overline{D}$ which is bounded and complete. By (1.9), $f^m : \overline{D} \to \overline{D}$ has a compact attractor L.

Consider $L \cap A \subset W = \cup \{f^{-i}(\overset{\circ}{A}) \mid i = 1, \ldots\}$; since $L \cap A$ is compact, there exists n such that $L \cap A \subset \cup \{f^{-i}(\overset{\circ}{A}) \mid i = 1, \ldots, n\}$ and $f^n(L \cap A) \subset \cup \{f^i(L \cap A) \mid i = 0, \ldots, n - 1\} = M$ which is compact, hence, $f(M) \subset M$. We claim that M is a compact attractor for $f : W \to W$.

In fact, if K is compact in W, as above, there exists s such that $f^{s+k}(K) \subset \cup \{f^{k+i}(A) \mid i = 0, \ldots, s - 1\} \subset D$; consider $K' = \cup \{f^{s+k+i}(K) \cup L \mid i = 1, \ldots\}$, it is compact since any open cover of K' contains a finite subcover of L which covers all but a finite number of $f^{s+k+i}(K)$, L being a compact attractor in $\overline{D}$; moreover, as above, there exists r such that $K'' = \cup \{f^i(K' \cap A) \mid i = 0, \ldots, r\}$ is compact and $f(K'') \subset K'' \subset W$; but $K' \cap \overset{\circ}{A} \subset K''$ and $f^{s+k}(K) \subset \cup \{f^{-i}(K' \cap \overset{\circ}{A}) \mid i = 1, \ldots\}$ so there exists ℓ such that $f^{\ell}(K) \subset K''$. Finally, let V be a neighbourhood of M then, since $L \cap \overset{\circ}{A} \subset M \subset V$ and $f(L \cap K'') \subset L \cap K''$, there exists t such that $L \cap K'' \subset f^{-t}(V)$; but $f^{-t}(V) \cup C(L \cap K'')$ is open and contains L thus there exists b such that $i > b$ implies $f^{\ell+i}(K) \subset f^i(K'') \subset f^{-t}(V)$; that is if $i > b + t + \ell$ then $f^i(K) \subset V$. □

THEOREM 4.6. *Let* P *be a cone in a Banach space such that* $\|x + y\| \geq \|x\|$ *for all* $x, y \in P$, *let* $f_1 : P \to P$ *be completely continuous and* $f_2 : P \to P$ *be homogeneously eventually condensing, eventually bounded and bounded on bounded subsets. Assume that the following conditions are satisfied:*

(4.6.1) *there exists an integer* n *such that for any completely continuous map* $f_3 : P \to P$, *we have that* $\gamma((f_2 + f_3)^n) < 1$;

(4.6.2) $\cup \{(f_1 + f_2)^i(B_{R_1}) \mid i = 0, \ldots\}$ *is bounded.*

Assume furthermore that conditions (4.1.1) *to* (4.1.3) *are satisfied for* $f = f_1 + f_2$, $A = B_{R_1}$ *and* $A = P \setminus \overset{\circ}{B}_{R_2}$; *then if* $R_1 > R_2$, *we have that ind* $(P, f, \overset{\circ}{B}_{R_1} \setminus B_{R_2}) = 1$ *and if* $R_2 > R_1$ *we have that* ind $(P, f, \overset{\circ}{B}_{R_2} \setminus B_{R_1})$ $= -1$, *and in both cases,* f *has a non trivial fixed point.*

Proof. By lemma 4.1, since f is homotopic to $f \circ \rho$, we get that $\Lambda(f : W \to W) = 1$, in both cases. By (4.6.2), (4.5), (3.1.4) and (3.1.2), for $A = B_{R_1}$, we have that

$$\text{ind } (P, f, \overset{\circ}{B}_{R_1}) = \text{ind } (P, f, W) = \Lambda(f : W \to W).$$

For $A = P \setminus \overset{\circ}{B}_{R_2}$, choose $R > R_2$ and consider the homotopy

$$h(x, t) = f_2(x) + ((1 - t) + t\phi(x)\psi(x))f_1(x)$$

for all $x \in P$ and all $t \in [0, 1]$, where

$$\psi(x) = \min \{1, R\,\|f_1(x)\|^{-1}\}$$

$$\phi(x) = \min \{1, \max \{0, (R - \|f_2(x)\|)(R - R_2)^{-1}\}\}.$$

We have that h is a homotopy between $f : W \to W$ and $g_1(x) = f_2(x) + \psi(x)\phi(x)f_1(x)$ since for all $x \in W$, either $h(x, t) = f(x)$ or $\|h(x, t)\| \geq \|f_2(x)\| \geq R_2$ that is, in all cases $h(x, t) \in W$. Thus $\Lambda(g_1 : W \to W) = 1$. Notice that since $\psi(x)\phi(x)f(x)$ is a completely continuous map, by (4.6.1), $\gamma(g_1^n) < 1$. Furthermore assume that $f_2^{\ell}(P)$ is bounded since $g_1(x) \in B_{2R} \cup \{f_2(x)\}$ for any $x \in P$, we get that

$$g_1^{\ell}(P) \subset B_{2R} \cup \ldots \cup f_2^{\ell-1}(B_{2R}) \cup f_2^{\ell}(P)$$

which is bounded since f_2 is bounded on bounded subsets of P. Thus by (4.5), $g_1 : W \to W$ has a compact attractor and by (3.1.4),

$$\Lambda(g_1 : W \to W) = \text{ind}\,(P, g_1, W) = \text{ind}\,(P, g_1, P \backslash B_{R_2}).$$

Similarly, $\text{ind}\,(P, g_1, P) = 1$. By (3.1.1), $\text{ind}\,(P, g_1, \overset{\circ}{B}_{R_2}) = 0$.

Finally, $H(x, t) = (h(x, t), t) = (f_2(x), 0) + (h(x, t) - f_2(x), t)$ is strongly compacting by (2.4), since $\text{Fix}(H)$ is compact: in fact, $\text{Fix}(H) \subset \text{Fix}(g_1 \times [0, 1])$, since if $h(x, t) = x \in B_{R_2}$ then $\|f_2(x)\| \leq R_2$ and $\phi(x) = 1$ and $\psi(x)\|f_1(x)\| \leq R_2$ and so $\psi(x) = 1$ that is $h(x, t) = f(x) = g_1(x) = x$. Thus, by (3.1.5), we have

$$\text{ind}\,(P, g_1, \overset{\circ}{B}_{R_2}) = \text{ind}\,(P, f, \overset{\circ}{B}_{R_2})$$

and by additivity, we get the conclusion. □

Finally let us give an example of function satisfying the conditions of theorem 4.6. Let $P = X$ of example 2.7, set $f_2(x) = L_{1/2,2}(x) = (L_{1/2}(x))^2$ then by (2.7), f_2 is homogeneously eventually condensing. Note that $L^2_{1/2}(x) = 0$ for all $x \in X$ and $\|L_{1/2}(x)\| \leq \|x\|$. Furthermore, if $f_3 : P \to P$ is completely continuous then

$$(f_2 + f_3)^2(x) = f_3 \circ (f_2 + f_3)(x) + (L_{1/2} \circ f_3(x) + L_{1/2}(L_{1/2}(x))^2)^2;$$

thus if $A \subset X$ is bounded it follows that $\gamma((f_2 + f_3)^2(A)) = 0$ since f_3 is compact, $f_2 + f_3$ is bounded on bounded subsets and $L_{1/2}(L_{1/2}(x))^2 = 0$ for all x. Set

$$f_1(x)(t) = y(t) + k \int_0^t x(s)x(1 - s)\,ds,$$

where $y \in X \backslash \{0\}$ and $k > 0$; by Arzela-Ascoli's theorem, f_1 is completely continuous. Let $R_2 = \frac{1}{2}\|y\| > 0$, note that $\|f_1 + f_2(x)\| \geq \|y\|$

$> \frac{1}{2}\|y\|$, thus conditions (4.1.1) to (4.1.3) are satisfied with $n = 1$ for $A = X \setminus B_{R_2}$. Assume that $\|y\| = 1$ and set $R_1 = 10$; then since

$$\|(f_1 + f_2)(x)\| \leq (1 + k)\|x\|^2 + \|y\|$$

and

$$\|(f_1 + f_2)^2(x)\| \leq \|y\| + \|y\|^2 + k((1 + 2k)\|x\|^4$$

$$+ 2(2 + k)\|x\|^2 \; \|y\| + \|y\|^2),$$

if $k < 2(10)^{-5}$ it follows that $(f_1 + f_2)^2(B_{10}) \subset B_{5/2}$ and $(f_1 + f_2)^3 (B_{10}) \subset B_9$; thus conditions (4.1.1) to (4.1.3) and (4.6.2) are satisfied and $f_1 + f_2$ has a fixed point of norm $1 \leq \|x\| < 10$.

REFERENCES

1. G. Darbo, "Punti uniti in transformazioni a codominio non compatto", Rend. Sem. Mat. Univ. Padova 24 (1955), 84-92.
2. G. Fournier, "Fixed Point Principles for cones of a linear normed space", Can. J. Math., Vol. XXXII, No. 6 (1980), 1372-1381.
3. G. Fournier and R. Fournier, "On the homotopy property of Nussbaum's fixed point index" to appear.
4. G. Fournier and H. O. Peitgen, "On some fixed point principles for cones in linear normed spaces", Math. Ann. 225 (1977), 205-218.
5. G. Fournier and H. O. Peitgen, "Leray Endomorphisms and cone maps", Ann. Scuola Normale Superiore Pisa, Serie IV, Vol. V, no. 1 (1978), 149-179.
6. M. A. Krasnosel'skii, "Fixed points of cone-compressing or cone-extending operators", Soviet Math. Dokl. 1 (1960), 1285-1288.
7. M. A. Krasnosel'skii, "Positive solutions of operator equations", (Groningen, Noordhoff, 1964).
8. K. Kuratowski, "Sur les espaces complets", Fund. Math., 15 (1930), 301-309.
9. K. Kuratowski, "Topologie I", Warsaw 1948, "Topologie II", Warsaw 1950.
10. R. D. Nussbaum, "The fixed point index for local condensing maps", Ann. Mat. Pura App. (IV), Vol. LXXXIX, 217-258.

11. R. D. Nussbaum, "Some asymptotic fixed point theorems", Trans. Amer. Math. Soc., 171 (1972), 349-375.

12. R. D. Nussbaum, "Periodic solutions of some nonlinear, autonomous functional differential equations, II", J. Diff. Eq. 14 (1973), 360-394.

13. R. D. Nussbaum, "Generalizing the fixed point index", Math. Ann., 228 (1977), no. 3, 259-278.

FIXED POINT THEOREMS FOR NONEXPANSIVE MAPPINGS IN CONVEX METRIC SPACES

M. D. Guay

University of Southern Maine
Portland, Maine
U. S. A.

K. L. Singh* and J. H. M. Whitfield†

Lakehead University
Thunder Bay, Ontario
Canada

Abstract

The main aim of the present paper is to prove the existence of fixed points for nonexpansive and equicontinuous mappings in convex and star-shaped metric spaces. Such spaces include Banach spaces and starshaped subset of Banach spaces, and our results generalize those of Belluce and Kirk, Chandler and Faulkner, Doston, Kirk and others.

1. Definitions and Preliminaries

Throughout this paper X and I will denote, respectively, a metric space X, with metric d and the unit interval $[0, 1]$.

The following definition was introduced by Takahashi [14].

DEFINITION 1.1. Let X be a metric space, a mapping $W : X \times X \times I \to X$ is said to be a convex structure on X if for all $x, y \in X$ and $\lambda \in I$ the following condition is satisfied

$$d(u, W(x, y, \lambda)) \leq \lambda d(u, x) + (1 - \lambda)d(u, y)$$

for all $u \in X$. We will call a metric space with a convex structure a convex metric space.

REMARK 1.2. If X is a Banach space then, as a metric space with $d(x,$

*Current affiliation: University of Minnesota-Duluth, Duluth, Minnesota, U.S.A.
†Reasearch supported in part by a grant from NSERC (Canada).

$y) = \|x - y\|$, the mapping $W : X \times X \times I \to X$ defined by $W(x, y, \lambda) = \lambda x + (1 - \lambda)y$ is a convex structure. Thus, a Banach space or any convex subset of a Banach space, is a convex metric space. There are many convex metric spaces which cannot be imbedded in any Banach space. The following two examples may be found in [14].

EXAMPLE 1.3. Let X be the family of closed intervals $[a_i, b_i]$ such that $0 \le a_i \le b_i \le 1$. For $I_i = [a_i, b_i]$, $I_j = [a_j, b_j]$ and $\lambda \in I$, we define a mapping W by $W(I_i, I_j, \lambda) = \{\lambda a_i + (1 - \lambda)a_j, \lambda b_i + (1 - \lambda)b_j\}$. Let the metric d on X be the Hausdorff distance, i.e.

$$d(I_i, I_j) = \sup_{a \in I} \{|\inf_{b \in I_i} \{|a - b|\} - \inf_{c \in I_1} \{|a - c|\}.$$

Then X is a convex metric space.

EXAMPLE 1.4. Let X be a linear space which is also a metric space with a metric d such that the following conditions are satisfied:

(1) For $x, y \in X$, $d(x, y) = d(x - y, 0)$;

(2) For $x, y \in X$, $\lambda \in I$, $d(\lambda x + (1 - \lambda)y, 0) \le \lambda d(x, 0) + (1 - \lambda)d(y, 0)$. Then X is a convex metric space.

REMARK 1.5. The mapping W is not continuous in general, however, if X is compact, then W is continuous [15, p. 64].

DEFINITION 1.6. Let X be a convex metric space. Let K be a nonempty subset of X. We say that K is <u>convex</u> if $W(x, y, \lambda)$ belongs to K for all $x, y \in K$ and $\lambda \in I$.

REMARK 1.7. The open sphere $S(x, r) = \{y \in X : d(x, y) < r\}$ and the closed sphere $\overline{S}(x, r) = \{y \in X : d(x, y) \le r\}$ are convex [14, Proposition 2]. Moreover, if $\{K_\alpha : \alpha \in \Delta\}$ is a family of convex subsets of X, then $\bigcap_{\alpha \in \Delta} K_\alpha$ is also convex [14, Proposition 1].

2. Mappings With Regular Orbital Diameters

We prove a fixed point theorem for nonexpansive mappings having regular orbital diameters, and thus generalize a result of Belluce and Kirk.

Let T be a mapping of a metric space X into itself and x be a point in X. $0(x)$, the orbit of x, is the set $\{T^n(x) : n \geq 0\}$ $(T^0(x) = x)$. The following notion is due to Belluce and Kirk [1, p. 141].

DEFINITION 2.1. Let X be a metric space and T a mapping of X into itself. Then for any point $x \in X$, T has <u>diminishing orbital diameter</u> at x if either $\delta(0((x)) = 0$ or there exists $n \geq 1$ such that $\delta(0(T^n(x)) < \delta(0(x))$, where $\delta(A)$ is the diameter of A.

Belluce and Kirk used this notion to obtain two very general fixed point theorems [1, Theorems 2 and 3]. Later Kirk used this notion in a number of papers (e.g. [9]). When X has certain convexity properties, this notion can be replaced by a more general one which links with the given metric space and given function.

Let K be a nonempty closed convex subset of a convex metric space X and let T be a mapping of K into itself. A subset L of K is said to be <u>T-invariant</u> if $T(L) \subset L$. For any point $x \in K$, we shall use $C(T, x)$ to denote the <u>smallest T-invariant</u> closed convex subset of K which contains x.

DEFINITION 2.2. Let X be a convex metric space and K be a closed convex subset of X. Let $T : K \to K$ be a mapping. For any $x \in K$, T has <u>regular orbital diameter</u> at x if either $\delta(C(T, x)) = 0$ or there exist points $y, z \in C(T, x)$ such that $\sup\{d(y, T^n z) : n \geq 0\} < \delta(C(T, x))$.

DEFINITION 2.3. Let B be a bounded set in a convex metric space X and $\delta(B)$ be its diameter, i.e. $\delta(B) = \sup\{d(x, y) : x, y \in B\}$. A convex set S of X is said to have <u>normal structure</u> if every bounded convex subset S_1 of S which contains more than one point, has a point which is non-diametral point of S_1, i.e. there is a point x of S_1 such that $\sup_{y \in S_1} d(x, y) < \delta(S_1)$.

REMARK 2.4. Every compact convex subset of a Banach space has normal structure [2]. So does every closed, bounded convex subset of a uniformly convex Banach space [3], [7]. Also, every compact convex subset of a convex metric space has normal structure [14, Proposition 5].

REMARK 2.5. It is clear that T has regular orbital diameter at a point $x \in K$ if either T has diminishing orbital diameter at x or $co(O(x))$ has normal structure, where co denotes the convex hull. However, the converse is not true in general. Indeed, let $X = \ell^{\infty}(2)$, i.e., the space of pairs $x = (a, b)$ with the norm $\|x\| = \max(|a|, |b|)$. Let T be the mapping defined as follows: $T(x) = T\{(a, b)\} = (|b|, -b)$. It is easy to see T is nonexpansive and T does not have diminishing orbital diameter at $x = (1, 1)$, but it has regular orbital diameter.

DEFINITION 2.6. A convex metric space X is said to have <u>property</u> (C) if every decreasing net of nonempty, closed, convex subsets of X has nonempty intersection.

It is a theorem of Smulian [6, p. 433] that a convex subset of a Banach space has property (C) when and only when it is weakly compact.

DEFINITION 2.7. A mapping $T : X \to X$ of a metric space into itself is called <u>nonexpansive</u> if $d(Tx, Ty) \leq d(x, y)$ for all $x, y \in X$.

THEOREM 2.8. Let X be a convex metric space having property (C) and K be a nonempty, closed, bounded and convex subset of X. Let $T : K \to K$ be a nonexpansive mapping. Suppose T has regular orbital diameter at each point of K. Then T has a fixed point in K.

<u>Proof</u>. Let G be the family of all nonempty closed and convex subsets of X which is mapped into itself by T. By property (C) and Zorn's Lemma, G has a minimal element D. If $\delta(D) = 0$, then the point in D is a fixed point of T. Suppose $\delta(D) > 0$. Let $x \in D$. By the minimality of D, $C(T, x) = D$. Since T has regular orbital diameter at x, there exist $y, z \in D$ such that $r = \sup\{d(y, T^n z) : n \geq 1\} < \delta(D)$. Let $E = \bigcup_{k=1}^{\infty} \bigcap_{n=k}^{\infty} B[T^n(z), r] \cap D$, where $B[x, r] = \{y \in X : d(x, y) \leq r\}$. Then E is nonempty $(y \in E)$, convex and T-invariant. It follows from the continuity of W, $\overline{E}$ is also a nonempty closed, convex T-invariant subset of D. By the minimality of D, $\overline{E} = D$. Consider $F = \bigcap_{u \in E} B[u, r] \cap D$. It follows from property (C), that $\phi \neq \bigcap_{n=1}^{\infty} \overline{co}(T^n(z)) \subset F$. Let $v \in F$. Then $d(Tv, Tu) \leq d(v, u) \leq r$ for all $u \in E$. So

$d(Tv, w) \leq r$ for all $w \in \overline{co}T(E)$. Again, $\overline{co}T(E)$ is a nonempty, closed convex T-invariant subset of D. So $\overline{co}T(E) = D$ and F is T-invariant. By the minimality of D, F = D. Hence, $\delta(D) = \delta(F) \leq r < \delta(D)$, a contradiction.

The method used in the above proof is similar to that of Belluce and Kirk [1]. Also, Kirk has used this method to obtain several different theorems along this direction. With the notion of regular orbital diameters, our Theorem 2.8 generalizes their two results [1, Theorems 2 and 3].

3. Starshaped Metric Spaces

In this section we prove fixed point theorems for nonexpansive mappings for compact star shaped metric spaces. The results of Doston and Kirk are obtained as corollaries.

DEFINITION. A metric space X is said to be <u>starshaped</u> if there exists an $x_0 \in X$ and a mapping $W : X \times \{x_0\} x I \to X$ such that, for each $x, y \in X$, $\lambda \in I$, we have $d(x, W(y, x_0, \lambda)) \leq \lambda d(x, y) + (1 - \lambda)d(x, x_0)$.

Starshaped metric spaces are generalizations of starshaped subsets of Banach spaces, where a subset K of a Banach space is starshaped if there exists an $x_0 \in K$ such that for each $x \in K$, $\lambda \in I$, $\lambda x + (1 - \lambda)x_0 \in K$. It is obvious that convex metric spaces are starshaped.

DEFINITION 3.2. A starshaped metric space X is said to satisfy <u>condition (I)</u> if for any $x, y \in X$, $\lambda \in I$, $d(W(x, x_0, \lambda), W(y, x_0, \lambda)) \leq \lambda d(x, y)$.

REMARK 3.3. Condition (I) is always satisfied in any normed linear space. Indeed, let $W(x, x_0, \lambda) = \lambda x + (1 - \lambda)x_0$ and $W(y, x_0, \lambda) = \lambda y + (1 - \lambda) x_0$. Then $\|W(x, x_0, \lambda) - W(y, x_0, \lambda)\| = \|\lambda x - \lambda x\| = \lambda\|x - y\|$.

THEOREM 3.4. Let X be a compact starshaped metric space satisfying condition (I). Let $T : X \to X$ be a nonexpansive mapping. Then T has a fixed point in X.

<u>Proof</u>. Let $x_0 \in K$. For $0 < k < 1$, define the mapping T_k as follows: $T_k(x) = W(Tx, x_0, k)$. Since X is starshaped, it follows that T_k maps X into itself. Moreover, T_k is a contraction. Indeed, $d(T_k$

$(x), T_k(y)) = d(W(Tx, x_0, k), W(Ty, x_0, k)) \leq kd(Tx, Ty) \leq kd(x, y)$ for all $x, y \in X$. It follows from Banach's contraction principle that there exists $x_k \in X$ such that $T_k(x_k) = x_k$. Now

$$d(x_k, T(x_k)) = d(T_k(x_k), T(x_k)) = d(T(x_k), W(T(x_k), x_0, k))$$

$$\leq kd(T(x_k), T(x_k)) + (1 - k)d(T(x_k), x_0)$$

$\to 0$ as $k \to 1$. Hence $d(x_k, T(x_k)) \to 0$. Since X is compact, $\{x_k\}$ has a convergent subsequence $\{x_{k_i}\}$ converging to some point $x \in X$. Then $d(x_{k_i}, T(x_{k_i})) \to 0$ and by the continuity of T, it follows that $T(x_{k_i}) \to T(x)$. So

$$d(x, T(x)) \leq d(x, x_{k_i}) + d(x_{k_i}, T(x_{k_i})) + d(T(x_{k_i}), T(x))$$

and taking the limit as $i \to \infty$, we have $x = T(x)$.

COROLLARY 3.5 [5, Theorem 1]. Let X be a Banach space and K be a nonempty compact starshaped subsed of X. Let $T : K \to K$ be a nonexpansive mapping. Then T has a fixed point in K.

DEFINITION 3.6. A convex metric space X is said to satisfy <u>condition (II)</u> if for all $x, y, z \in X$ and $\lambda \in I$

$$d(W(x, z, \lambda), W(y, z, \lambda)) \leq \lambda d(x, y).$$

THEOREM 3.7. Let X be a convex metric space satisfying condition (II). Let K be a closed bounded convex subset of X. Let M be a compact subset of K. Suppose $T : K \to K$ is nonexpansive mapping such that for each $x \in K$,

(i) $\overline{co}\{T^n(x)\} \cap M \neq \phi$

(ii) $d(y, Ty) \leq d(x, Tx)$ if $y \in co\{T^n(x)\}$.

Then there exists a point $x \in M$ such that $Tx = x$.

<u>Proof</u>. Let x_0 be a fixed element of K. Define $T_k : K \to K$ as follows: $T_k(x) = W(Tx, x_0, k)$, $0 < k < 1$. As in the proof of Theorem

3.4, there exists $x_k \in X$ such that $T_k(x_k) = x_k$. Therefore,

$$d(x_k, T(x_k)) = d(T_k(x_k), T(x_k)) \leq kd(T(_k)T(x_k)) + (1 - k)d(T(x_k), x_0).$$

Let $\varepsilon > 0$. Since K is bounded, we may choose k sufficiently near 1 so that $d(x_k, T(x_k)) < \varepsilon/3$.

By (i) there exists an element $z \in \overline{co}\{T^m(x_k)\}$ such that $d(z, m) < \varepsilon/3$ for some $m \in M$.

Therefore,

$$d(m, T(m)) \leq d(m, z) + d(z, T(m)) + d(T(z), T(m))$$

$$\leq 2d(z, m) + d(z, T(m)) < \varepsilon.$$

Hence, $\inf\{d(m, T(m)) : m \in M\} = 0$. Since M is compact and T is continuous this infimum is achieved at some point $x \in M$, which is fixed under T.

COROLLARY 3.8 [10, Corollary 2.1]. Let X be a Banach space. Let K be a closed convex subset of X and let M be a compact subset of K. Suppose $T : K \to K$ is a nonexpansive mapping such that for each $x \in K$,

(i) $\overline{co}\{T^n(x)\} \cap M \neq \phi$ and

(ii) $||y - Ty|| \leq ||x - Tx||$ if $y \in co\{T^n(x)\}$.

Then there is a point $x \in M$ such that $Tx = x$.

4. Attractors For Compact Sets

Fixed point theorems are proved for nonexpansive and equicontinuous mappings on convex metric spaces with attractors. These results include those of Chandler and Faulkner and others.

DEFINITION 4.1. Let X be a convex metric space. We say that X has property (W) if the convex hull of any precompact subset of X is precompact.

It is well known that locally convex Hausdorff topological linear spaces (in particular, Banach spaces) have property (W). Also, strongly convex metric spaces with stable convex structure have property (W) [12,

Theorem 3.4].

DEFINITION 4.2. Let X be a metric space and $G : X \to X$ be a family of functions. G is said to be equicontinuous on X if for each $\varepsilon > 0$, there is a $\delta(\varepsilon) > 0$ such that for all $x, y \in X$ if $d(x, y) < \delta$, then $d(Tx, Ty) < \varepsilon$ for each $T \in G$.

For the proof of our next result, we need to recall the following.

LEMMA 4.3 [8, p. 531(ii).] Let X be a compact metric space and $T : X \to X$ with $\{T^n : n \geq 1\}$ equicontinuous. Then there exists a retraction $r : X \to C$, where $C = \cap\{T^n(X) : n \geq 1\}$.

LEMMA 4.4 [4, Lemma 2]. Let T be a nonexpansive self-map of a compact metric space X. Then there exists a nonexpansive retraction $r : X \to C$.

DEFINITION 4.5. Let X be a topological space and $T : X \to X$ a map. A nonempty subset $M \subseteq X$ is called an attractor for compact sets under T if (i) M is compact and $T(M) \subseteq M$, (ii) for any compact set $G \subseteq X$ and an open set 0 containing M, there exists an $N = N(G, 0)$ such that $T^n(G) \subseteq 0$ for all $n \geq N$.

THEOREM 4.6. Let X be a complete convex metric space which satisfies condition (I) and has property (W). Let C be a closed starshaped subset of X. Let $T : C \to C$ be a nonexpansive mapping. If there exists $M \subseteq C$, an attractor for compact sets, then T has a fixed point in C.

Proof. Let y be the starpoint of C and define the closed starhull of M to be $L = c\ell\{W(x, y, t) : t \in I, x \in M\}$. M is compact, so $\overline{co}[M \cup \{y\}]$ is also compact. But $L \subseteq \overline{co}[M \cup \{y\}]$, so L is compact. Let $D = \bigcup_{n=0}^{\infty} T^n(L)$, where $T^0(L) = L$. Clearly, D is T-invariant. We shall show that D is totally bounded. Given $\varepsilon > 0$, we consider the $\varepsilon/2$ open neighbourhood of M which we denote by $M_{\varepsilon/2}$. Let $\alpha > 0$ and consider $M_\alpha = \{x \in C : d(x, M) < \alpha\}$. It follows from the definition of an attractor for compact sets under T, that there exists a positive integer N such that $n \geq N$ implies that $T^n(L) \subseteq M_{\varepsilon/2}$. Since M is compact, it has a finite $\varepsilon/2$-net implying that $M_{\varepsilon/2}$ has a finite ε-net.

Since $\bigcup_{n=0}^{N} T^n(L)$ is compact, $\bigcup_{n=0}^{\infty} T^n(L)$ has a finite ε-net. Since ε was arbitrary and D is complete, we have that D is compact.

Now, let $R = \bigcap_{n=0}^{\infty} T^n(D)$. Then R is nonempty compact, $T(R) = T(\bigcap_{n=0}^{\infty} T^n(D)) = \bigcap_{n=0}^{\infty} T^{n+1}(D) = R$. So by the definition of M, $R \subseteq M$. By Lemma 4.4 there exists a nonexpansive retraction $r : D \to R$. Note that the map $F = Tor$ maps L continuously into itself, which is compact and starshaped. It follows from Theorem 3.4 that there exists $y_0 \in L$ such that $Fy_0 = y_0$. As $ry_0 \in R$ and R is T-invariant, the equation $y_0 = Fy_0 = T(ry_0)$ shows that $y_0 \in R$. Thus, $y_0 = ry_0$, since r is an identity map on R. Thus, $y_0 = T(ry_0) = Ty_0$. This completes the proof.

Using the remark following definition 3.1, we have

COROLLARY 4.7 [4, Theorem 2]. Let C be a closed, starshaped subset of a Banach space X and $T : C \to C$ nonexpansive. If there exists $M \subseteq C$, an attractor for compact sets, then T has a fixed point in C.

DEFINITION 4.8. Let X be a convex metric space and $T : X \to X$ be a mapping. T satisfies the <u>fixed point property</u> if T has a fixed point in every compact convex subset which it leaves invariant.

REMARK 4.9. It follows from Schauder-Tychonoff theorem that for a continuous mapping T on a locally convex Hausdorff topological linear space has the fixed point property.

THEOREM 4.10. Let X be a complete convex metric space with property (W). Let C be a closed convex subset of X. Let T be a self mapping of C satisfying

(i) there exists $M \subseteq C$ an attractor for compact sets under T, and

(ii) the family $\{T^n\}_{n=1}^{\infty}$ is equicontinuous.

Furthermore, if T has fixed point property, then T has a fixed point in C.

<u>Proof.</u> Let $L = \overline{co}[M]$. Then L is nonempty compact convex subset

of C, not necessarily T-invariant. Let $D = \overline{\bigcup_{n=0}^{\infty} T^n(L)}$ $(T^0(L) = L)$.

We shall now show that D is T-invariant and compact. Since $T[T^n(L)] = T^{n+1}(L)$ for every $n \geq 0$, it follows that $\bigcup_{n=0}^{\infty} T^n(L)$ is T-invariant. The continuity of T implies that D is T-invariant.

Following the proof of Theorem 4.7, it follows that D is compact. Let $R = \bigcap_{n=0}^{\infty} T^n(D)$. As in the proof of Theorem 4.6, R is nonempty compact and $R \subseteq M$. By Lemma 4.3, there exists a retraction $r : D \to R$. Note that the map F = Tor maps L continuously into itself, which is compact and convex. It follows from the hypothesis that there exists $y_0 \in L$ such that $Fy_0 = y_0$. But since $r(y_0) \in R$ and R is T-invariant, we have $y_0 \in R$, which implies, since r is an identity on R, that $r(y_0) = y_0$. Thus, $y_0 = T(ry_0) = Ty_0$. This completes the proof.

Using Remark 4.7, we have the following corollary to Theorem 4.11.

COROLLARY 4.11. Let C be a closed convex subset of a Banach space X and $T : C \to C$ be a mapping satisfying

(i) there exists $M \subset C$ which is an attractor for compact sets under T, and

(ii) the family $\{T^n\}_{n=1}^{\infty}$ is equicontinuous (relative to the metric induced by the norm). Then T has a fixed point in C.

REFERENCES

1. L. P. Belluce and W. A. Kirk, "Fixed-point theorems for certain classes of nonexpansive mappings", Proc. Amer. Math. Soc. 20 (1969), 141-146.

2. M. S. Brodskii and D. P. Milman, "On the center of convex sets (Russian), Dokl. Akad. Nauk SSR 59 (1948), 837-840.

3. F. E. Browder, "Nonexpansive nonlinear operators in a Banach space", Proc. Nat. Acad. Sci. 54 (1965), 1041-1044.

4. E. Chandler and G. Faulkner, "Fixed points in nonconvex domains", Proc. Amer. Math. Soc. 80 (1980), 835-838.

5. W. G. Doston, Jr., "Fixed-point theorems for nonexpansive mappings on star-shaped subsets of Banach spaces", J. London Math. Soc. 2 (1972), 408-410.

6. N. Dunford and J. T. Schwartz, "Linear operators, Part I", Interscience, New York, 1958.

7. M. Edelstein, "A theorem on fixed points under isometries", Amer. Math. Monthly 70 (1963), 298-300.

8. L. Janos, "On representation of self mappings", Proc. Amer. Math. Soc. 26 (1970), 529-533.

9. W. A. Kirk, "On mappings with diminishing orbital diameters", J. Lon. Math. Soc. 44 (1969), 107-111.

10. W. A. Kirk, "Nonexpansive mappings and the weak closure of sequence of iterates", Duke Math. J. 36 (1969), 639-645.

11. W. Takahashi, "A convexity in metric space and nonexpansive mappings I", Kodai Math. Sem. Rep. 22 (1970), 142-149.

12. L. A. Talman, "Fixed points for condensing multifunctions in metric spaces with convex structure", Kodai Math. Sem. Rep. 29 (1977), 62-70.

AN ANTIPODAL COINCIDENCE THEOREM FOR FAMILIES OF COMPACT PERTURBATIONS

J. Jaworowski

Indiana University
Bloomington, Indiana
U. S. A.

Introduction

Suppose that V is a Banach space, $S \subset V$ is the unit sphere and $\{f_t : S \to V, t \in T\}$ is a continuous family of compact perturbations of the inclusion map $S \subset V$, parametrized over a compact space T, which map S into a subspace $V_k \subset V$ of codimension k. By this we mean that the map $f : S \times T \to V$ $(x, t) \to f_t x$, is continuous, $f(S \times T) \subset V_k$ and $f = \iota + \phi$, where $\iota : (x, t) \to x$ and $\phi : S \times T \to V$ is a compact map. By the antipodal coincidence set of f we mean $A_f = \{x \in S \mid f(x, t) = f(-x, t)\}$. We study the size of A_f or of the set $\hat{A}_f = \hat{A}_{f/\sim}$, obtained from A_f by the antipodal identification $(x, t) \sim (-x, t)$. Specifically, we will show that for each $i \geq 0$, there exists an injective map

$$H^i(T; \mathbb{Z}/_2) \to H^{i+k}(\hat{A}_f; \mathbb{Z}/_2).$$

The construction is based on the results of [3], where the antipodal coincidence set was studied for fibre preserving maps of finite dimensional sphere bundles into vector space bundles. It seems plausible that the results of this paper can also be extended to a more general setting of fibre preserving maps of Banach sphere bundles. We intend to study this problem in a subsequent paper.

Our result is also a generalization of a theorem of Granas [1].

2. Main Result

We will be using the Alexander-Spanier cohomology theory (we will need the continuity property of cohomology) with coefficients in $\mathbb{Z}/_2$. efficient group will be suppressed from the notation.

Let $\alpha : Z \to Z$ be a fixed point free involution of a space Z. The characteristic class of α will be denoted by $u(\alpha)$; it is an element of $H^1(\hat{Z})$ where $\hat{Z}$ denotes the orbit space, $\hat{Z} = Z/\alpha$. In this paper, α will usually denote either the antipodal map $a : S \to S$, or more generally, $a \times 1_T : S \times T \to S \times T$, $a(x, t) = (-x, t)$, where S is a unit sphere in a Banach space; or the restriction $a_Y : Y \to T$ of a to some symmetric (i.e., invariant under α) subset $Y \subset S \times T$. Then $u(a \times 1_T) = u(a) \times 1 \in H^1(S \times T) = H^1(\hat{S} \times T)$, where 1 is the non-zero constant class in $H^0(T)$, and $u(a_y) = u(a \times 1_T)|\hat{Y} \in H^1(\hat{Y})$, the restriction of $u(a \times 1_T)$ to $\hat{Y}$. In either case, of $\alpha = a : S \to S$, or $\alpha = a \times 1_T : S \times T \to S \times T$, the characteristic class will be simply denoted by $u = u(a) \in H^1(\hat{S})$, or $u = u(a \times 1_Y) \in H^1(\hat{S} \times T)$, respectively.

If $f : S \times T \to V$ is a map, the restriction of $a \times 1_T$ to the antipodal coincidence set A_f will be denoted by a_f, and its characteristic class by $u_f = u(a_f) = u \mid \hat{A}_f$.

We denote by p the projection $S \times T \to T$ and by $\hat{p}$ the induced map $\hat{p} : \hat{S} \times T \to T$; and by $p_f : A_f \to T$ and $\hat{p}_f : \hat{A}_f \to T$ the restrictions of p and $\hat{p}$, respectively.

THEOREM. Let V be a Banach space, let $S \subset V$ be its unit sphere, let V_k be a subspace of V of codimension k and let $f : S \times T \to V$ be a compact perturbation of the map $\iota : S \times T \to S \subset V$, $(x, t) \to x$, such that $f(S \times T) \subset V_k$. Then the cohomology map

$$e^{i,k} : H^i(T) \to H^{i+k}(\hat{A}_f)$$

$$t \to (\hat{p}_f^* t) \cup u_f^k$$

is injective.

Proof. We will first prove two lemmas. A proof of similar assertions, in a more general setting, is contained in the proof of Theorem 2.1 of [2], but we give a proof in our case for the sake of completeness.

LEMMA 1. Let $f : S \times T \to V$ be as in the Theorem. Then, for any neighborhood U of A_f in $S \times T$, there exists an $\varepsilon > 0$ such that for each $(x, t) \in U$, $\|f(x, t) - f(-x, t)\| < \varepsilon$ implies that $x \in U$.

Proof of Lemma 1. Assuming the contrary, we would be able to find a sequence of points $(x_n, t_n) \notin U$ such that

$$\|f(x_n, t_n) - t_n) - f(-x_n, t_n)\| < \frac{1}{n}. \tag{1}$$

Let $\phi : S \times T \to V$ be a compact map such that $f(x, t) = x + \phi(x, t)$ for each $(x, t) \in S \times T$. We have $f(x, t) - f(-x, t) = 2x + \phi(x, t) - \phi(-x, t)$, and we can assume that the sequence $y_n = \phi(x_n, t_n) - \phi(-x_n, t_n)$ converges to a point $y_o \in V$. By (1) it follows that the sequence $x_n = \frac{1}{2}(f(x_n, t_n) - f(-x_n, t_n) - y_n)$ is converging to $x_o = -\frac{y_n}{2}$. Since T is compact, we can assume that the sequence $\{t_n\}$ is converging to $t_o \in T$. Therefore, the sequence $\{(x_n, t_n)\}$ is converging to $(x_o, t_o) \notin A_f$; but, by the continuity, $f(x_o, t_o) - f(-x_o, t_o) = 0$, which is impossible.

Recall that a map $\psi : X \to V$ is said to be finite-dimensional if $\psi(X)$ is contained in a finite dimensional subspace of V. For any compact map $\phi : X \to V$ and for any $\varepsilon > 0$, there exists a finite-dimensional compact map $\psi : X \to V$ such that $\|\phi - \psi\| < \varepsilon$ (compare [1] or [2]). Thus, any compact perturbation $f = \iota + \phi$ of $\iota : S \times V \to V$ can, for any $\varepsilon > 0$, be ε-approximated by a compact finite-dimensional perturbation $g = \iota + \psi$ of ι.

LEMMA 2. Let $f : S \times T \to V$ and $V_k \subset V$ be as in the Theorem and let U be a neighborhood of A_f. Then there exists a compact finite dimensional perturbation $g : S \times T \to V$ of the map $\iota : S \times T \to S \subset V$, $g = \iota + \psi$, where $\psi : S \times T \to V$ is compact and finite-dimensional, such that $g(S \times T) \subset V_k$ and $A_g \subset U$.

Proof of Lemma 2. Choose $\varepsilon > 0$ as in Lemma 1. Let π be a projection of V onto a subspace W^k complementary to V_k and let θ be a compact finite dimensional η-approximation to ϕ, where $\eta = \text{Min}\{1, \frac{1}{2\|1 - \pi\|}\}$. Let $\psi = \pi \circ \psi + (1 - \pi) \circ \theta$ and $g = \iota + \psi$. Then $f - g = \phi - \psi = (1 - \pi) \circ (\phi - \theta)$; $\pi \circ g = 0$, hence $g(S \times T) \subset V_k$. Suppose $(x, t) \in A_g$. Then $\|f(x, t) - f(-x, t)\| \leq \|f(x, t) - g(x, t)\| + \|f(-x, t) - g(-x, t)\| \leq \|\phi(-x, t) - \psi(-x, t)\| + \|\phi(-x, t) - \psi(-x, t)\| \leq 2\|1 - \pi\| \cdot \|\phi - \theta\| < \varepsilon$. By Lemma 1, $x \in U$. Therefore, $A_g \subset U$.

Proof of the Theorem. Let $t \in H^i(T)$ and suppose that $e^{i,k}t = 0$, i.e., $[(\hat{p}^*t) \cup u^k] \mid \hat{A}_f = 0$. Then there is a symmetric (i.e., invariant under the involution $(x, t) \to (-x, t)$) neighborhood U of A_f in $S \times T$ such that $[(\hat{p}^*t) \cup u^k] \mid \hat{U} = 0$. By using Lemma 2, choose a compact finite dimensional perturbation g of f such that $g(S \times T) \subset V_k$ and $A_g \subset U$. Suppose that Y^m is a finite dimensional subspace of V such that $g(S \times T) \subset Y^m$ and W^k be a k-dimensional subspace of V complementary to V_k. Let R^n be the n-dimensional subspace of V spanned by Y^m and W^k.

Let $g_1 = g \mid (S \cap R^n) \times T$. Then g_1 maps $(S \cap R^n) \times T \cong S^{n-1} \times T$ into a subspace of R^n of codimension k. By Theorem 2 of [3], $e^{i,k} : H^i(T) \to H^{i+k}(\hat{A}_{g_1})$ is injective.

But $A_{g_1} \subset A_g \subset U$ and

$$e^{i,k}t = [(\hat{p}^*t) \cup u^k] \mid A_{g_1} = \{[\hat{p}^*t) \cup u^k] \mid \hat{U}\} \mid A_g = 0.$$

Hence, $t = 0$, which proves that $e^{i,k} : H^i(T) \to H^{i+k}(\hat{A}_f)$ is injective.

COROLLARY. Under the assumption of the Theorem the topological (covering) dimension of A_f is $\dim \hat{A}_f = \dim A_f \geq d(T) + k$, where $d(T)$ is the cohomology of T.

REFERENCES

1. A. Granas, "An extension to functional spaces of Borsuk-Ulam Theorem on antipodes", Bull. Acad. Pol. Sci., 10 (1962), 81-86.

2. P. Holm and E. H. Spanier, "Involutions and Fredholm maps", Topology 10 (1971), 203-218.

3. J. Jaworowski, "Fibre preserving maps of sphere bundles into vector space bundles", Proc. of the Fixed Point Theory Workshop, Sherbrooke, 1980; Lecture Notes in Mathematics, Vol. 886, Springer-Verlag, 1981.

REMARKS ON SET-VALUED MAPPINGS IN BANACH SPACES*

W. A. Kirk

University of Iowa
Iowa City, Iowa
U.S.A.

We illustrate here how a 'minimum condition' initially used in the study of fixed point theory for mappings which are not self-mappings of their domains (see [4], [6]), and then used extensively in the study of mappings satisfying local metric conditions (see, e.g., [5], [7]), also provides a criterion which guarantees the existence of fixed points for certain set-valued mappings. Our first theorem, which is for set-valued local contractions, is an analog of Proposition 1 of [5]. We then use the global version of this result to obtain a new fixed point theorem for set-valued nonexpansive pseudo-contractions.

Let E be a Banach space. For a subset K of E, let $\mathcal{B}(K)$ denote the collection of all nonempty bounded closed subsets of K endowed with the usual Hausdorff metric H. (Thus for $A, B \in \mathcal{B}(K)$,

$$H(A, B) = \inf\{r : A \subset N_r(B) \text{ and } B \subset N_r(A)\},$$

where $N_r(S) = \{x \in K : \mathrm{dist}(x, S) < r\}$.)

We shall use the following notation: For $S \subset E$, let $|S| = \inf\{\|x\| : x \in S\}$, and for $D, K \subset E$, let $\partial_K D$ denote the relative boundary of D in K. In the theorem below we use the term <u>local contraction</u> in the strong sense — $U : S \subset E \to \mathcal{B}(E)$ is a local contraction

*Research supported by the National Science Foundation under grant MCS80-01604.

if there exists $\alpha \in (0, 1)$ such that for each $x \in S$, $H(U(u), U(v)) \leq \alpha\|u - v\|$ for all $u, v \in S$ sufficiently near x.

THEOREM 1. Let E be a Banach space, let K be a closed convex subset of E, and let D be an open subset of E with $D \cap K \neq \phi$. Suppose $U : \overline{D \cap K} \to \mathcal{B}(K)$ is a continuous mapping which is a local contraction on $D \cap K$, and suppose there exists $z \in D \cap K$ such that

$$|z - U(z)| \leq |x - U(x)| \quad \text{for all} \quad x \in \partial_K D. \tag{*}$$

Then there exists $x \in D \cap K$ such that $x \in U(x)$.

Our second theorem concerns mappings $T : S \subset E \to \mathcal{B}(E)$ which are nonexpansive in the sense $H(T(u), T(v)) \leq \|u - v\|$, $u, v \in S$, and which are also pseudo-contractions, thus satisfying in addition the condition

$$(\lambda - 1)\|u - v\| \leq |(\lambda I - T)(u) - (\lambda I - T)(v)|$$

for all $u, v \in S$, $\lambda > 1$. It is known ([3]) that this latter condition implies accretiveness of the mapping $A = I - T$ in the sense that for each $x, y \in S$, $u \in A(x)$, $v \in A(y)$, it follows that $\langle u - v, j\rangle \geq 0$ for all $j \in E^*$ satisfying $\|j\|^2 = \|x - y\|^2 = \langle x - y, j\rangle$. In the context of Theorem 2, pseudo-contractiveness is a strong assumption, although for single-valued mappings it is an assumption weaker than that of nonexpansiveness (cf. [1]). However the fixed point theory for set-valued nonexpansive mappings is not yet well developed, and it is by no means clear whether or not in the set-valued case the theorem would remain true in absence of an assumption of pseudo-contractive type. (We point out, however, that a set-valued nonexpansive mapping of a bounded closed convex subset K of a uniformly convex space E whose values are nonempty compact subsets of K is known to always have a fixed point without further assumptions (see [8]; cf. [2]).

THEOREM 2. Let E be a Banach space, let K be a closed convex subset of E, and let D be a bounded open subset of E for which $D \cap K \neq \phi$. Suppose $T : \overline{D \cap K} \to \mathcal{B}(K)$ is a nonexpansive pseudo-contraction and suppose there exists $z \in D \cap K$, such that

$$|z - T(z)| < |x - T(x)| \quad \text{for all} \quad x \in \partial_K D. \tag{*}$$

Then there exists a nonexpansive mapping $f : \overline{D \cap K} \to D \cap K$ whose fixed points are also fixed points of T. In particular, if $\overline{D \cap K}$ has the fixed point property with respect to nonexpansive (point-valued) mappings, then there exists $x \in \overline{D \cap K}$ such that $x \in T(x)$.

We remark that if $K \subset D$ in the above theorem, then (*) is satisfied vacuously, $T : K \to \mathcal{B}(K)$, and this theorem then reduces to a special case of Theorem 2.5 of [1], where T is assumed to be only lipschitzian (rather than nonexpansive). We do not know if such a weakening is possible in Theorem 2.

We shall require the following well-known facts (cf. [9]) which follow quickly from the definition of the Hausdorff metric.

LEMMA 1. If $A, B \in \mathcal{B}(E)$ and if $x \in A$, then for each $\varepsilon > 0$ there exists $y \in B$ such that $\|x - y\| \leq H(A, B) + \varepsilon$.

LEMMA 2. Suppose $\{A_n\}$ is a sequence in $\mathcal{B}(E)$ and suppose $\lim_{n\to\infty} H(A_n, A) = 0$ for $A \in \mathcal{B}(E)$. Then if $x_n \in A_n$ and if $\lim_{n\to\infty}\|x_n - x\| = 0$, it follows that $x \in A$.

Proof of Theorem 1. Clearly we may suppose $z \notin U(z)$; hence $\inf\{|x - U(x)| : x \in \partial_K D\} > 0$. Let $\alpha \in (0, 1)$ be a uniform local Lipschitz constant for U. We first observe that:

$$\begin{aligned}&\text{if } \{w_n\} \subset U(z) \text{ satisfies } \|z - w_n\| \to |z - U(z)| \text{ as } n \to \infty,\\ &\text{then for } n \text{ sufficiently large, } \operatorname{seg}[z, w_n) \subset D.\end{aligned} \tag{1}$$

To see this, suppose for each n there exists $m_n \in \operatorname{seg}[z, w_n]$ such that $m_n \in \partial_K D$ with $\operatorname{seg}[z, m_n) \subset D$. Then it follows that U is a global contraction on $\operatorname{seg}[z, m_n]$ so by (*) and contractiveness:

$$\begin{aligned}|z - U(z)| &\leq |m_n - U(m_n)|\\ &\leq \|m_n - w_n\| + H(U(m_n), U(z))\\ &\leq \|m_n - w_n\| + \alpha\|m_n - z\|\\ &= \|z - w_n\| - (1 - \alpha)\|m_n - z\|\\ &\leq \|z - w_n\| - (1 - \alpha)|z - \partial_K D|.\end{aligned}$$

Upon letting $n \to \infty$ one obtains a contradiction, and this proves (1).

Now choose $z_1 \in U(z) \cap D$ so that $seg[z, z_1) \subset D \cap K$ and

$$((1 + \alpha)/2)\|z - z_1\| \leq |z - U(z)|.$$

By Lemma 1 there exists $w \in U(z_1)$ such that

$$\begin{aligned}\|z_1 - w\| &\leq H(U(z), U(z_1)) + ((1 - \alpha)/2)\|z - z_1\| \\ &\leq \alpha\|z - z_1\| + ((1 - \alpha)/2\|z - z_1\| \\ &= ((1 + \alpha)/2\|z - z_1\| \\ &\leq |z - U(z)|.\end{aligned}$$

In particular, $|z_1 - U(z_1)| \leq |z - U(z)|$, so in view of (1) we may suppose w to have been chosen so that $seg[z_1, w) \subset D \cap K$. Letting $\beta = (1 + \alpha)/2$ and $z_2 = w$, we have:

$$\begin{aligned}&|z_1 - U(z_1)| \leq |z - U(z)|; \\ &\|z_1 - z_2\| \leq \beta\|z - z_1\| .\end{aligned}$$

A straightforward induction argument yields existence of a sequence $\{z_n\}$ in $D \cap K$ such that for each n,

$$\begin{aligned}&|z_n - U(z_n)| \leq |z_{n-1} - U(z_{n-1})|; \\ &\|z_n - z_{n+1}\| \leq \beta^n\|z - z_1\| .\end{aligned}$$

Since $\beta \in (0, 1)$, $\{z_n\}$ is a Cauchy sequence so there exists $z \in \overline{D \cap K}$ such that $z_n \to z$ as $n \to \infty$. Since U is continuous on $\overline{D \cap K}$, $H(U(z_n), U(z)) \to 0$ as $n \to \infty$, and since $z_n \in U(z_{n-1})$, Lemma 2 implies $z \in U(z)$. By assumption $\inf\{|x - U(x)| : x \in \partial_K D\} > 0$, so $z \in D \cap K$, completing the proof.

<u>Proof of Theorem 2</u>. (This argument is patterned after one originally used in [6]; cf. also, [1, Theorem 2.5].) As before we may suppose $z \in T(z)$ and $\inf\{|x - T(x)| : x \in \partial_K D\} > 0$. Since D is bounded it is possible to choose $\alpha \in (0, 1)$ so near 1 that

$$\alpha|z - T(z)| + (1 - \alpha)\|z - y\| < \inf\{\alpha|x - T(x)| - (1 - \alpha)\|x - y\| : x \in \partial_K D\}.$$

For each $y \in \overline{D \cap K}$, define $U_y : \overline{D \cap K} \to \mathcal{B}(K)$ by

$$U_y(x) = (1 - \alpha)y + \alpha T(x).$$

Then

$$\begin{aligned} |z - U_y(z)| &= |z - (1 - \alpha)y - \alpha T(z)| \\ &\leq \alpha|z - T(z)| + (1 - \alpha)\|z - y\| \\ &< \inf\{\alpha|x - T(x)| - (1 - \alpha)\|x - y\| : x \in \partial_K D\} \\ &\leq \inf\{|\alpha(x - T(x)) + (1 - \alpha)(x - y)| : x \in \partial_K D\} \\ &= \inf\{|x - U_y(x)| : x \in \partial_K D\}. \end{aligned}$$

Thus U_y is a set-valued contraction mapping which satisfies (*), so by Theorem 1 there exists for each $y \in \overline{D \cap K}$ a point $z_y \in D \cap K$ such that $z_y \in U_y(z_y)$. Now let $u, v \in \overline{D \cap K}$, with $z_u \in U_u(z_u)$ and $z_v \in U_v(z_v)$. Since T is a pseudo-contraction, for each $\lambda > 1$,

$$(\lambda - 1\|z_u - z_v\| \leq |\lambda(z_u - z_v) - (T(z_u) - T(z_v))|$$

and setting $\lambda = 1/\alpha$,

$$(1 - \alpha)\|z_u - z_v\| \leq |z_u - z_v - \alpha(T(z_u) - T(z_v))|.$$

But since $z_u \in (1 - \alpha)u + \alpha T(z_u)$ and $z_v \in (1 - \alpha)v + \alpha T(z_v)$, it follows that

$$|z_u - z_v - \alpha(T(z_u) - T(z_v))| \leq (1 - \alpha)\|u - v\| .$$

In particular, $u = v$ implies $z_u = z_v$ and so the mapping $f : u \to z_u$ is a single-valued nonexpansive mapping of $\overline{D \cap K} \to D \cap K$.

REFERENCES

1. D. Downing and W. O. Ray, "Some remarks on set-valued mappings", Nonlinear Anal. TMA (to appear).

2. K. Goebel, "On a fixed point theorem for multivalued nonexpansive mappings", Annales Univ. Maria Curie - Sklodowska 29 (1975), 69-72.

3. T. Kato, "Nonlinear semigroups and evolution equations", J. Math. Soc. Japan 19 (1967), 508-520.

4. W. A. Kirk, "On zeros of accretive operators in uniformly convex spaces", Bollettino Un. Mat. Ital. (5), 17-A(1980), 249-253.

5. W. A. Kirk, "Locally nonexpansive mappings in Banach spaces", Proc. Workshop on Fixed Point Theory (Univ. Sherbrooke, 1980), to appear.

6. W. A. Kirk and R. Schöneberg, "Some results on pseudo-contractive mappings", Pacific J. Math. 71 (1977), 89-100.

7. W. A. Kirk and R. Schöneberg, "Mapping theorems for local expansions in metric and Banach spaces", J. Math. Anal. Appl. 72 (1979), 114-121.

8. T. C. Lim, "A fixed point theorem for multivalued nonexpansive mappings in a uniformly convex Banach space", Bull. Amer. Math. Soc. 80 (1974), 1123-1126.

9. S. B. Nadler, Jr., "Multivalued contraction mappings", Pacific J. Math. 30 (1969), 475-480.

FIXED POINTS AND CLOSE-TO-NORMAL STRUCTURE IN LOCALLY CONVEX SPACES*

S. A. Naimpally, K. L. Singh,† and J. H. M. Whitfield

Lakehead University
Thunder Bay, Ontario
Canada

Abstract

In this paper, we prove fixed and common fixed point theorems for non-expansive, generalized nonexpansive and semigroups of nonexpansive mappings in locally convex spaces. These results extend those of Browder, Goebel, Kirk, Shimi, Lim, Reich and others. Also, we give necessary and sufficient conditions for a convex subset of a locally convex space to have close-to-normal structure.

1. Normal and Close-to-normal Structure

In this section, we prove some results for normal and close-to-normal structure. These extend the results of Veit and Theip. Also in the next section, application is made of these concepts to obtain fixed points.

Throughout this paper E will denote a Hausdorff, locally convex linear topological space and Q a (fixed) family of continuous seminorms which generate the topology of E. For any $p \in Q$ and $A \subseteq E$, $\delta_p(A)$ will denote the p-diameter of A, that is, $\delta_p(A) = \sup\{p(x - y) : x, y \in A\}$, coA will denote the convex hull of A, $B_p(x, r)(B_p[x, r])$ will denote the open (closed) ball with centre x and radius r. For $A \subseteq E$, $d_p(x, A) = \inf\{p(x - y) : y \in A\}$.

DEFINITION 1.1. A convex subset of K of E is said to have <u>normal structure</u> if for every closed bounded convex subset W of K such that

*This research supported in part by Grants from NSERC (Canada).

† Current affiliation: University of Minnesota - Duluth, Duluth, Minnesota, U. S. A.

$\delta_p(W) > 0$ there exists $x \in W$ such that $\delta_p(W - x) < \delta_p(W)$ for each $p \in Q$.

DEFINITION 1.2. Let K be a nonempty subset of E and let $\{W_\alpha : \alpha \in A\}$ be a decreasing net of nonempty bounded subsets of K. For $x \in K$, $\alpha \in A$, and $p \in Q$, define

$$\begin{aligned} r_{p,\alpha}(x) &= \sup\{p(x - y) : y \in W_\alpha\} \\ r_p(x) &= \inf\{r_{p,\beta}(x) : \beta \in A\} \\ r_p &= \inf\{r_p(x) : x \in K\} \\ M &= \{x \in K : r_p(x) = r_p\}. \end{aligned} \tag{I}$$

The set M and the number r_p will be called <u>asymptotic centre</u> and the <u>asymptotic radius</u>, respectively, of $\{W_\alpha : \alpha \in A\}$ in K with respect to p. Also we will use the notation $ar(\{W_\alpha : \alpha \in A\}, K)$ for the asymptotic radius of $\{W_\alpha : \alpha \in A\}$ with respect to K.

DEFINITION 1.3. A convex set K of E is said to have <u>asymptotic normal structure</u> with respect to Q, if given any bounded, closed convex subset W of K which contains more than one point, given any decreasing net of nonempty subsets $\{W_\alpha : \alpha \in A\}$ of W and given any $p \in Q$ such that $\delta_p(W) > 0$, the asymptotic centre of $\{W_\alpha : \alpha \in A\}$ in W with respect to p is a proper subset of W.

DEFINITION 1.4. A convex subset K of E is said to have <u>close-to-normal structure</u> if for any bounded, closed, convex subset H of K with $\delta_p(H) > 0$ there exists $x \in H$ such that $p(x - y) < \delta_p(H)$ for all $y \in H$.

DEFINITION 1.5. A <u>diametral sequence</u> is a nonconstant bounded sequence $\{x_n\}_{n=1}^{\infty}$ such that for some $p \in Q$, $\delta_p(\{x_n\}) > 0$ and there exists an integer N such that $d_p(x_{n+1})co(x_1, x_2, \ldots, x_n)) = \delta_p(\{x_n\}_{n=1}^{\infty}$ for all $n > N$.

The following Lemma is proved in [13].

LEMMA 1.1. For a convex subset K of E, the following are equivalent.

(i) K does not contain a bounded sequence $\{x_n\}_{n\geq 1}$ such that, for some $p \in Q$ with $\delta_p(\{x_n\}) > 0$,

$$d_p(x_{n+1}, co(x_1, x_2, \ldots, x_n)) \to \delta_p(x_n)$$

as $n \to \infty$.

(ii) K has normal structure.

(iii) K has asymptotic normal structure.

Using Lemma 1.1, we have the following.

LEMMA 1.2. A convex subset K of E has normal structure if and only if it possesses the following property: (R) For each nonconstant bounded sequence $\{x_n\}$ in K there exists $p \in Q$ such that the function $g(x) = \limsup_n p(x_n - x)$ is not constant in $co\{x_n\}$.

Proof. Suppose K has normal structure. Let $\{x_n\}$ be a nonconstant bounded sequence in K and let $W = co\{x_n\}$. Define a decreasing net $\{W_k : k = 1, 2, \ldots\}$ of subsets of W by $W_k = \{x_n : n \geq k\}$. For any $x \in W$, $r_p(x) = \limsup_n p(x_n - x) = g(x)$, which by Lemma 1.1. is not constant in W.

Conversely, suppose K has property (R). Let $\{x_n\}$ be a nonconstant bounded sequence in K. By property (R), there exists a point $x \in W = co\{x_n\}$ such that $\limsup p(x_n - x) < \delta_p(\{x_n\})$. For sufficiently large n, $x \in co(x_1, x_2, \ldots, x_{n-1})$ and thus $\limsup_n d_p(x_n, co(x_1, x_2, \ldots, x_{n-1}) \leq \limsup_n p(x_n - x) < \delta_p(\{x_n\})$. So, $\{x_n\}$ is a nondiametral sequence, hence by Lemma 1.1, K has normal structure.

REMARK 1.1. It is clear that if K has normal structure (asmyptotic normal structure), then K has close-to-normal structure. However, the converse is not true, in general, as can be seen from the following example.

EXAMPLE 1.1. [2, p. 439] Let H be a Hilbert space and B the space isomorphic to H with norm defined by (for $x \in H$)

$$\|x\| = \sup\{\tfrac{1}{2}\|x\|_H, |x_n|\}.$$

Let $K = \{x : \|x\|_H \leq 1$ and $x_i \geq 0$ for all $i\}$. Then K does not have normal structure, however, it follows from [21, Theorem 2] that K has close-to-normal structure. Also, the answer to the problem [21, (iii), p. 355], "Does every weakly compact convex subset of a Banach space have close-to-normal structure?" is no [20, Example 1.3].

THEOREM 1.1. A convex subset K of E has close-to-normal structure if and only if it contains no diametral sequence.

Proof. Suppose K contains a diametral sequence $\{x_n\}_{n=1}^{\infty}$. Let $K_0 = co(\{x_n\}_{n=1}^{\infty}) \subset K$. If $x_0 \in K_0$, then $x_0 = \sum_{i=1}^{q} \alpha_i x_i$, $\alpha_i \geq 0$, $\forall i = 1, 2, \dots q$, $\sum_{i=1}^{q} \alpha_i = 1$ and $x_0 \in co(x_1, x_2, \dots, x_{n-1})$ for all $n > q$. Since $\{x_n\}_{n=1}^{\infty}$ is a diametral sequence, there exists $p \in Q$ and an integer N such that $d_p(x_{n+1}, co(x_1, x_2, \dots, x_n)) = \delta_p(\{x_n\}_{n=1}^{\infty}) > 0$, $\forall n > N$. Then $\delta_p(\{x_n\}_{n=1}^{\infty}) \geq p(x_0 - x_m) \geq \delta_p(\{x_n\}_{n=1}^{\infty})$, $\forall m > q$, $m > N$. Let $y_0 = x_{q+N} \in K_0$, then $p(x_0 - y_0) = \delta_p(K_0) = \delta_p(\{x_n\}_{n=1}^{\infty})$. This shows that K does not have close-to-normal structure.

Conversely, suppose K does not have close-to-normal structure. Then K contains a bounded convex subset H such that $\beta = \delta_p(H) > 0$ and, for each $x \in H$, there is $y \in H$ such that $p(x - y) = \beta$. Choose $x_1, x_2 \in H$ such that $p(x_1 - x_2) = \beta$. When $\{x_1, x_2, \dots, x_n\} \subset H$ have been chosen , we take $x_{n+1} \in H$ such that $p(y_n - y_{n+1}) = \beta$, where $y_n = \frac{1}{n} \sum_{i=1}^{n} x_i \in H$. Proceeding in this way, we get a sequence $\{x_n\}_{n=1}^{\infty} \subset K$.

Let $x \in co(x_1, x_2, \dots, x_n)$ be arbitrary, $x = \sum_{i=1}^{n} \alpha_i x_i$, $\alpha_i \geq 0$, $\forall i = 1, 2, \dots, n$, $\sum_{i=1}^{n} \alpha_i = 1$. For $\alpha = \max\{\alpha_1, \alpha_2, \dots, \alpha_n\}$ we have

$$y_n = \sum_{i=1}^{n} \frac{\alpha_i x_i}{n\alpha} - \sum_{i=1}^{n} \frac{\alpha_i x_i}{n\alpha} + \sum_{i=1}^{n} \frac{x_i}{n}$$

$$= \frac{x}{n} + \sum_{i=1}^{n} (\frac{1}{n} - \frac{i}{n})x_i;$$

$$\frac{1}{n\alpha} + \sum_{i=1}^{n} (\frac{1}{n} - \frac{\alpha_i}{n\alpha}) = 1 \quad \text{and} \quad \frac{1}{n} - \frac{\alpha_i}{n\alpha} \geq 0, \ \forall i = 1, 2, \ldots, n.$$

Then

$$\beta = p(y_n - x_{n+1}) = p(\frac{x}{n\alpha} + \sum_{i=1}^{n} (\frac{1}{n} - \frac{\alpha_i}{n\alpha})x_i - x_{n+1})$$

$$\leq \frac{1}{n\alpha} p(x - x_{n+1}) + \sum_{i=1}^{n} (\frac{1}{n} - \frac{\alpha_i}{n\alpha}) p(x_i - x_{n+1})$$

$$\leq \frac{1}{n\alpha} p(x - x_{n+1}) + \beta(1 - \frac{1}{n\alpha}).$$

Hence

$$\frac{\beta}{n\alpha} \leq \frac{1}{n\alpha} \ p(x - x_{n+1})$$

implies that $p(x - x_{n+1}) = \beta$. Since $x \in co(x_1, x_2, \ldots, x_n)$ was arbitrary, it follows that $d_p(x_{n+1}, co(x_1, x_2, \ldots, x_n)) = \inf p(x - x_{n+1}) = \beta$, $\forall n$, $x \in co(x_1, x_2, \ldots, x_n)$. Thus $\{x_n\}_{n=1}^{\infty}$ is a diametral sequence in K.

THEOREM 1.2. A convex subset K of E has close-to-normal structure if and only if it does not contain a sequence $\{x_n\}_{n=1}^{\infty}$ such that for some $c > 0$, $p(x_n - x_m) = c$, $p(x_{m+1} - \overline{x}_n) = c$, for all $n \geq 1$, $m \geq 1$, where

$$\overline{x}_n = \frac{1}{n} \sum_{i=1}^{n} x_i.$$

<u>Proof</u>. Suppose K does not have close-to-normal structure. Then there is a bounded convex subset H of K such that $\delta_p(H) > 0$ and for every $x \in H$ there is a $y \in H$ such that $p(x - y) = \delta_p(H)$. By induction we construct a nonconstant sequence $\{x_n\}_{n=1}^{\infty} \subset H$ as follows: Take x_1, $x_2 \in H$ such that $p(x_1 - x_2) = \delta_p(H)$. Suppose $x_1, x_2, \ldots, x_n \in H$ has

been constructed with the properties that

$$p(x_i - x_k) = \delta_p(H),\ \forall i,\ k = 1, 2, \ldots, n$$

and

$$p(x_{k+1} - \bar{x}_k) = \delta_p(H),\ \forall k = 1, 2, \ldots, n - 1.$$

We choose $x_{n+1} \in H$ such that $p(x_{n+1} - \bar{x}_n) = \delta_p(H)$.

We show that for this x_{n+1}, we have $p(x_{n+1} - x_i) = \delta_p(H)$, $\forall i = 1, 2, \ldots, n$. Since $p(x_{n+1} - x_n) = \delta_p(H)$,

$$\delta_p(H) = \frac{n\delta_p(H)}{n} \geq \sum_{i=1}^{n} p(x_{n+1} - x_i) \geq p(x_{n+1} - \bar{x}_n) = \delta_p(H).$$

From this it follows that $\frac{1}{n}\sum_{i=1}^{n} p(x_{n+1} - x_i) = \delta_p(H)$. Hence, $p(x_{n+1} - x_i) = \delta_p(H)$, $\forall i = 1, 2, \ldots, n$. So the sequence $\{x_n\}_{n=1}^{\infty} \subset H$ satisfies the condition of Theorem 1.2 with $c = \delta_p(H)$.

Conversely, suppose K contains a sequence $\{x_n\}_{n=1}^{\infty}$ satisfying the condition of Theorem 1.2. Let $x \in co(x_1, x_2, \ldots, x_n)$. Then $x = \sum_{i=1}^{n} \alpha_i x_i$, $\alpha_i \geq 0$, $\forall i = 1, 2, \ldots, n$, $\sum_{i=1}^{n} \alpha_i = 1$. Let $\alpha = \max\{\alpha_1, \alpha_2, \ldots, \alpha_n\}$, $\gamma_0 = n\alpha$, $\gamma_i = \alpha_i - \alpha$, $i = 1, 2, \ldots, n$. We have $0 \leq \gamma_0 \leq n$; $\gamma_i \leq 0$, $\forall i = 1, 2, \ldots, n$ and $\sum_{i=1}^{n} \gamma_i = 1 - \gamma_0$. We can write x as $x = \sum_{i=1}^{n} (\alpha_i - \alpha + \alpha)x_i = n\alpha \sum_{i=1}^{n} \frac{x_i}{n} + \sum_{i=1}^{n} (\alpha_i - \alpha)x_i = \gamma_0\bar{x}_n + \sum_{i=1}^{n} \gamma_i x_i$. Hence, $p(x_{n+1} - x) \leq \sum_{i=1}^{n} \alpha_i p(x_{n+1} - x_i) = c$ and $p(x_{n+1} - x) = p(x_{n+1} - \gamma_0\bar{x}_n - \sum_{i=1}^{n} \gamma_i x_i) =$

$p(\gamma_0 x_{n+1} - \gamma_0\bar{x}_n + (1 - \gamma_0)x_{n+1} - \sum_{i=1}^{n} \gamma_i x_i) = p(\gamma_0(x_{n+1} - \bar{x}_n) + \sum_{i=1}^{n}$

$\gamma_i(x_{n+1} - x_i)) \geq p(\gamma_0(x_{n+1} - \overline{x}_n)) - \sum_{i=1}^{n} p(x_{n+1} - x_i) = \gamma_0 p(x_{n+1} - \overline{x}_n) -$

$\sum_{i=1}^{n} \gamma_i p(x_{n+1} - x_i) = \gamma_0 c - \sum_{i=1}^{n} \gamma_i c = \gamma_0 c + (1 - \gamma_0)c = c.$ It follws that $p(x_{n+1} - x) = c$ for all n, and for all $x \in co(x_1, x_2, \ldots, x_n)$. Hence,

$$d_p(x_{n+1}, co(x_1, x_2, \ldots, x_n) = \delta_p(\{x_n\}_{n=1}^{\infty}).$$

Thus $\{x_n\}_{n=1}^{\infty}$ is a diametral sequence in K and hence, K does not have close-to-normal structure by Theorem 1.1. This completes the proof.

Immediate consequences of Theorems 1.1 and 1.2 we have the following.

COROLLARY 1.1. [20, Proposition 1] A convex subset K of a Banach space E has close-to-normal structure if and only if it contains no diametral sequence.

COROLLARY 1.2. [20, Proposition 2] A convex subset K of a Banach space E has close-to-normal structure if and only if it does not contain a sequence $\{x_n\}_{n=1}^{\infty}$ such that for some $c > 0$, $\|x_n - x_m\| = c$, $\|x_{n+1} - \overline{x}_n\| = c$, for all $n \geq 1$, $m \geq 1$, where $\overline{x}_n = \frac{1}{n}\sum_{i=1}^{n} x_i$.

2. Fixed Points

Using the results obtained in section 1, we prove a fixed point theorem for nonexpansive type mappings. As corollaries, we obtain results of Browder, Kannan, Goebel, Kirk and Shimi.

LEMMA 2.1. Let K be a weakly compact convex subset of E. Let $T : K \to K$ be a mapping satisfying the condition

$$\text{(K)} \quad p(Tx - Ty) \leq \frac{1}{2}[p(x - Tx) + p(y - Ty)]$$

for all $x, y \in K$ and for all $p \in Q$. Then for each p, the set $\{p(x - Tx) : x \in K\}$ has a smallest element.

Proof. Let $r > 0$. Consider the set K_r defined by $K_r = \{y \in K : p(y - Ty) \leq r\}$. Let $I = \{r \in [0, \infty) : K_r \neq \phi\}$. Since K is bounded, $I \neq \phi$. It is sufficient to show that I has a smaller element or, equivalently, $\bigcap_{r \in I} K_r \neq \phi$. For each $r \in I$, let H_r be the closed convex hull of $T(K_r)$. Since for any r, s in I with $r < s$, $K_r \subseteq K_s$, the family $F = \{H_r : r \in I\}$ has the finite intersection property. Since each H_r is closed and convex, it is weakly closed. So by weak compactness of K, F has a nonempty intersection. It suffices to show that $H_r \subseteq K_r$ for each $r \in I$. Let $r \in I$, $y \in H_r$, $\varepsilon > 0$. Then there exist $a_1, a_2, \ldots a_n$ with $a_i \geq 0$, $\sum_{i=1}^{n} a_i = 1$ and $y_1, y_2, \ldots, y_n \in K_r$ such that $p(y - \sum_{i=1}^{n} a_i Ty_i) < \varepsilon$. Thus

$$p(y - Ty) \leq p(y - \sum_{i=1}^{n} a_i Ty_i) + p(\sum_{i=1}^{n} a_i Ty_i - Ty)$$

$$< \varepsilon + \sum_{i=1}^{n} a_i p(Ty_i - Ty)$$

$$\leq \varepsilon + \sum_{i=1}^{n} \frac{a_i}{2} [p(y_i - Ty_i) + p(y - Ty)]$$

$$< \varepsilon + \frac{1}{2} [r + p(y - Ty)].$$

So, $p(y - Ty) \leq 2\varepsilon + r$. Since ε was arbitrary, $p(y - Ty) \leq r$, that is $y \in K_r$. Hence $H_r \subseteq K_r$.

LEMMA 2.2. Let K be a nonempty weakly compact convex subset of E. Let $T : K \to K$ be a continuous mapping satisfying condition (K).

(i) Suppose that T has no fixed point. Then there is a T-invariant closed convex subset H of K such that $\delta_p(H) > 0$ and $p(x - Tx) = \delta_p(H)$ for all $x \in H$.

(ii) Suppose that for any closed T-invariant subset H of K with $\delta_p(H) > 0$ there exists $x \in H$ for which $p(x - y) < \delta_p(H)$, $y \in H$. Then T has a unique fixed point. In particular, T has a unique fixed point if H has close-to-normal structure.

Proof. (i) By Zorn's Lemma there exists a minimal nonempty T-invariand closed, convex subset H of K. Since H is a closed convex subset of a weakly compact set K, H is weakly compact. By Lemma 2.1, there exists $x \in H$ such that $p(x - Tx) = \inf\{p(y - Ty) : y \in H\}$. Let $r_0 = p(x - Tx)$. By hypothesis $r_0 > 0$. Let $K_{r_0} = \{y \in K : p(y - Ty) \leq r\}$ and let H_{r_0} be the convex hull of $T(K_{r_0})$. As in the proof of Lemma 2.1, it follows that $H_{r_0} \subseteq K_{r_0}$. So $T(H_{r_0}) \subseteq T(K_{r_0}) \subset \overline{co}T(K_{r_0}) = H_{r_0}$. By the minimality of H, $H_{r_0} = H$ and therefore $K_{r_0} = H$. By the choice of r_0, $p(y - Ty) = r_0$ for all $y \in H$. Now for any $y, z \in K_{r_0}$, $p(Ty - Tz) \leq \frac{1}{2}[p(y - Ty) + p(z - Tz)] = r_0$. So, $\delta_p(T(K_{r_0})) \leq r_0$. Therefore, $\delta_p(H) = \delta_p(H_{r_0}) = \delta_p(T(K_{r_0})) \leq r_0$. Thus, $\delta_p(H) = r_0$. Hence $p(y - Ty) = \delta_p(H)$ for all $y \in K$.

(ii) Follows from (i).

LEMMA 2.3. The following are equivalent.

(i) Every weakly compact convex subset of E has close-to-normal structure.

(ii) Every map satisfying the condition of Lemma 2.2 on a weakly compact subset of E has a unique fixed point.

Proof. (i) $\Rightarrow$ (ii) follows from Lemma 2.2.

(ii) $\Rightarrow$ (i). Suppose to the contrary, that there exists a weakly compact convex subset of E which does not have close to normal structure. Then K contains a weakly compact convex subset H such that $\delta_p(H) > 0$ and for any $x \in H$, $p(x - Tx) = \delta_p(H)$. Then T is a map satisfying condition (K) which has no fixed point.

The simple proofs of following lemmas will be omitted.

LEMMA 2.4. Let $T : E \to E$ be a mapping satisfying the following condition: there exist $a, b, c \geq 0$, with $a + 2b + 2c = 1$ such that

$$(S)\quad p(Tx - Ty) \leq ap(x - y) + b[p(x - Tx) + p(y - Ty)]$$

$$+ c[p(x - Tx) + p(y - Ty)]$$

for all $x, y \in E$, for all $p \in Q$. Then for any $x \in E$, $p(Tx - T^2x) \leq p(x - Tx)$.

LEMMA 2.5. If $b > 0$ in (S), then there is a number $k < 2$ such that for each $x \in E$, $p(Tx - T^3x) \leq kp(x - Tx)$.

LEMMA 2.6. If $b > 0$ and $c > 0$, then there is a number $m < 1$ such that for each $x \in E$, $p(T^2x - T^3x) \leq mp(x - Tx)$.

LEMMA 2.7. Suppose E is complete in the following sense: If $p(x_n - x_m) \to 0$ for each p, then there exists x such that $x_n \to x$. Let $T : E \to E$ be a mapping (not necessarily continuous) satisfying condition (S) with $b > 0$, $c > 0$. Then T has a unique fixed point.

Now we state and prove a generalization of Theorem in [6].

THEOREM 2.1. Let K be a nonempty weakly compact convex subset of E with normal structure. Let $T : K \to K$ be a continuous mapping satisfying

$$(P)\quad p(Tx - Ty) \leq a_1p(x - y) + a_2p(x - Tx) + a_3p(y - Ty)$$

$$+ a_4p(x - Tx) + a_5p(y - Tx)$$

for all $x, y \in E$, for all $p \in Q$ and $a_i \geq 0$, $\sum_{i=1}^{5} a_i = 1$. Then T has a fixed point in K. (A mapping satisfying condition (P) will be a generalized nonexpansive mapping.)

<u>Proof</u>. By interchanging the roles of x and y we see that the condition (P) can be written as

$$p(Tx - Ty) \leq ap(x - y) + b[p(x - Tx) + p(y - Ty)]$$

$$+ c[p(x - Ty) + p(y - Tx)],$$

where $a, b, c \geq 0$ and $a + 2b + 2c = 1$.

It is enough to prove the theorem for the case $b = 0$ or $c = 0$ only, for otherwise if $b > 0$, $c > 0$, we can apply Lemma 2.7, since K is complete.

By the weak compactness of K and Zorn's Lemma, there exists a subset C of K which is minimal in the family of all nonempty closed and convex subsets of K which are invariant under T. We show that C contains exactly one element.

CASE 1. $b = 0$. Let x_0 be an arbitrary point in C and $x_n = T^n x_0 (n = 1, 2, \ldots)$. Then, either $\{x_n\}$ is constant implying $\{x_0\} = C$ or $\{x_n\}$ is nonconstant and, hence by Lemma 1.2, there is a number $r > 0$ such that the set $M = \{y \in C : \limsup p(x_n - y) \leq r\}$ is nonempty closed convex subset of C and $M \neq C$. If $y \in M$

$$p(Ty - T^n x_0) \leq ap(y - T^{n-1}x_0) + c[p(y - T^n x_0) + p(T^{n-1}x_0 - Ty)]$$

and thus $\limsup\limits_n p(Ty - x_n) \leq \frac{a + c}{1 - c} \lim\limits_n p(y - x_n) \leq r$. So, $T(M) \subseteq M$ and this contradicts the minimality of C. Hence, $C = \{x_0\}$.

CASE 2. $c = 0$. In the view of Theorem 4.6 [19] and Lemma 2.3 the cases $c = b = 0$ and $c = a = 0$ need not be considered. So we may assume $c = 0$, $a > 0$, $b > 0$.

Let $d = \inf\{p(x - Tx) \quad x \in C\}$ and $\varepsilon > 0$. Then there is a point $x \in C$ such that $p(x - Tx) < d + \varepsilon$. Let $y = \frac{T^2x + T^3x}{2}$. Then $y \in C$. Now

$$p(y - Ty) \leq \frac{1}{2}[p(T^2x - Ty) + p(T^3x - Ty)]$$

$$\leq \frac{1}{2}[ap(Tx - y) + b[p(Tx - T^3x) + p(y - Ty)]]$$

$$+ \frac{1}{2}[ap(T^2x - y) + b[p(T^2x - T^3x) + p(y - Ty)]].$$

Hence

$$(1 - b)p(y - Ty) \leq \frac{1}{2}[ap(Tx - y) + bp(Tx - T^2x)]$$

$$+ \frac{1}{2}[ap(T^2x - y) + bp(T^2x - T^3x)].$$

$$p(Tx - y) = p(Tx - \frac{T^2x + T^3x}{2}) \leq \frac{1}{2}[p(Tx - T^2x) + p(Tx - T^3x)],$$

$$p(T^2x - y) = p(T^2x - \frac{T^2x + T^3x}{2}) = \frac{1}{2}p(T^2x - T^3x).$$

Thus

$$(1 - b)p(y - Ty) \leq \frac{1}{2}\{[\frac{a}{2} p(Tx - T^2x) + p(Tx - T^3x)]$$

$$+ bp(Tx - T^2x)\} + \frac{1}{2}[\frac{a}{2} p(T^2x - T^3x) + bp(T^2x - T^3x)].$$

Now using Lemmas 2.4 and 2.5, we have

$$(1 - b)p(y - Ty) \leq \frac{1}{2}\{\frac{a}{2}[p(x - Tx) + kp(x - Tx)] + bp(T^2x - T^3x)\}$$

$$+ \frac{1}{2}[\frac{a}{2} p(x - Tx) + bp(x - Tx)]$$

$$\leq (\frac{a}{2} + \frac{k}{4} a + b)p(x - Tx).$$

So, $(1 - b)d \leq (1 - b)p(y - Ty) \leq (\frac{a}{2} + \frac{k}{4} a + b)(d + \varepsilon)$. Letting $\varepsilon \to 0$ and assuming $d > 0$, we see that, since $a > 0$, $(1 - b)d < (a + b)d$, and this contradicts $a + 2b = 1$. Thus, $d = 0$ and hence, there is a sequence $\{x_n\}$ in C such that $p(x_n - Tx_n) \to 0$. If $\{x_n\}$ is constant, it defines a fixed point of T. Otherwise, by Lemma 1.2, there is a number $r > 0$ such that the set $M = \{y \in C : \limsup_n p(x_n - y) \leq r\}$ is a nonempty closed convex subset of C and $M \neq C$. For each $y \in M$ and each n,

$$p(Ty - x_n) \leq p(Ty - Tx_n) + p(Tx_n - x_n)$$

$$\leq ap(y - x_n) + b[p(y - Ty) + p(x_n - Tx_n)] + p(Tx_n - x_n)$$

$$\leq ap(y - x_n) + b[p(y - x_n) + p(x_n - Ty)] + (1 + b)p(x_n - Tx_n).$$

Since $\lim_n p(Tx_n - x_n) = 0$, we have $\limsup_n p(Ty - Tx_n) \leq \frac{a + b}{1 - b} \limsup_n$

$p(y - x_n) \le r$. So, $T(M) \subseteq M$, a contradiction to the minimality of C. Hence, $C = \{x_0\}$ and the theorem is proved.

COROLLARY 2.1. [6] Let E be a uniformly convex Banach space, K a non-empty, bounded, closed and convex subset of E, and $T : K \to K$ a continuous mapping satisfying for $x, y \in K$,

$$\|Tx - Ty\| \le a_1\|x - y\| + a_2\|x - Tx\| + a_3\|y - Ty\|$$

$$+ a_4\|x - Ty\| + a_5\|y - Tx\| ,$$

where $a_i \ge 0$, $\sum_{i=1}^{5} a_i \le 1$. Then T has a fixed point in K.

Proof. K being closed, bounded convex subset of a uniformly convex Banach space is weakly compact. Moreover, K has normal structure [4], [5].

Taking $a_1 = 1$ and $a_2 = a_3 = a_4 = a_5 = 0$, we also get Theorem 1 [4] as a corollary of Theorem 2.1.

3. Fixed Points For Nonexpansive And Generalized Contraction Mappings

Our main aim here is to extend some results of Kirk and Reich from the Banach space setting to locally convex spaces.

DEFINITION 3.1. A mapping $T : E \to E$ is called nonexpansive if for all $x, y \in E$ and $p \in Q$, $p(Tx - Ty) \le p(x - y)$.

LEMMA 3.1. Let H and K be two closed subsets of E such that $H \cap K \neq \phi$. If H is convex, then $\partial_H(K) = \phi$ iff $H \subseteq K$, here $\partial_H(K)$ denotes the relative boundary of $H \cap K$ in H.

Proof. If $H \subseteq K$, then $H \setminus K = \phi$ and therefore, $\partial_H(K) = \phi$. Suppose $\partial_H(K) = \phi$. We show that $H \subseteq K$. Suppose $H - K \neq \phi$. Take $x \in H \cap K$ and $y \in H - K$. Since H is convex, the segment $\{\alpha x + (1 - \alpha) y\} \subseteq H$. Let $\gamma = \inf\{\alpha : \alpha x + (1 - \alpha)y \in K\}$. Then there is a sequence $\{\alpha_n\}$ such that $z_n = \alpha_n x + (1 - \alpha_n)y \in K$ for all n. Since K is closed, $\lim_{n\to\infty} z_n = \lim_n (\alpha_n x + (1 - \alpha_n)y) \in K$. So $\gamma > 0$. Choose an arbitrary $\varepsilon > 0$. Take $\eta \in (0, \gamma)$ such that $\gamma - \eta < \dfrac{\varepsilon}{p(x - y)}$ for some $p \in Q$. Then $\eta x + (1 - \eta)y \in S_\varepsilon(\gamma x + (1 - \gamma)y) \cap (H - K)$. Hence, $\gamma x + (1 - \gamma)y \in \partial_H(K)$, showing $\partial_H(K) \neq \phi$, a contradiction.

THEOREM 3.1. Let H be a closed, convex subset of E, and K be a nonempty weakly compact convex subset of H which possesses normal structure. If $T : K \to H$ is nonexpansive, and if $T : \partial_H(K) \to K$, then T has a fixed point.

Proof. Let G be the family of all closed convex subsets of H such that for $F \in G$, $F \cap K \neq \phi$ and $T : F \cap K \to K$. G is nonempty, because $H \in G$. Let $\{C_\alpha\}$ be a descending chain of subsets of G and $C = \cap C_\alpha$. Note that $C \cap K$ is nonempty, since each of the sets $C_\alpha \cap K$ is a nonempty weakly compact subset of E. Also, since $T : C_\alpha \cap K \to C_\alpha$ for each α, clearly $T : C \cap K \to C$. Since C is closed and convex, $C \in G$ and therefore by Zorn's Lemma, G has a minimal element.

Let D be such a minimal element. Let us first observe that $\partial_D(K) \neq \phi$ for otherwise, it will follow from Lemma 3.1 that $D \subseteq K$ and $T : D \cap K \to D$ would imply $T : D \to D$. The result will then follow from Theorem 4.6 [19].

We shall show that $D \cap K$ is a singleton. Suppose not, that is $D \cap K$ consists of at least two points. Then $r = \delta_p(D \cap K) > 0$. Since K has normal structure, there exists a point $x \in D \cap K$ and $p \in Q$ such that $\sup\{p(x - y) : y \in D \cap K\} = \alpha < r$.

Let $L = \{z \in E : D \cap K \subseteq B[z, \alpha]\}$.

It is easily seen that L is closed and convex and since $x \in D \cap L$, $(D \cap L) \cap K \neq \phi$. Also, there exist points $u, v \in D \cap K$ such that $p(u - v) > \alpha$. Such points cannot be elements of L, therefore, $L \cap D$ is a proper subset of D. We claim that $D \cap L \in G$. As observed $(D \cap L) \cap K \neq \phi$, it is enough to show that $T : (D \cap L) \cap K \to D \cap L$.

Let $w \in (L \cap D) \cap K$. Let $M = B[T(w), \alpha] \cap D$. If $M \in G$, then since $M \subset D$ and D is minimal, $M = D$. This implies $D \cap K \subset D \subset B[T(w), \alpha]$ and hence, $T(w) \in L$. Since $T : D \cap K \to D$, this in turn implies $T(w) \in D \cap L$. Therefore, $T : (D \cap L) \cap K \to D \cap L$, if $M \in G$ for every $w \in (D \cap L) \cap K$.

We complete the proof by showing $M \in G$. Suppose $x \in M \cap K$. Then

$x \in D \cap K$, so that $p(x - w) \leq \alpha$, because $w \in L$. Hence, $p(Tx - Tw) \leq p(x - w) \leq \alpha$ and $T(x) \in B[T(w), \alpha]$. But $x \in M \cap K$ also implies $x \in D \cap K$ and hence, $T(x) \in D$. Therefore, $T(x) \in B[T(w), \alpha] \cap D = M$ that is $T : M \cap K \to M$.

Finally, since $\partial_D(k) \neq \phi$, it follows that $M \cap K \neq \phi$. Indeed, let $y \in \partial_D(K)$, then $y \in D \cap K$ and $p(y - w) \leq \alpha$, which implies $p(Ty - Tw) \leq p(y - w) \leq \alpha$ and therefore, $T(y) \in M$. But also $\partial_D(K) \subset \partial_H(K)$ implies $T(y) \in K$, hence $T(y) \in M \cap K$ and $M \cap K \neq \phi$.

The above theorem reduces to Theorem 4.6 [19] by taking $H = K$. A more interesting consequence of the above theorem is when $H = E$.

COROLLARY 3.1. Let K be a nonempty weakly compact convex subset of E with normal structure. If $T : K \to E$ is nonexpansive, and if T maps the boundary of K into K, then T has a fixed point in K.

Let K be a nonempty subset of a real linear space E. If $x \in K$, we put (after Halpern [7]), $I_K(x) = \{z \in E : z = x + c(y - x)$ for some $y \in K$ and $c \geq 0\}$. We will be interested in two conditions which may be satisfied by $T : K \to E$.

(1) $T(y) \in I_K(y)$ for each $y \in bdg(K)$, the set of bounding points of K.

(2) For some $w \in int(K)$, $T(y) - w \neq m(y - w)$ for all $y \in bdy(K)$ and $m > 1$, here $bdy(K)$ denotes the boundary of K.

REMARK 3.1. If G is an open subset of E and $T : X \to E$ satisfies (1), where $X = c\ell(G)$. Then given any $w \in G$, it also satisfies (2) [16].

THEOREM 3.2. Let E be semireflexive and K be a nonempty, closed, bounded convex subset of E with nonempty interior. Suppose K has normal structure. Let $T : K \to E$ be a nonexpansive mapping satisfying (1), then T has a fixed point.

Proof. By (1), for each $x \in K$, there exists a number $c(x) \geq 0$ and a point $g(x) \in K$ such that $T(x) = x + c(x)(g(x) - x)$.

Since K is convex, by changing c and g, if necessary we may assume that $c(x) \geq 1$. Let D, the set ordered by inclusion be defined as follows: $D = \{C : C$ is nonempty closed convex subset of K such that $g(x) \in C$ for all $x \in C\}$.

By an application of Zorn's Lemma, we obtain a minimal element L

of D. We claim that L consists of a single element. Suppose not, since K has normal structure there exists a point $y \in L$ such that $r = \sup \{p(x - y) : x \in L\} < \delta_p(L)$.

Let $H = \cap\{B[z, r] : z \in L\}$. H is nonempty, since $y \in H$. Clearly, H is closed and convex. Let $M = H \cap L$. Then M is a proper subset of L. We shall show that $M \in D$.

Let $u \in M$ and consider $S = B[T(u), r] \cap L$. Fix a point $v \in L$. Define the function $g : L \to E$ by $g(w) = v + t(T(w) - v)$ for all $W \in L$. g is a t-set contraction for $0 \leq t < 1$ [18]. Using Remark 3.1 and Theorem 2 [15], it follows that g has a fixed point. Since $L \subseteq B[u, r]$, it follows that $\inf\{p(w - Tu) : w \in L\} \leq r$. The weak lower semicontinuity of the seminorm p implies that S is nonempty. $g(s) \in S$ for each $s \in S$, because $T(L) \subset B[T(u), r]$.

Thus, $S \in D$, so that $S = L$. This means $T(u) \in H$, consequently, $g(u) \in M$, and $M \in D$. But this contradicts the minimality of L. Thus, L consists of a single point, which is a fixed point of T.

DEFINITION 3.2. Let K be a nonempty subset of E. A mapping $T : K \to E$ is called a <u>generalized contraction</u> if for each $x \in K$, there is a number $k_p(x) < 1$ such that $p(Ty - Tx) \leq k_p(x)p(x - y)$ for each $y \in K$ and $p \in Q$.

THEOREM 3.3. Let E be semireflexive and K be a nonempty closed, bounded convex subset of E with nonempty interior. If $T : K \to E$ is a generalized contraction satisfying (1), then T has a fixed point in K.

Proof. As in the proof of Theorem 3.2, we get a minimal subset L with a positive diameter d. Fix a point $w \in L$ and let $S = B[T(w), k_p(z)d] \cap L$. It follows from Theorem 2 [15] that S is nonempty, because $T(L) \subseteq B[T(w), k_p(z)d]$. Thus, $S = L$, and $T(w)$ is a nondiametral point of L. Now we may proceed exactly as in the proof of Theorem 3.2 with $y = T(w)$.

4. Fixed Points For Reversible Semigroup of Nonexpansive Mappings

We prove a common fixed point theorem for families of nonexpansive mappings which generalizes a corrected version of Theorem 10 [14].

THEOREM 4.1. Let H be a bounded closed convex subset of E with normal structure. Let K be a weakly compact subset of H. Let S be a left reversible topological semigroup of self nonexpansive mappings of H satisfying the following conditions:

(i) For each $a \in H$, the mapping $s \to s(a)$ from S into H is continuous;

(ii) For each $x \in H$ and $s \in S$, $[\overline{co}\ s\,S(x)] \cap K \neq \phi$. Then S has a common fixed point.

Proof. Let G be a subset of H minimal with respect to being nonempty, closed, convex, mapped into itself by each member of S and having a nonempty intersection with K. Using the fact that closed convex sets are weakly closed, applying Zorn's Lemma and the weak compactness of K, the existence of such G is guaranteed.

If G consists of a single point, then this is the desired fixed point. Suppose G consists of more than one point. Let $p \in Q$ be such that $\delta_p(G) > 0$ and consider S to be directed by the partial ordering defined by $s_1 \geq s_2$ if and only if $s_1 S \subseteq \overline{s_2 S}$, whenever $s_1, s_2 \in S$. The left reversibility of S guarantees that S becomes a directed set under $\geq$. Pick $x_0 \in G$ and define $W(x_0) = \{W_s = \overline{sS(x_0)} : s \in S\}$. Let $r = ar(W(x_0), G)$. Since H possesses normal structure, there exists $x \in G$ such that $r(x) = r_1 > r$. Here $r(\cdot)$ is understood to be relative to the decreasing net $W(x_0)$. If r_2 is such that $r_1 > r_2 > r$, then the set $L = \{x \in G : r(x) \leq r_2\}$ is nonempty proper closed convex subset of G. Let $s_1 \in S$, $x \in L$ and $\varepsilon > 0$. There exists $t \in S$ such that $tS(x) \subseteq W_t \subseteq B[r_2 + \varepsilon, x]$. Since s_1 is nonexpansive, we have $s_1 t\,S(x) \subseteq B[r_2 + \varepsilon, s_1(x)]$, so that $W_{s_1 t} \subseteq B[r_2 + \varepsilon, s_1(x)]$. Since $\{W_s : s \in S\}$ is decreasing, $W_s \subseteq B[r_2 + \varepsilon, s_1(x)]$ for all $s \geq s_1 t$. Hence, $r(s_1(x)) \leq r_2 + \varepsilon$ and $s_1(x) \in L$. Since $\varepsilon > 0$ is arbitrary, it follows that L is invariant under S. Let $x \in L$. Since L is closed, convex and S-invariant, we have $\overline{co}\ s\,S(x) \subseteq L$. Therefore, $L \cap K \neq \phi$, by condition (ii). This contradicts the minimality of G.

COROLLARY 4.1. Let H be a closed, bounded convex subset of a Banach space E with normal structure. Let K be a weakly compact subset of H. Let S be a left reversible topological semigroup of self nonexpansive mappings of H satisfying the following conditions:

(i) For each $a \in H$, the mapping $s \to s(a)$ from S into H is continuous;

(ii) For each $x \in H$ and $s \in S$, $[\overline{co}\ s\ (x)] \cap K \neq \phi$.

Then S has a common fixed point.

Corollary 4.1 as stated in [14, Theorem 10] is not true without added assumption of normal structure on H. In fact, taking $H = K$, it follows from [1] that even a single nonexpansive mapping does not have a fixed point.

REFERENCES

1. Dale Alspach, "A fixed point free nonexpansive map", preprint.
2. L. P. Belluce, W. A. Kirk and E. F. Steiner, "Normal structures in Banach spaces", Pacific J. Math. 26 (1968), 443-440.
3. M. S. Brodskii and D. P. Milman, "On the center of a convex set", Dokl. Akad. Nauk, SSSR 59 (1948), 837-840.
4. F. E. Browder, "Nonexpansive nonlinear operators in a Banach space", Proc. Nat. Acad. Sci. U.S.A. 54 (1965), 1041-1044.
5. M. Edelstein, "A theorem on fixed points under isometries", Amer. Math. Monthly 70 (1963), 298-300.
6. K. Goebel, W. A. Kirk and T. N. Shimi, "A fixed point theorem in uniformly convex spaces", Boll. Un. Mat. Ital. (4), (1973), 67-75.
7. Benjamin Halpern, "Fixed point theorems for set-valued maps in infinite dimensional spaces", Math. Ann. 189 (1970), 87-98.
8. R. Kannan, "Some results on fixed points III", Fund. Math. 70 (1971), 169-177.
9. R. Kannan, "Some results on fixed points IV", Fund. Math. 74 (1972), 181-187.
10. W. A. Kirk, "Nonexpansive mappings and the weak closure of sequence of iterates", Duke Math. J. 36 (1969), 639-645.
11. W. A. Kirk, "Mappings of generalized contractive type", J. Math. Anal. Appl. 27 (1969), 409-412.
12. W. A. Kirk, "Fixed point theorems for nonlinear nonexpansive and generalized contraction mappings", Pacific J. Math. 38 (1971), 89-94.
13. T. C. Lim, "Characterizations of normal structure", Proc. Amer. Math. Soc. 43 (1974), 313-319.
14. T. C. Lim, "On asymptotic centers and fixed points of nonexpansive mappings", Canad. J.Math. 32 (1980), 421-430.
15. Simeon Reich, "A fixed point theorem in locally convex spaces". Bull. Cal. Math. Soc. 63 (1971), 199-200.
16. Simeon Reich, "Fixed points in locally convex spaces", Math. Z. 125 (1972), 17-31.
17. Simeon Reich, "Fixed points of nonexpansive functions", J. London Math. Soc. (2) 7 (1973), 5-10.

18. B. N. Sadovskii, "Limit-compact and condensing operators", Russian Math. Surveys, 27 (1972), 85-156.

19. K. L. Singh, "Fixed points and the sequence of iterates", Ph.D. Dissertation, Texas A & M University (1980).

20. Ho Duc Veit and Nguyen Theip, "A note on close to normal structure", Comment. Math. Univ. Carolinae 20 (1979), 29-36.

21. C. S. Wong, "Close-to-normal structure and its applications", J. Functional Analysis 16 (1974), 353-358.

SOME FIXED POINT THEOREMS FOR GENERALIZED NONEXPANSIVE MAPPINGS

B. E. Rhoades

Indiana University
Bloomington, Indiana
U.S.A.

Let X be a metric space, T a selfmap of X. Many fixed point theorems have been proved for contractive type mappings satisfying conditions of the form

$$\|Tx - Ty\| \leq km(x, y) \tag{1}$$

when $0 \leq k < 1$ and $m(x, y)$ is some subset of the collection: $\|x - y\|$, $\|x - Tx\|$, $\|y - Ty\|$, $\|x - Ty\|$, $\|y - Tx\|$. (See [10], for example, for a comprehensive treatment of such theorems.) In each case, the contractive definition is strong enough to guarantee the existence and uniqueness of the fixed point in any metric space in which there is an x such that some subsequence of $\{T^n x\}$ has a limit point.

If one allows $k = 1$ in [11], then additional hypothesis must be added, either to the space X, or to the contractive definition, or both, in order to guarantee the existence of a fixed point.

Let K be a nonempty subset of a metric space X, T a selfmap of K. T will be called a generalized nonexpansive (G) mapping if, for each $x, y \in K$,

$$\|Tx - Ty\| \leq \max\{\|x - y\|, \|x - Tx\|, \|y - Ty\|, \|x - Ty\|, \|y - Tx\|\} \tag{2}$$

Massa and Roux [7] have written a very fine paper dealing with G mappings in uniformly convex spaces. Their paper surveys the known literature on G mappings as of 1978, and contains a number of examples to demonstrate that the conditions in their theorem cannot be substantially weakened. As noted in their paper, (2) is equivalent to

$$\|Tx - Ty\| \le a(x, y)\|x - y\| + b(x, y)\|x - Tx\| \tag{3}$$

$$+ b(y, x)\|y - Ty\| + c(x, y)\|x - Ty\| + c(y, x)\|y - Tx\|,$$

where a, b, c are nonnegative real valued mappings from $K \times K$ satisfying $(a + b + c)(x, y) + (b + c)(y, x) \le 1$ for all $x, y \in K$.

For completeness, we quote their main result.

THEOREM 1. Let X be a uniformly convex metric space, K a closed, convex subset of X, T a selfmap of K satisfying (3). If, in addition,

$$\sup_{x,y\in K} (b(x, y) + c(y, x)) < 1 \tag{4}$$

and

$$\operatorname{Inf}_{\|x\| \le r} \|x - Tx\| = 0 \quad \text{for some } r < \infty, \tag{5}$$

then T has a fixed point in K.

They show by example that, if (4) is not satisfied, then there exists a compact set K and a selfmap satisfying (3), which is also asymptotically regular, i.e. $\lim \|T^n x - T^{n+1}x\| = 0$, but such that T has no fixed points in K. We shall therefore be mainly interested in maps which satisfy conditions (3) and (4).

In [12], a mapping is called Kannan-type if T satisfies

$$\|Tx - Ty\| \le b(x, y)\|x - Tx\| + b(y, x)\|y - Ty\|$$

for all $x, y \in K$, where $b : K \times K \to [0, 1]$ such that $b(x, y) + b(y, x) \le 1$ and $b(x, y) \to 1$ implies $\max\{\|x - Tx\|, \|y - Ty\|\} = 0$ or ∞. For theorems involving maps of this type, including the special case in which $b \equiv \frac{1}{2}$, the reader may consult [5], [9], [13] or [14].

There are maps which satisfy conditions, in addition to (3) and (4), and which possess a unique fixed point. But such maps are not generalizations of nonexpansive maps. Two such examples occur in [6] and [11].

THEOREM 2. Let X be a complete reflexive space, K a closed, bounded, convex subset of X, T a selfmap of K satisfying (3) and (4). Then T has a fixed point in K.

In light of Theorem 1, it is sufficient to show that (5) is satisfied. Since K is convex, it is starlike. Let k_0 be a star-center for K. For each $t \in (0, 1)$, define $S_t(x) = tTx + (1 - t)k_0$. Then $\|S_t(x) - S_t(y)\| = t\|Tx - Ty\| \leq tm(x, y)$, where $m(x, y) = \max \{\|x - y\|, \|x - Tx\|, \|y - Ty\|, \|x - Ty\|, \|y - Tx\|\}$. From [3], S_t has a unique fixed point in K for each t. Call it u_t. Let $\{t_j\}$ be any sequence in (0, 1) with limit 1. Since X is reflexive, and K is bounded, $\{u_{t_j}\}$ has a weakly convergent subsequence with limit $u_0 \in X$. Since K is closed, $u_0 \in K$. For notational convenience, denote this weakly convergent subsequence by $\{u_j\}$. Then S_{t_j} has u_j as a fixed point. We may write $Tu_j - u_j = t_j Tu_j + (1 - t_j)k_0 - u_j + (1 - t_j)(Tu_j - k_0) = (1 - t_j)(Tu_j - k_0)$. Therefore, $\lim_j \|Tu_j - u_j\| = 0$. Since K is bounded, (5) is satisfied, and T has a fixed point in K.

A closed convex subset K of a real Hilbert space H is said to have the fixed point property for a family $\mathcal{F}$ of mapping if, for each $T \in \mathcal{F}$, $T : K \to K$, T has a fixed point in K. Ray [8] has shown that K has the fixed point property for nonexpansive maps if and only if K is bounded. We shall now extend this result to P_2 and G mappings. A P_2 mapping satisfies the condition

$$\|Tx - Ty\|^2 \leq \|x - y\|^2 + k\|(I - T)x - (I - T)y\|^2$$

for each $x, y \in K$, $0 \leq k < 1$. Clearly, P_2 and G mappings are generalizations of nonexpansive mappings. We shall first show that P_2 and G are independent.

Since $T \in P_2$ implies T is continuous simply pick a discontinuous $T \in G$. Examples abound. For instance, one can use one of those in [7].

For $T \in P_2$ but $T \notin G$, use the T of [10, p. 165], i.e., $Tx = 1 - x/a$, $0 \le x \le a$, $Tx = 0$ for $a < x \le 1$, where $1/3 < a < 1/2$. For $0 \le x < a(1 - a) < a < y < 1 - x/a < 1$, $|Tx - Ty| = |1 - x/a| > \max\{1 - (1 + a)x/a, y\} = m(x, y)$. We shall now show that $T \in P_2$ for $1/3 < k < 1/2$. Clearly, $T \in P_2$ for $a < x < y \le 1$. For $0 \le x < y \le a$, $|Tx - Ty| = |x - y|/a$, and $|(I - T)x - (I - T)y| = (a + 1)|x - y|/a$, so that $T \in P_2$ for $k = (1 - a)/(1 + a)$. (Note that $1/3 < a < 2/3$ implies $1/3 < k < 1/2$.) For $0 \le x \le a < y \le 1$, $T \in P_2$ implies

$$(1 - x/a)^2 \le (x - y)^2 + \frac{(1-a)}{1+a}\left[y + 1 - \frac{(a+1)x}{a}\right]^2$$

Rewriting the above inequality, we have

$$1 - \frac{2x}{a} \le -2xy + y^2 + \frac{(1-a)}{1+a}\left[(y + 1)^2 - 2\frac{(a+1)(y+1)x}{a}\right],$$

$$\frac{2a}{1+a} \le -2\left(1 + \frac{(1-a)}{a}\right)xy + 2x + \frac{2y^2}{1+a} + \frac{2y(1-a)}{1+a},$$

$$2a^2 \le -2(1 + a)xy + 2xa(1 + a) + 2ay^2 + 2ay(1 - a).$$

It is sufficient to show that $f(x, y) \ge 0$ over the domain $D = \{(x, y) |\ 0 \le x \le a,\ a \le y \le 1$, where

$$f(x, y) = ay^2 + a(1 - a)y + a(1 + a)x - (1 + a)xy - a^2.$$

Using calculus, it can be shown that f has no interior extreme values, and assumes its minimum of 0 on the boundary. Therefore $T \in P_2$.

THEOREM 3. Let K be a closed, convex subset of a real Hilbert space. Then K has the fixed point property for P_2 or G mappings if and only if K is bounded.

The necessity follows by using the same example as in [8] since P_2 and G include nonexpansive mappings as special cases. Suppose K is bounded. If $T \in P_2$, then T has a fixed point in K by [2, Theorem 12]. If $T \in G$, then T has a fixed point in K by Theorem 2.

For arbitrary metric spaces, the following is true.

THEOREM 4. Let X be a metric space, T a selfmap of X satisfying (3) and (4). If, in addition,

(i) T is asymptotically regular

and

(ii) there exists an $x \in X$ such that $\{T^n x\}$ has a convergent subsequence with limit $y \in X$, then y is a fixed point of T and $\lim T^n x = y$.

From (ii) there exists a subsequence $\{k_n\}$ of the natural numbers such that $\lim_n T^{k_n} x = y$. Using the triangular inequality,

$$\|Ty - y\| \le \|Ty - T^{k_n+1} x\| + \|T^{k_n+1} x - T^{k_n} x\| + \|T^{k_n} x - y\| . \tag{6}$$

From (3),

$$\|Ty - T^{k_n+1} x\| \le a\|y - T^{k_n} x\| + b\|y - Ty\| + b'\|T^{k_n} x - T^{k_n+1} x\| \tag{7}$$

$$+ c\|y - T^{k_n+1} x\| + c'\|T^{k_n} x - Ty\| ,$$

where a, b, b', c, c' are evaluated at $(y, T^{k_n} x)$, and $b'(x, y) = b(y, x)$, $c'(x, y) = c(y, x)$.

Since $\|y - T^{k_n+1} x\| \le \|y - T^{k_n} x\| + \|T^{k_n} x - T^{k_n+1} x\|$, and $\|T^{k_n} x - Ty\| \le \|T^{k_n} x - T^{k_n+1} x\| + \|T^{k_n+1} x - Ty\|$, we have, substituting in (7).

$$(1 - c')\| Ty - T^{k_n+1} x\| \le (a + c)\| y - T^{k_n} x\| + b\|y - Ty\| \tag{8}$$

$$+ (b' + c + c')\|T^{k_n} x - T^{k_n+1} x\| .$$

Substituting (8) into (6),

$$(1 - \frac{b}{1-c'}) \|Ty - y\| \le (1 + \frac{a+c}{1-c'}) \|y - T^{k_n} x\|$$

$$+ (1 + \frac{b'+c+c'}{1-c'}) \|T^{k_n} x - T^{k_n+1} x\| .$$

Taking the limit as $n \to \infty$ yields $y = Ty$ since, from (4), $\sup(b + c') < 1$ and T is asymptotically regular.

Theorem 4 contains theorems of [1] and [4] as special cases.

REFERENCES

1. L. P. Belluce and W. A. Kirk, "Fixed Point theorems for certain classes of nonexpansive mappings", Proc. Amer. Math. Soc. 20 (1969), 141-146.
2. F. Browder and W. V. Petryshyn, "Construction of fixed points of nonlinear mappings in Hilbert space", J. Math Anal. Appl. 20 (1967), 197-228.
3. L. B. Ciric, "A generalization of Banach's contraction principle", Proc. Amer. Math. Soc. 45 (1974), 267-273.
4. G. Emmanuele, "Fixed Point theorems for generalized nonexpansive mappings in complete metric spaces", preprint.
5. A. A. Gillespie and B. B. Williams, "Some Theorems on Fixed Points in Lipschitz - and Kannan-type Mappings", J. Math. Anal. Appl. 74 (1980), 382-387.
6. M. Gregus, Jr., "A Fixed Point Theorem in Banach Space", Boll. UMI 17A (1980), 193-198.
7. S. Massa and D. Roux, "A Fixed Point Theorem for Generalized Non-expansive Mappings", Boll. UMI 15-A (1978), 624-634.
8. W. O. Ray, "The fixed point property and unbounded sets in Hilbert space", Trans. Amer. Math. Soc. 258 (1980), 531-537.
9. S. Reich, "Remarks on fixed points", Atti. Acad. Naz. Lincei, 52 (1972), 689-697.
10. B. E. Rhoades, "A comparison of various definitions of contractive mappings", Trans. Amer. Math. Soc. 226 (1977), 257-290.
11. B. E. Rhoades, "A Generalization of a Fixed Point Theorem of Bogin", Math. Seminar Notes 6 (1978), 1-7.
12. D. Roux and C. Zanco, "Kannan maps in normed spaces", Atti Acad. Nax. Lincei 65 (1978), 252-258.
13. P. Soardi, "Su un problema di punto di S. Reich", Boll. UMI 4 (1971), 841-845.
14. P. Soardi, "Struttura quasi-normale e teoremi di punto unito", Trieste Univ. Inst. Mat. Rend. 4 (1972), 105-113.

FIXED POINT SETS OF HOMOTOPIES

H. Schirmer

Carleton University
Ottawa, Ontario
Canada

In recent years, several authors have been interested in the realization of a subset A of a topological space X as the fixed point set $\text{Fix } f = \{x \in X \mid f(x) = x\}$ of a continuous selfmap of X under minimal assumptions on A. If X is Hausdorff, then A is necessarily closed, and in order to avoid questions concerning the fixed point property, it has also been assumed that A is non-empty. In many cases these assumptions suffice. X is called a space with the <u>complete invariance property</u> (CIP) if every closed and non-empty subset A of X is the fixed point set of a map $f : X \to X$, and it is known that the class of spaces with the CIP includes all convex subsets of a normed linear space, compact topological manifolds, locally finite simplicial complexes with the weak topology, locally compact metrizable topological groups and 1-dimensional Peano continua. A survey of these results, which are due to Boju Jiang (Po-Chu Chiang), J. R. Martin, S. B. Nadler, Jr., H. Robbins, H. Schirmer, E. D. Tymchatyn and L. E. Ward, Jr., and references can be found in [4].

A similar problem can be considered for homotopies $H : X \times I \to X$, where $I = [0, 1]$ is the unit interval. The fixed point set of a homotopy H is defined by

$$\text{Fix } H = \{(x, t) \in X \times I \mid H(x, t) = x\},$$

and hence we are interested in minimal assumptions on a set $K \subset X \times I$ under which K can be realized as the fixed point set of a homotopy of

X. Again K is closed if X is Hausdorff, and we have also assumed that $K \cap (X \times t) \neq \emptyset$ for every $t \in I$. But in general, this is not enough, as the continuity of H implies conditions on the change of $K \cap (X \times t)$ with varying t. To clarify the situation, define that a set $A \subset X \times 0$ is joined tc $B \subset X \times 1$ by a continuum (i.e. a compact and connected space) $C \subset X \times I$ if $C \cap (X \times 0) \subset A$ and $C \cap (X \times 1) \subset B$. Then the following proposition is an easy consequence of a rather complex theorem by F. Browder (see [1], Theorem 1).

PROPOSITION. Let X be a compact ANR and $H : X \times I \to X$ a homotopy from $f = H \mid X \times 0$ to $g = H \mid X \times 1$. If the fixed point index of the map $f : X \to X$ is non-zero, then there exists a continuum C in Fix H joining Fix f and Fix g.

The proposition shows that in order to realize the closed set $K \subset X \times I$ as the fixed point set of a homotopy it is generally necessary to assume that K contains a continuum joining $X \times 0$ and $X \times 1$. We can prove a theorem which shows that these assumptions on K are sufficient for certain polyhedra. Recall that a local cut point (or separating point) is a point x which has a connected neighbourhood N so that $N - x$ is not connected.

THEOREM. Let P be a compact and connected polyhedron without local cut points, and $K \subset P \times I$ a closed set which contains a continuum C joining $P \times 0$ and $P \times 1$. Then there exists a homotopy $H : P \times I \to P$ with Fix $H = K$.

The assumptions on P in the Theorem are e.g. satisfied by those polyhedra for which it is presently known that the Nielsen number $N(f)$ of a selfmap f can be realized by a map homotopic to f (see [2], Main Theorem) and by all compact triangulable manifolds of dimension ≥ 2. Nevertheless, the class of spaces for which we have proved the Theorem is more restrictive than the class of spaces which are known to have the CIP, even if attention is confined to polyhedra. The reason is that we have considered only the case where H can be obtained via path fields, and hence $H \mid P \times 0$ is homotopic to the identify map of P. But there is little doubt that the Theorem holds for a much wider class of spaces, and a combination of the techniques from our proof with those of [3] will possibly suffice to establish it for all locally finite complexes with the weak topology.

A proof of the Theorem will appear in Pacific J. of Mathematics.

REFERENCES

1. F. E. Browder, "On continuity of fixed points under deformations of continuous mappings", Summa Brasil. Math. 4 (1960), 183-191.

2. Boju Jiang (Po-chu Chiang), "On the least number of fixed points", Amer. J. Math. 102 (1980), 749-763.

3. Boju Jiang and H. Schirmer, "Fixed point sets of continuous self-maps of polyhedra", to appear in the Proceedings of the Conference on Fixed Point Theory, Sherbrooke 1980, Springer Lecture Notes.

4. Boju Jiang, "Fixed point sets of continuous selfmaps", to appear in the Proceedings of the Conference on Fixed Point Theory, Sherbrooke 1980, Springer Lecture Notes.

A THEOREM ON THE MINIMIZATION OF A CONDENSING MULTIFUNCTION AND FIXED POINTS*

V. M. Sehgal

University of Wyoming
Laramie, Wyoming
U.S.A.

S. P. Singh *

Memorial University of Newfoundland
St. John's, Newfoundland
Canada

Abstract

Let G be subset of a locally convex separated topological vector space E with $\text{int}(G) \neq \phi$, $\text{cl}(G)$ convex and quasi-complete. Let $F : \text{cl}(G) \to 2^E$ be a continuous, condensing multifunction with compact and convex values and a bounded range. It is shown here that either f has a fixed point or for each $w \in \text{int}(G)$, there exists a $u = u(w) \in \delta(\text{cl}(G))$ such that

$$p(fu - u) = \inf\{p(x - y) \mid x \in f(u),\ y \in \text{cl}(G)\},$$

where p is the Minkowski's functional of the set $(\text{cl}(G) - w)$. Several fixed point theorems are obtained as consequences of this result.

Let G be a nonempty convex and compact subset of a locally convex separated topological vector space E. A well-known result of Ky Fan [1] states that for any continuous mapping $F : G \to E$, there exists

*This work was supported in part by NSERC and Scientific Affairs Division, NATO.

a $u \in G$ and a continuous seminorm p on E such that

$$p(fu - u) = \min\{p(fu - x) \mid x \in G\}. \tag{1}$$

In this paper, we investigate (1) for condensing multifunction with convex but not necessarily compact domain and give sufficient conditions for (1) to hold for such functions. The main result is used to derive some fixed point theorems for multifunctions. The terminology of this paper follows after Su and Sehgal [3].

Let X and Y be topological spaces. A mapping $f : X \to 2^Y$ is (a) upper semicontinuous (u.s.c.) if for each closed set B of Y, the set $f^{-1}(B) = \{x \mid f(x) \cap B \neq \phi\}$ is a closed subset of X, (b) lower semicontinuous (l.s.c.) if $f^{-1}(U)$ is an open subset of X for each U open in Y, and (c) continuous if f is both u.s.c. and l.s.c.

Throughout, let E denote a locally convex separated topological vector space and Δ a base of absolutely convex neighborhoods of the origin. A subset X of E is totally bounded if for each $U \in \Delta$, there exists a finite subset $F \subseteq X$ such that $X \subseteq F + U$. For $A \subseteq E$, let $M(A) = \{U \in \Delta \mid A \subseteq X + U$ for some totally bounded subset X of $E\}$. Note that for $A \subseteq E$, $M(Co(A)) = M(A)$, where $Co(A)$ denotes the convex hull of A.

Let $G \subseteq E$. A multifunction $f : G \to 2^E$ is called condensing (see Himmelberg, Porter, van Vleck [2]), if, for each bounded but not totally bounded subset B of G,

$$M(B) \subseteq M(f(B)) \text{ but } M(B) \neq M(f(B)),$$

where $f(B) = \{f(x) \mid x \in B\}$.

For a subset S of E, let $\delta(S)$ denote the boundary of S, $cl(S)$ the closure of S and $int(S)$ the interior of S. Further, let $cc(E)$ denote the non-empty convex and compact subsets of E. A closed subset S of E is quasi-complete if its closed bounded subsets are complete.

The following result is proved in [3] and is used in the proof of main result.

THEOREM 1. Let X be a convex, quasi-complete subset of E and $f : X \to cc(E)$ be an u.s.c. condensing multi-function. If $f(X)$ is bounded and $f(x) \cap X \neq \phi$ for each $x \in X$, then f has a fixed point.

For $G \subseteq E$ with $cl(G)$ convex and $int(G) \neq \phi$, let $P = \{p(w) \mid p(w)$ is the Minkowski's functional of $cl((G) - w)$; $w \in int(G)\}$. Also, for subsets A and B of E and $p \in P$, let

$$p(A - B) = \inf \{p(x - y) \ / \ x \in A, \ y \in B\}.$$

The main result is:

THEOREM 2. Let G be a subset of E with $int(G) \neq \phi$, $cl(G)$ convex and quasi complete. If $f : cl(G) \rightarrow cc(E)$ is a continuous, condensing multifunction with a bounded range, then for each $w \in int(G)$, there exists a $u = u(w) \in cl(G)$ such that

$$p(fu - u) = p(f(u) - cl(G)), \quad \text{where} \quad p = p(w).$$

Further, if $p(fu - w) \leq 1$, then $u \in f(u)$ and if $p(f(u) - w) > 1$, then $u \in \delta(cl(G))$.

<u>Proof</u>. Let $w \in int\ (G)$ and $p = p(w)$. Define a mapping

$$F : cl(G) \rightarrow [0, 1]$$

by

$$F(x) = (\max\{1, p(f(x) - w)\})^{-1}.$$

Note that for any $x \in cl(G)$, $0 < F(x) \leq 1$. Let

$$g : cl(G) \rightarrow cc(E)$$

be defined by

$$g(x) = \{z \in f(x) \mid p(z - w) = p(f(x) - w)\}.$$

By Theorem 1, there exists a $u \in cl(G)$ with $u \in F(u)g(u) + (1 - F(u))w$. Clearly $u \in f(u)$ if $p(fu - w) \leq 1$. If $p(fu - w) > 1$, choose a $z \in g(u)$ with $u = F(u)z + (1 - F(u))w$. It can be shown that $p(fu - u) \leq p(z - u) = p(fu - w) - 1$. Consequently, $p(fu - u) = p(fu - cl(G))$.

Furthermore, in this case, $p(u - w) = F(u)p(z - w) = F(u)p(f(u) - w) = 1$. This implies that $u \in \delta(cl(G))$.

For a subset A of E and an $x \in A$, let

$$I_A(x) = \{x + c(z - x) \mid c > 0 \text{ and } z \in A\}.$$

As an application of Theorem 2, we have the following fixed point results.

COROLLARY 1. Let G be a subset of E with $int(G) \neq \phi$, $cl(G)$ convex and quasi-complete. Let $f : cl(G) \to cc(E)$ be a continuous, condensing multifunction with a bounded range. If f satisfies any one of the following conditions:

(a) for each $u \in \delta(cl(G))$, $f(u) \cap cl(G) \neq \phi$

(b) there exists a $p = p(w) \in P$ such that if $p(f(u) - u) > 0$ for $u \in \delta(cl(G))$, then $p(f(u) - y) < p(f(u) - u)$ for some $y \in I_{cl(G)}(u)$.

(c) for each $u \in \delta(cl(G))$, $f(u) \cap cl(I_{cl(G)}(u)) \neq \phi$

(d) there exists a $p = p(w) \in P$ such that

$$\liminf_{t \to 0} \frac{p((1 - t)u + tfu) - cl(G))}{t} = 0 \quad \text{for each } u \in \delta(cl(G)).$$

Then f has a fixed point. Details will appear elsewhere.

REFERENCES

1. Ky Fan, "Extensions of Two Fixed Point Theorems of F. E. Browder", Math. Z., 112 (1969), 234-240.

2. C. J. Himmelberg, J. R. Porter and F. S. van Vleck, "Fixed Point Theorems for Condensing Multifunctions", Proc. Amer. Math. Soc. 23 (1969), 635-641.

3. C. H. Su and V. M. Sehgal, "Some Fixed Point Theorems for Condensing Multifunctions in Locally Convex Spaces", Proc. Amer. Math. Soc. 50 (1975), 150-154.

REMARKS ON THE EXAMPLE OF ALSPACH

R. Sine

University of Rhode Island
Kingston, Rhode Island
U.S.A.

Dale Alspach has recently given an example of a fixed point free nonexpansive map of a weakly compact convex set. His example and a slight variant of it are used to resolve some other questions related to nonexpansive maps.

If (M, d) is a metric space, we define a nonexpansive map $T : M \to M$ by the condition $d(Tx, Ty) \leq d(x, y)$. A major problem for some time has been whether a nonexpansive map of K into itself must have a fixed point if K is a nonempty weakly compact convex set in a B-space [15, p. 36]. Dale Alspach [1] has recently settled the problem negatively by giving a fixed point free isometry in L_1 [0, 1] (thus showing that L_1 is not just another pretty space).

Some time ago, Ralph De Marr [7] gave an example in which the fixed point set was not convex. (Also, see [15, p. 57]). Thus the norm closed fixed point set might fail to be weakly closed. Recently Gideon Schechtman [12] using methods similar to Alspach's, produced an example which does show (as W. A. Kirk pointed out to the present author) that Fix(T) can fail to be weakly closed.

In the linear setting the fixed point set is a weakly closed subspace and (if the B-space is reflexive) there is a nonexpansive linear projection mapping onto the fixed point set. It is then reasonable, in the nonlinear setting, to hope for a nonexpansive retraction mapping onto the fixed point set. Ronald Bruck [3] has given a wide array of conditions under which this is the case. Other conditions have been given in [11] and [13]. Bruck [4, p. 68] has also given an example of a linear positive nonexpansive map T of C[0, 1] for which the fixed point set is not the range of a nonexpansive retraction mapping. However, in this example Fix(T) is the range of a linear projection of norm two. Implicit in Alspach's example is the fact that Fix(T) need not even be connected. To make this explicit we give a brief description of the example.

Let Ω be the unit square. Consider the mapping $\phi : \Omega \to \Omega$ given by

$$(x, y) = \begin{cases} (1/2x,\ 2y) \bmod 1 & 0 \le y < 1/2 \\ (1/2(x+1),\ 2y) \bmod 1 & 1/2 \le y < 1. \end{cases}$$

This is the "Baker's transformation" of ergodic theory [9, p.9]. For a popular graphical representation of this mapping see [2, p.9] or [10, p.5]. This is clearly a (2 dimensional Lebesgue) measure preserving transformation. It is less clear that this transformation is ergodic but this has been known for many decades. (See [9, p. 29].) We now use (Ω, ϕ) to induce a transformation of the order interval $J = \{0 \le f \le 1\}$ in $L_1[0, 1]$. Each f in J defines a set A in Ω (the subgraph of f). (Map A by ϕ to obtain a second set $B = \phi(A)$ which is also a subgraph.) We set Tf equal to that function g of which B is the subgraph. It is then immediate that the induced map T on J is isometric. Since ϕ is ergodic on Ω (the only invariant sets have measure zero or one) we see that the only invariant functions for T in J have integral zero or one. Thus $\mathrm{Fix}(T) = \{0, 1\}$ and the fixed point set is not connected.

REMARK 1. Whenever the fixed point set consists of just two points it is clear that the set of points halfway between will be convex invariant and fixed point free. This set is the example given by Alspach (it is the set of Tchebyshev centers of the order interval and is weakly compact as it is weakly closed in the weakly compact order interval).

REMARK 2. A convex set K is said to have the <u>hereditary fixed point</u>

property for nonexpansive maps if each nonempty closed bounded convex subset of K has the fixed point property for nonexpansive maps. It has been open whether a convex set would have the hereditary fixed point property if it had the fixed point property. We can now answer this in the negative by embedding Alspach's example in L_∞ and then extending to the enveloping order interval in L_∞. This order interval in L_∞ has the fixed point property for all nonexpansive maps. See [14] and [16] for this L_∞ fixed point theorem and [11] for the embedding and extension technique.

Ronald Bruck has studied the ω-limit set for a nonexpansive map in [5] and [6]. He observed that the set had the virtue of being nonempty but that it was not at all clear that the weak ω-limit set would be invariant. He has given conditions under which it would be invariant. Alspach's example shows that it can fail to be invariant in certain cases. For this let f be the constant 1/2 function. We claim that the weak ω-limit set for $\{T^n f\}$ is just the singleton set $\{f\}$. Since J is weakly compact it is sufficient to consider sequences. Suppose $T^n f \to g$ [weakly] for n running through N in Z_+. Consider

$$\int hT^n f\, dx \to \int hg\, dx$$

where h is taken to be a continuous function. We see (by the nature of the functions $T^n f$) that

$$\int hT^n f\, dx \to 1/2 \int h dx.$$

Thus $g = f$ as the continuous functions are w^* dense in $L_\infty[0, 1]$. Since the only possible weak cluster point of the orbit of f is $\{f\}$, we conclude that the singleton set $\{f\}$ is the weak ω-limit set.

In 1964 Michael Edelstein [8] proved that a nonexpansive map T of a nonempty weakly compact convex set K in a strictly convex B-space has a fixed point whenever T has a nonempty (strong) ω-limit set. Thus it might be conjectured that the existence of a nonempty ω-limit set would imply a fixed point even without strict convexity. A less bold conjecture would be that a precompact orbit suffices. If all orbits are precompact and K is weakly compact, this is the case and is easily established by standard arguments. If all orbits are precompact and K is an order interval in a sup norm space, then Fix(T) is nonempty and for some sup norm sequence spaces a single precompact orbit is sufficient [14].

We next give a slight modification of Alspach's example to show there is a fixed point free isometry of an order interval in $L_1[0, 1]$ which has a point of period two. The description is easy but we must do a bit more to establish the properties.

Let $\Theta : \Omega \to \Omega$ be given by $\Theta(x, y) = (x, 1 - y)$ so that Θ, and hence $\theta\,\phi$, is a measure preserving map of the unit square Ω. Then $\theta\,\phi$ induces an isometry S of the order interval J in $L_1[0, 1]$. It is clear that $S(0) = 1$ and $S(1) = 0$ so that 0 is a point of J of period two. To show that S is fixed point free we will argue that $\theta\,\phi$ is ergodic. Now there is a representation of (Ω, ϕ) as the shift operator $\hat{\phi}$ on the space of zero-one bilateral sequences. Any set A in this sequence space can be approximated by a set B which is finite dimensional (only a finite number of coordinates are constrained). For such a set B we have $\mu(\hat{\phi}^n B \cap B) = \mu(\hat{\phi}^n B)\mu(B)$ for large n (because when the sets of constrained coordinates are disjoint this is a product set). It follows that $(\mu(A))^2 = \mu(A)$ for any invariant set A. Thus μA is zero or one. Now the effect of composing with Θ on what we already have is to conjugate half of the sequential representation. That is, Θ induces a map $\hat{\Theta}$ on sequences so that

$$\hat{\Theta}(\ldots a_{-1}, a_0, a_1, a_2 \ldots) = (\ldots a_{-1}, a_0, a_1 + 1, a_2 + 1, \ldots)$$

where the arithmetic is carried out mod 1 of course. Thus a sufficiently high iterate of $\theta\,\phi$ will still shift the constrained coordinates to a disjoint set of coordinates. So the same argument for ergodicity applies. Since $\theta\,\phi$ is ergodic, it follows that ST is fixed point free.

REMARK. The question has been raised of whether the fixed point property for nonexpansive maps holds for order intervals in $L_1[0, 1]$ as it does for $L_\infty[0, 1]$. Our modification of Alspach's example shows that it does not.

REFERENCES

1. Dale Alspach, "A fixed point free nonexpansive map", preprint.
2. V. I. Arnold and A. Avez, "Ergodic Proglems of Classical Mechanics", W. A. Benjamin Inc., 1968.

3. Ronald Bruck, Jr., "Properties of fixed-point set of nonexpansive mappings in Banach spaces", Trans. Amer. Math. Soc. 179 (1973), 251-262.

4. Ronald Bruck, Jr., "A common fixed point theorem for a commuting family of nonexpansive mappings", Pac. J. Math. 53 (1974), 59-71.

5. Ronald Bruck, Jr., "On the almost-convergence of iterates of a nonexpansive mapping in Hilbert space and the structure of the weak ω-limit set", Israel J. Math. 29 (1978), 1-16.

6. Ronald Bruck, Jr., "On the convex approximation property and the asymptotic behavior of nonlinear contractions in Banach spaces", preprint.

7. Ralph De Marr, "Common fixed-points for commuting contraction mappings", Pacific J. Math., 13 (1963), 1139-41.

8. Michael Edelstein, "On non-expansive mappings of Banach spaces", Proc. Camb. Phil. So., 60 (1964), 439-447.

9. Paul Halmos, "Lectures on Ergodic Theory", Chelsea, 1956.

10. Donald Ornstein, "Ergodic Theory, Randomness and Dynamical Systems", Yale Univ. Press, 1974.

11. William Ray and Robert Sine, "Nonexpansive mappings with precompact orbits", preprint.

12. Gideon Schechtman, "Some remarks on commuting families of nonexpansive operators", preprint.

13. Robert Sine, "Structure and asymptotic behavior of abstract dynamical systems", Nonlinear Anal. 2 (1978), 119-127.

14. Robert Sine, "On nonlinear contraction semigroups in sup norm spaces", Nonlinear Anal. 3 (1979), 885-890.

15. D. R. Smart, "Fixed Point Theorems", Cambridge Univ. Press, 1974.

16. P. M. Soardi, "Existence of fixed points of nonexpansive mappings in certain Banach lattices", Proc. Amer. Math. Soc. 73 (1979), 25-29.

SEQUENCE OF ITERATES FOR NONEXPANSIVE TYPE MAPPINGS IN BANACH SPACES

K. L. Singh* and J. H. M. Whitfield†

Lakehead University,
Thunder Bay, Ontario
Canada

Abstract

For mappings $T : C \to C$, C a nonempty convex subset of a Banach space, we define a condition (α) which simultaneously generalizes several known nonexpansive type mappings. In this paper we investigate strong and weak convergence of sequences of iterates of convex combinations of iterates of such mappings.

1. Preliminaries

Throughout this paper we will study mappings which satisfy condition (α) (see (8) below) which provides a simultaneous generalization of the types of mappings defined in (1) - (7). Let $T : X \to X$ where X is a metric space with metric d (generally X is taken to be a Banach space and the metric is that induced by the norm), then T is said to be:

1. Contraction: There exists a number k, $0 \leq k < 1$ such that, for each $x, y \in X$, $d(Tx, Ty) \leq kd(x, y)$.
2. Contractive: For each $x, y \in X$, $x \neq y$, $d(Tx, Ty) < d(x, y)$.
3. Nonexpansive: For each $x, y \in X$, $d(Tx, Ty) \leq d(x, y)$.
4. Condition (K): There exists a number k, $0 < k < \frac{1}{2}$ such that, for each $x, y \in X$, $d(Tx, Ty) \leq k[d(x, Tx) + d(y, Ty)]$.
5. Condition (Y): There exist non-negative numbers a, b, c satisfying $a + b + c < 1$ such that, for each $x, y \in X$, $d(Tx, Ty) \leq ad(x, y) + bd(x, Tx) + cd(y, Ty)$.

*Current affiliation: University of Minnesota - Duluth, Duluth, Minnesota, U. S. A.

†This author's research is supported in part by an NSERC (Canada) Grant.

6. <u>Condition (H-R)</u>: if there exist non-negative numbers q, r, s satisfying $q + 2r + 2s < 1$ such that, for all $x, y \in X$,

$$d(Tx, Ty) \leq qd(x, y) + r[d(x, Tx) + d(y, Ty)] + s[d(x, Ty) + d(y, Tx)].$$

7. <u>Condition (β)</u>: if for all $x, y \in X$, $d(Tx, Ty) \leq \max\{d(x, y), \frac{1}{2}[d(x, Tx) + d(y, Ty)], \frac{1}{2}[d(x, Ty) + d(y, Tx)]\}$.

8. <u>Condition (α)</u>: if there exists a constant k, $0 \leq k < 1$ such that, for all $x, y \in X$, $d(Tx, Ty) \leq \max\{d(x, y), \frac{1}{2}[d(x, Tx) + d(y, Ty)], kd(x, Ty), kd(y, Tx)\}$.

The following implications exist among the above mappings. A detailed discussion of various types of contractive mappings may be found in [29].

(1) $\Rightarrow$ (2) $\Rightarrow$ (3) $\Rightarrow$ (8), 8 $\not\Rightarrow$ (n) for $n = 1, 2, 3$.

(3) $\not\Rightarrow$ (2) $\not\Rightarrow$ (1).

(4) and (n) are independent for $n = 1, 2, 3$.

(4) $\Rightarrow$ (5) $\Rightarrow$ (6) $\Rightarrow$ (7) $\Rightarrow$ (8).

(8) $\not\Rightarrow$ (7) $\not\Rightarrow$ (6) $\not\Rightarrow$ (5) $\not\Rightarrow$ (4).

EXAMPLE. Let $M_1 = \{\frac{m}{n} : m = 0, 1, 3, 9; \ldots, n = 1, 4, \ldots, 3k + 1\}$. $M_2 = \{\frac{m}{n} : m = 1, 3, 9, 27; n = 2, 5, \ldots, 3k + 2\}$.

Let $M = M_1 \cup M_2$ with the usual metric. Define $T : M \to M$ by

$$T(x) = \begin{cases} \frac{3x}{4} & \text{for } x \in M_1 \\ \frac{x}{2} & \text{for } x \in M_2. \end{cases}$$

T satisfies condition (α). Indeed, if both x and y are in M_1 or in M_2, then $|Tx - Ty| \leq \frac{3}{4}|x - y|$. Now suppose $x \in M_1$ and $y \in M_2$. Then for $x > \frac{2}{3}y$, we have $|Tx - Ty| = \frac{3}{4}(x - \frac{2}{3}y) \leq \frac{3}{4}(x - \frac{y}{2}) = \frac{3}{4}|x - Ty|$, for $x < \frac{2}{3}y$, $|Tx - Ty| = \frac{3}{4}(\frac{2}{3}y - x) \leq \frac{3}{4}|y - x|$. Interchanging the roles of x and y, we have $|Tx - Ty| \leq \frac{3}{4}|y - Tx|$. Letting $k = \frac{4}{5}$, we see that (α) is satisfied. We show that (β) is not satisfied. Take $x = 1$, $y = \frac{3}{5}$. Then $d(Tx, Ty) = \frac{9}{20} \not\leq \max\{\frac{2}{5}, \frac{11}{40}, \frac{17}{40}\} = \max\{d(x, y), \frac{1}{2}[d(x, Tx) + d(y, Ty)], \frac{1}{2}[d(x, Ty) + d(y, Tx)]\}$.

In this section we prove two theorems for mappings satisfying condition (α). These extend the results of Browder and Petryshyn, Kannan, Kirk, Massabo, Ray and Chatterjee, Ray and Rhoades, Schaefer, Singh and others.

THEOREM 1.1. Let X be a Banach space and let C be a nonempty convex subset of X. Let $T : C \to C$ be a mapping satisfying condition (α). Define the mapping $S : C \to C$ by

$$S = \alpha_0 I + \alpha_1 T + \ldots + \alpha_m T^m,$$

where $\alpha_i \geq 0$, $\alpha_1 > 0$ and $\sum_{i=1}^{m} \alpha_i = 1$. Then $Sx = x$ if and only if $Tx = x$.

<u>Proof</u>. Clearly, $Tx = x$ implies $Sx = x$. To show the converse, assume $Sx = x$. Let $x_0 = x$, $x_1 = Tx_0$, $x_2 = T^2 x_0, \ldots, x_m = T^m x_0 = Tx_{m-1}$. Let $\delta = \max \| x_p - x_q \|$, $p, q = 0, 1, \ldots, m$ and $p_0 = \min\{p : p > q$ for some q and $\delta = \| x_p - x_q \| \}$. Then $\delta = \| x_{p_0} - x_q \|$ implies $q = 0$. For if $q > 0$, then $p_0 > 1$ and

$$\delta = \| x_{p_0} - x_q \| = \| Tx_{p_0 - 1} - Tx_{q-1} \|$$

$$\leq \max\{\| x_{p_0-1} - x_{q-1} \|, \tfrac{1}{2}[\| x_{q-1} - Tx_{q-1} \| + \| x_{p_0-1} - Tx_{p_0-1} \|],$$
$$k\| x_{p_0-1} - Tx_{q-1} \|, k\| x_{q-1} - Tx_{p_0-1} \|$$

$$= \max\{\| x_{p_0-1} - x_{q-1} \|, \tfrac{1}{2}[\| x_{q-1} - x_q \| + \| x_{p_0-1} - x_{p_0} \|],$$
$$k\| x_{p_0-1} - x_q \|, k\| x_{q-1} - x_{p_0} \| \}$$

$$< \delta,$$

which is a contradiction.

Since $\alpha_1 > 0$, $\alpha_0 < 1$ and $Sx = x$, we get $x = \sum_{i=1}^{m} \beta_i Tx^i$, where

$\beta_i = \dfrac{\alpha_i}{1 - \alpha_0}$. Now if $\beta_1 = 1$, then $\alpha_i = 0$, $i \geq 2$ and $S = \alpha_0 I + \alpha_1 T$, $\alpha_0 + \alpha_1 = 1$, in which case $Tx = x$ is easily shown.

Suppose, then that $\beta_1 < 1$. Now $\sum_{i=1}^{m} \beta_i = 1$ and $\beta_i \geq 0$. Set $\gamma = \dfrac{\beta_i}{1 - \beta_1}$ and $y = \sum_{i=2}^{m} \gamma_i T^i x$. Thus, $x = \beta_1 Tx + (1 - \beta_1)y$. By the convexity of C, $y \in C$, since $\sum_{i=2}^{m} \gamma_i = 1$ and $\gamma_i \geq 0$. Consequently,

$$||T^{p_0}x - y|| = ||\sum_{i=2}^{m} \gamma_i (T^{p_0}x - T^i x)||$$

$$\leq \sum_{i=2}^{m} \gamma_i ||T^{p_0}x - T^i x||$$

$$= \sum_{i=2}^{m} \gamma_i ||x_p - x_i|| \leq \delta.$$

Thus, $\delta = ||x_{p_0} - x|| = ||T^{p_0}x - \beta_1 Tx + (1 - \beta_1)y||$

$$\leq \beta_1 ||T^{p_0}x - Tx|| + (1 - \beta_1)\, ||T^{p_0}x - y||$$

$$< \beta_1 \delta + (1 - \beta_1)\delta = \delta.$$

This is a contradiction, and $x = Tx$.

REMARK 1.1. Theorem 1 [18], Theorem 4 [17], Lemma 4.1 [20] and Lemma 1 [27] are special cases of Theorem 1.1.

DEFINITION 1.1. A normed linear space X is said to be <u>uniformly convex</u> if given $\varepsilon > 0$ there exists $\delta(\varepsilon) > 0$ such that $||x - y|| \geq \varepsilon$ for $||x|| \leq 1$, and $||y|| \leq 1$ implies $||\frac{x + y}{2}|| \leq 1 - \delta(\varepsilon)$.

DEFINITION 1.2. A mapping T from a Banach space X into itself is said to be <u>asymptotically regular</u> if $T^{n+1}x - T^n x$ converges to 0 as $n \to \infty$ for all $x \in X$.

Uniform convexity of the space is not enough to ensure asymptotic regularily even for nonexpansive mappings. For example, let $X = R$, define $T : X \to X$ by $Tx = -x$, then clearly T is nonexpansive, but T is not asymptotically regular. However, in a uniformly convex Banach space, a convex combination of mappings satisfying condition (α) turns out to be asymptotically regular as is seen from the following.

THEOREM 1.2. Let X be a uniformly convex Banach space and C be a nonempty convex subset of X. Let $T : C \to C$ be a mapping satisfying condition (α) and having nonempty fixed point set. Let $S : C \to C$ be a mapping as defined in Theorem 1.1. Then S is asymptotically regular.

<u>Proof.</u> It follows from Theorem 1.1 that $F(S) = F(T)$. Let $y \in F(T)$. Set $x_n = S^n x_0$, $x_0 \in C$. Now $\|x_{n+1} - y\| = \|Sx_n - y\| \leq \sum_{i=0}^{m} \alpha_i \|T^i x_n - y\|$. Using condition (α) and $y \in F(T)$ we obtain $\|T^i x_n - y\| \leq \|T^{i-1} x_n - y\| \leq \dots \leq \|x_n - y\|$. So $\|x_{n+1} - y\| \leq \sum_{i=0}^{m} \alpha_i \|T^i x_n - y\| \leq \|x_n - y\|$. Therefore, $\{\|x_n - y\|\}$ is a nonincreasing sequence; and we let $\lim_{n\to\infty} \|x_n - y\| = d$. If $d = 0$, the theorem is proved. Suppose $d > 0$. Now $x_{n+1} - y = Sx_n - y = \alpha_0(x_n - y) + (1 - \alpha_0)z_n$, where $z_n = \frac{1}{1 - \alpha_0} \sum_{1}^{m} \alpha_i (T^i x_n - y)$. Since $\|T^i x_n - y\| \leq \|x_n - y\|$ and $\sum \alpha_i = 1$, we have $\lim_{n\to\infty} \sup \|z_n\| \leq d$. Also $\lim \|x_n - y\| = d = \lim \|x_{n+1} - y\|$. Because X is uniformly convex, $\lim_{n\to\infty}(x_n - y - z_n) = 0$. But $x_{n+1} - x_n = (1 - \alpha_0)(x_n - y - z_n)$, and so $\lim_{n\to\infty}(x_{n+1} - x_n) = 0$, completing the proof.

REMARK 1.2. Special cases of Theorem 1.2 are Theorem 5 [17], Theorem 2 [18], Theorem 1 [27] and Theorem 3.1 [26] for the mapping S.

REMARK 1.3. For $\alpha_0 = \lambda$, $\alpha_1 = 1 - \lambda$, $\alpha_2 = \alpha_3 = \dots = \alpha_m = 0$, special cases of Theorem 1.2 are Theorem 5 [3], Theorem 1.1 [32], Theorem 1 [35], Schaefer [30] and others.

REMARK 1.4. The Remark following Theorem 3.1 [26] is not correct, as there are quasi-nonexpansive mappings which do not satisfy (3.1.1) [26]. However, any mapping satisfying (3.1.1) [26] with at least one fixed point is quasi-nonexpansive.

2. Strong Convergence Of The Iterates $S^n x$

In general, even in Hilbert space, for nonexpansive mappings (in fact, for contractive mappings) we cannot have that $T^n x \to x_0$ for every $x \in X$ and $Tx_0 = x_0$. Indeed, let $X = \ell_2 = \{(x_1, x_2, \ldots, x_n \ldots), x_i$ real for each $i = 1, 2, \ldots$ and $\sum_{n=1}^{\infty} |x_n|^2 < \infty\}$. For $x \in X$, $\|x\| = (\sum_{n=1}^{\infty} |x_n|^2)^{1/2}$. If $x = (x_1, x_2, \ldots, x_n \ldots) \in X$. Define $T : X \to X$ by $T(x) = (0, \alpha_1 x_1, \alpha_2 x_2, \ldots, \alpha_n x_n, \ldots)$, where $\alpha_i = 1 - \frac{1}{i^2}$, $i = 2, 3, \ldots$. Clearly, T is contractive, with fixed point $(0, 0, \ldots 0, \ldots)$. Now, let $x = (1, 0, \ldots, 0, \ldots) \in X$, then $T^n x = (0, 0, \ldots, \prod_{i=1}^{n+1} \alpha_i, 0 \ldots)$ for $n = 1, 2, \ldots$. Thus, $\|T^n x\| = \frac{n+2}{2(n+1)} \to \frac{1}{2}$ as $n \to \infty$ and hence, $T^n x$ does not converge to 0.

Also, an example due to Genel and Lindenstrauss [10] shows that even in Hilbert space $T^n_{1/2}(x)$ does not converge to a fixed point of a non-expansive mapping T where $T_\lambda = \lambda I + (1 - \lambda)T$, $0 < \lambda < 1$. However, with some additional assumptions, we can get the strong convergence of sequences of iterates $S^n x$.

DEFINITION 2.1. Let X be a real Banach space and D be a bounded subset of X. The <u>measure</u> of <u>noncompactness</u> of D denoted by $\gamma(D)$, is defined as follows: $\gamma(D) = \inf\{\varepsilon > 0 : D$ can be covered by a finite number of subsets of diameter $< \varepsilon\}$. We will use the following properties of γ.

1. $\gamma(D) = \gamma(\overline{D})$, where $\overline{D}$ is the closure of D.
2. $\gamma(C \cup D) = \max\{\gamma(C), \lambda(D)\}$.

3. $C \subset D$ implies $\gamma(C) \leq \gamma(D)$.

4. $\gamma(a(D)) = |a|\gamma(D)$ for any real number a.

The proofs of these along with other properties of γ may be found in [21].

Closely related to the measure of noncompactness is the concept of k-set contraction.

DEFINITION 2.2. Let X be a real Banach space. A continuous mapping $T : X \to X$ is said to be a <u>k-set contraction</u> if for any bounded subset of X we have $\gamma(T(D)) \leq k\gamma(D)$. T is said to be <u>densifying</u> if for any bounded subset D of X such that $\gamma(D) \neq 0$, we have $\gamma(T(D)) < \gamma(D)$.

DEFINITION 2.3. A Banach space X is said to be <u>strictly convex</u> if $\|\gamma x + (1 - \gamma)y\| < 1$ for all γ, $0 < \gamma < 1$ and all $x, y \in X$ such that $\|x\| = \|y\| = 1$.

THEOREM 2.1. Let X be a strictly convex Banach space and C be a nonempty, closed, bounded convex subset of X. Let $T : C \to C$ be a densifying mapping satisfying condition (α). Let $S : C \to C$ be the mapping as defined in Theorem 1.1. Then for any $x \in C$, the sequence of iterates $\{S^n x\}_{n=1}^{\infty}$ converges to a fixed point of T.

<u>Proof</u>. Since T is densifying on C, S is also densifying. Indeed, the continuity of S follows from that of T and I. It remains to show that for any bounded subset B of C with $\gamma(B) \neq 0$, we must have $\gamma(S(B)) < \gamma(B)$. By the definition of S, we have $S(B) \subset \alpha_0 B + \alpha_1 T(B) + \alpha_2 T^2(B) + \ldots + \alpha_m T^m(B)$. Hence, by properties (3) and (4) of γ, $\gamma(S(B)) \leq \alpha_0\gamma(B) + \alpha_1\gamma(T(B)) + \alpha_2\gamma(T^2(B)) + \ldots + \alpha_m\gamma(T^m(B)) < \alpha_0\gamma(B) + \alpha_1\gamma(B) + \alpha_2\gamma(B) + \ldots + \alpha_m\gamma(B) = \gamma(B)$.

Since S is densifying, we conclude from [23, Proposition 1] that $F(S)$, and hence from Theorem 1.1, $F(T)$ is nonempty. Let $D = \bigcup_{n=0}^{\infty} S^n x$. Then $S(D) = \bigcup_{n=1}^{\infty} S^n x$. Then $S(D) \subset D$ and $D = \{x\} \cup S(D)$. Let $\overline{D}$ be the closure of D, it follows from the continuity of S, $S(\overline{D}) \subset \overline{S(D)}$, that $\overline{D}$ is invariant under S.

We need to show that $\overline{D}$ is compact. It is enough to show that $\gamma(\overline{D}) = 0$. Using property (1) it suffices to show that $\gamma(D) = 0$. Suppose not, i.e. $\gamma(D) > 0$. Since $D = \{x\} \cup S(D)$, $\gamma(D) = \max\{\gamma\{x\}, \gamma(S(D))\}$ (by property (2)) $= \max\{0, \gamma(S(D))\} = \gamma(S(D)) < \gamma(D)$, a contradiction. Hence, $\gamma(D) = 0$. Thus $\overline{D}$ is compact. Hence, the sequence of iterates $\{S^n x\}$ has a convergent subsequence.

The result will follow from [7, Corollary 3.1] if we can show that for each $x \notin F(S)$ and $u \in F(S)$, (hence, $x \notin F(T)$ and $u \in F(T)$, we have $||Sx - u|| < ||x - u||$. Now for $x \notin F(T)$, $u \in F(T)$ and any $p > 1$, using condition (α), we have

(1.) $||T^p x - u|| \le ||T^{p-1}x - u|| \le \ldots \le ||x - u||$. Let $z = \frac{1}{1 - \alpha_0}[\alpha_1 T(x - u) + \alpha_2(T^2 x - u) + \ldots + \alpha_m(T^m x - u)]$. If $x \notin F(T)$ by Theorem 1.1, we have $z \neq x - u$ and, by (1), we get

(2.) $$||z|| \le \frac{1}{1 - \alpha_0}[\alpha_1||Tx - u|| + \alpha_2||T^2x - u|| + \ldots + \alpha_m||T^m x - u||]$$
$$\le \frac{1}{1 - \alpha_0}[\alpha_1 + \alpha_2 + \ldots + \alpha_m]\,||x - u|| = ||x - u||.$$ Now

$$||Sx - u|| = ||Sx - Su|| = ||\alpha_0(x - u) + \alpha_1(Tx - u) + \alpha_2(T^2x - u) + \ldots + \alpha_m(T^m x - u)||$$

$$= ||\alpha_0(x - u) + (1 - \alpha_0)z||$$

$$\le ||x - u||\,\left|\left|\frac{\alpha_0(x - u)}{||x - u||} + (1 - \alpha_0)\frac{z}{||x - u||}\right|\right| .$$

If the strict inequality holds in (2), then we have

$$||Sx - u|| \le ||x - u||\ \left|\left|\frac{\alpha_0||x - u||}{||x - u||} + (1 - \alpha_0)\frac{z}{||x - u||}\right|\right|$$

$$< ||x - u||\,(\alpha_0 + (1 - \alpha_0)) = ||x - u|| .$$

If the equality sign holds in (2), since the two vectors $\frac{x - u}{||x - u||}$ and $\frac{z}{||x - u||}$ are different, and $\left|\left|\frac{x - u}{||x - u||}\right|\right| = \left|\left|\frac{z}{||x - u||}\right|\right| = 1$, using the strict convexity of X we obtain $\left\| \frac{\alpha_0(x - u)}{\|x - u\|} + (1 - \alpha_0) \frac{z}{\|x - u\|} \right\| < 1$. Therefore, $\|Sx - u\| < \|x - u\|$ and the theorem is proven.

REMARK 2.1. Special cases of Theorem 2.1 for mapping S are Theorem 4.1 [20] and Theorem 6 [27].

REMARK 2.2. For $\alpha_0 = \lambda$, $\alpha_1 = 1 - \lambda$, $\alpha_2 = \alpha_3 = \ldots = \alpha_m = 0$, special cases of Theorem 2.1 are Theorem 4.2 [32], Theorem 1 [34], Theorem 2 [35], Theorem 6 [23], Theorem 1.5 [25], Theorem 4 [7], Theorem [12], Theorem [13], Theorem 2.1 [20], results of [1], [9], [30] and others.

REMARK 2.3. Theorem 1.5 [25] as stated is incorrect without the added assumption of strict convexity.

THEOREM 2.2. Let X be a uniformly convex Banach space, C a closed convex subset of X, T a self mapping of C which satisfies condition (α), S as defined in Theorem 1.1. Suppose that the set of fixed points of T is nonempty. Let us assume further that S satisfies the following condition:

(A) I - S maps bounded closed sets into closed sets.

Then, for any $x_0 \in C$, the sequence $\{S^n x_0\}$ converges to a fixed point of T in C.

Proof. If y is a fixed point of S, it follows that $\|S^{n+1}x_0 - y\| \leq \|S^n x_0 - y\|$, $n = 1, 2, \ldots$. So the sequence $\{S^n x\}$ is bounded. Let D be the strong closure of $\{S^n x_0\}$. By condition (A), it follows that $(I - S)D$ is closed. By Theorem 1.2, S is asymptotically regular over C so that $\lim\|(I - S)\, S^n x_0\| = 0$ and it lies in the strong closure of $(I - S)D$. So there exists $z \in D$ such that $(I - S)z = 0$. But this implies that either $z = S^n x_0$ for some n, or there exists a sequence $\{S^{n(j)}x_0\}$ converging to z. Since z is a fixed point of S, we can then conclude that, in either case, the whole sequence $\{S^n x_0\}$ converges to z.

REMARK 2.4. Special cases of Theorem 2.2 for mapping S are Theorem 2 [27], Theorem 6 [17].

REMARK 2.5. For $\alpha_0 = \lambda$, $\alpha_1 = 1 - \lambda$, $\alpha_2 = \alpha_3 = \ldots = \alpha_m = 0$, special cases of Theorem 2.3 are Corollary 4.3 [32], Corollary 2.1 [24] and Theorem 2.8 [22].

REMARK 2.6. The assumption (2.1) in Corollary 2.1 [24] is superfluous, as it is known [2, Theorem 2] that in a uniformly convex Banach space for T nonexpansive, $I - T$ maps bounded closed sets into closed sets.

DEFINITION 2.4. A mapping T from a Banach space X into itself is said to be demicompact if every bounded sequence $\{x_n\}$, such that $\{(I - T)x_n\}$ converges strongly contains a convergent subsequence.

DEFINITION 2.5. A mapping $T : X \to Y$ from a Banach space X into a Banach space Y is called <u>compact</u> if it is continuous and maps bounded sets into relatively compact sets.

REMARK 2.7. It is known that any compact mapping is demicompact [22, p. 39]. However, the following example demonstrates the existence of a demicompact mapping which is not compact.

EXAMPLE 2.1. Let $B = \{e_1, e_2, \ldots, e_n \ldots\}$ be the usual orthonormal basis for ℓ_2. Define $T : B \to B$ by $T(e_i) = e_{i+1}$. Then T is continuous (in fact, an isometry) but not compact. However, T is demicompact. Indeed, if $\{e_i\}_{i=1}^{\infty}$ is a bounded sequence in B such that $e_i - Te_i$ converges, then $\{e_i\}_{i=1}^{\infty}$ must be finite.

The following example shows that there is no connection between the demicompactness and the continuity of a mapping.

EXAMPLE 2.2. Let $X = [0, 1]$ with the usual metric. Define $T : X \to X$ as $Tx = x/2$ if $x \neq 0$ and $T(0) = 1$. Then T is not continuous. However, T is demicompact. Indeed, if $\{x_n\}$ is a bounded sequence in X such that $x_n - Tx_n \to 0$ as $n \to \infty$, then from Bolzano-Weierstrass theorem it follows that $\{x_n\}$ has a convergent subsequence.

Mappings which are simultaneously demicompact and continuous have an important topological property expressed in the following.

LEMMA 2.1 [22, Lemma 2.1]. Let $T : X \to X$ be a continuous demicompact mapping of a Banach space into itself. Then the mapping $I - T$ maps closed bounded sets into closed sets.

It follows from Lemma 2.1 that Theorem 2.3 remains true if S is demicompact and T is continuous.

COROLLARY 2.1. Let X be a uniformly convex Banach space, C a closed convex subset of X, T a self mapping of C which satisfies condition (α), S as defined in Theorem 1.1. Suppose that the set of fixed points of T is nonempty and T is compact. Then for any $x_0 \in C$, the sequence $\{S^n x_0\}$ converges to a fixed point of T in C.

Proof. It is enough to show that S is demicompact. Let $\{x_n\} \subset C$ be such that $x_n - Sx_n$ converges strongly to y. Since T is compact, $\{Tx_n\}$ contains a convergent subsequence $\{Tx_{n_j}\}$. Let $u = \lim_j Tx_{n_j}$. Since T is continuous, $T^2 x_{n_j} \to Tu$, $T^3 x_{n_j} \to T^2 u$, etc. Thus, $y = \lim(x_n - Sx_n)$ $= \lim_j (x_{n_j} - Sx_{n_j}) = \lim_j (1 - \alpha) x_{n_j} - \lim_i \sum_{i=1}^{m} \alpha_i T^i x_{n_j}$. Hence, we conclude that $\{x_{n_j}\}$ is convergent. Thus, S is demicompact.

REMARK 2.8. Special cases of Corollary 2.1, for mapping S are Theorem 3 [27], Theorem 7 [17].

REMARK 2.9. For $\alpha_0 = \lambda$, $\alpha_1 = 1 - \lambda$, $\alpha_2 = \alpha_3 = \dots \alpha_m = 0$, special cases of Corollary 2.1 are results of [19], [30].

Let D be a closed subset of a Banach space and $T : D \to X$. Let $\{\alpha_n\}$, $\{\beta_n\}$ be two sequences such that $0 \le \alpha_n, \beta_n \le 1$ for all n. Let $x_1 \in D$ be such that x_{n+1} is defined by

$$x_{n+1} = (1 - \alpha_n) x_n + \alpha_n T y_n \tag{1}$$

for each n, where $y_n = \beta_n T x_n + (1 - \beta_n) x_n$. We will call x_{n+1} defined by (1) generalized Ishikawa iterate. Note that if $0 \le \alpha_n \le \beta_n \le 1$ for all n, $\lim_{n\to\infty} \beta_n = 0$ and $\sum \alpha_n \beta_n = \infty$, then $\{x_n\}$ is Ishikawa process. If $\beta_n = 0$, $0 < \alpha_n < 1$ and $\sum \alpha_n$ diverges, then $\{x_n\}$ is a normal Mann process as discussed in Doston [8]. Also, if $\beta_n = 0$, $\alpha_n = 1$ for each n, then $\{x_n\}$ is the sequence of iterates of x_1.

DEFINITION 2.6. Let X be a Banach space. $T : X \to X$ is called a <u>quasi-nonexpansive</u> mapping if for all $x \in X$ and $y \in F(T)$, we have $\|Tx - y\| \leq \|x - y\|$.

All the mappings considered in this paper are quasi-nonexpansive, but not conversely. Indeed, if T satisfies condition (α), then for $x \in X$ and $y \in F(T)$, we have $\|Tx - y\| \leq \max\{\|x - y\|, \frac{1}{2}\|x - Tx\|, k\|x - y\|, k\|Tx - y\|\} \leq \|x - y\|$. For the converse, let $X = R$ and define $T : X \to X$ by $Tx = \frac{2}{3} x \sin 1/x$ if $x \neq 0$ and $T(0) = 0$. Then 0 is the only fixed point of T. In fact, $Tx = x$ implies $\sin 1/x = \frac{3}{2}$ which is impossible. Also, $|Tx - 0| = |\frac{2}{3} x \sin 1/x| \leq \frac{2}{3}|x| = \frac{2}{3}|x - y|$. Thus, T is quasi-nonexpansive. Let $x = \frac{2}{\pi}$ and $y = \frac{2}{3\pi}$. Then $|Tx - Ty| = \frac{16}{9\pi} > \max\{\frac{4}{3\pi}, \frac{2}{3\pi}, k \cdot \frac{14}{9\pi}, k \cdot \frac{2}{3\pi}\} = \max\{|x - y|, \frac{1}{2}[|x - Tx| + |y - Ty|], k|x - Ty|, k|y - Tx|\}$. Therefore, T does not satisfy condition (α).

THEOREM 2.3. Let D be a closed subset of a Banach space X and let $T : D \to X$ be a quasi-nonexpansive mapping. Suppose that $x_0 \in D$ is such that (1) yields $\{x_n\}$ either as a sequence of generalized Ishikaw iterates or a sequence of iterates. If $\lim d(x_n, F) = 0$, then $\{x_n\}$ converges to a fixed point of T.

<u>Proof</u>. It suffices to discuss the case when x_n does not belong to F for any n. If $\{x_n\}$ yields a sequence of generalized Ishikawa iterates, then for $y \in F$ we have

$$\|x_{n+1} - y\| = \|(1 - \alpha_n)x_n + \alpha_n Ty_n - y\| \leq (1 - \alpha_n)\|x_n - y\| + \alpha_n\|y_n - y\|.$$

$$\|y_n - y\| = \|\beta_n Tx_n + (1 - \beta_n)x_n - y\| \leq \beta_n\|Tx_n - y\| + (1 - \beta_n)\|x_n - y\| \leq \beta_n\|x_n - y\| + (1 - \beta_n)\|x_n - y\| = \|x_n - y\|.$$

Thus, in either case $\|x_{n+1} - y\| \leq \|x_n - y\|$. Choose $\varepsilon > 0$. Then there exists an n_0 such that $d(x_n, F) < \varepsilon/2$ for $n \geq n_0$. Hence, if $m, n \geq n_0$, for some $z \in F$, $\|x_n - x_m\| \leq \|x_n - z\| + \|z - x_m\| \leq 2\|x_{n_0} - z\| < \varepsilon$.

This shows that $\{x_n\}$ is a Cauchy sequence in D and hence, there exists $v \in D$ with $\lim\|x_n - v\| = 0$. From $\lim d(x_n, F) = 0$, it follows that $v \in \overline{F} = F$ (since F is closed). Again for any $z \in F$, $\|v - Tv\| \leq \|v - z\| + \|z - Tv\| \leq 2\|v - z\|$ implies $\|v - Tv\| \leq 2d(v, F) = 0$. This completes the proof.

REMARK 2.10. We first note that $\sum\alpha_n\beta_n = \infty$ is not used in the above proof. Also, it is evident that if $\alpha_n = 1$, $\beta_n = 0$, then the theorem holds for complete metric spaces. If $\beta_n = 0$, we get a result of Das, Singh and Watson [5] as a corollary of Theorem 2.3.

REMARK 2.11. The special case, namely $\alpha_n = 1$, $\beta_n = 0$ for each n, of Theorem 2.3 is a generalization of Theorem 1.2 [24] since T is not assumed to be continuous.

REMARK 3. Observe that in the proof of Theorem 2, Ishikawa [15] first shows that $\lim d(x_n, F) = 0$. Since as already remarked, a nonexpansive map with nonempty fixed points set is necessarily quasi-nonexpansive, we see that the hypothesis of Ishikawa's theorem imply those above. Hence, Theorem 2 [15] is a corollary of Theorem 2.3.

3. Weak Convergence Of The Sequence Of Iterates

DEFINITION 3.1. A mapping T of a Banach space X into itself is said to be <u>demiclosed</u> if for any sequence $\{x_n\}$ such that $x_n \to x$ weakly and $Tx_n \to y$ then $y = Tx$.

EXAMPLE 3.1 [2, Theorem 2]. If X is uniformly convex and T is nonexpansive, then $I - T$ is demiclosed.

THEOREM 3.1. Let X be a uniformly convex Banach space. Let C be a closed, bounded and convex subset of X. Let $T : C \to C$ be a mapping satisfying condition (α) with at most one fixed point $y \in C$. Let S be as defined in Theorem 1.1. Suppose furthermore that $I - S$ is demiclosed. Then for any $x_0 \in C$, the sequence $\{S^n x_0\}$ converges weakly to

$y \in C$.

Proof. Since by hypothesis $I - S$ is demiclosed, it follows that if $\{u_j\}$ converges weakly to $u_0 \in C$ and $(I - S)(u_j)$ converges strongly to w, then $(I - S)u_0 = w$. Now let $x_n = S^n x_0$, $n = 1, 2, \ldots$ and suppose $\{x_{n_i}\}$ converges weakly to u_0. By Theorem 1.2, S is asymptotically regular, so $\lim_{i\to\infty}(I - S)(x_{n_i}) = \lim_{i\to\infty}(S^n x_0 - S^{n+1}x_0) = 0$, and thus the demiclosedness of $I - S$ implies $(I - S)u_0 = 0$. Thus, u_0 is a fixed point of S. By Theorem 1.1, the fixed points of S and T coincide. Therefore, u_0 is the unique fixed point of T and every weakly convergent subsequence of $\{x_n\}$ converges weakly to u_0. If $\{x_n\}$ does not converge weakly to u_0, then there exists a weak neighbourhood W of u_0 and a subsequence $\{x_{n_k}\}$ of $\{x_n\}$ with the property that $x_{n_k} \notin W$, $k = 1, 2, \ldots$. By the reflexivity of X and boundedness of $\{x_n\}$, we conclude that some subsequence of $\{x_{n_k}\}$ converges weakly and by what we have shown, this weakly convergent subsequence must converge to u_0. This implies terms of the sequence $\{x_{n_k}\}$ must lie in W, a contradiction. Therefore, $\{S^n x_0\}$ converges weakly to u_0.

LEMMA 3.1. Let H be a strictly convex Banach space and C be a convex subset of H. Let $T : C \to C$ be a mapping satisfying condition (α), with nonempty fixed point set. Let $S : C \to C$ be a mapping as defined in Theorem 1.1. Then the fixed point set of T and hence of S is convex.

Proof. We may assume that F consists of more than one point, otherwise the result is proved. Let $x, y \in F$, it is enough to show that $z = \lambda x + (1 - \lambda)y \in F$, for $0 < \lambda < 1$. Since T satisfies condition (α), we have $\|Tz - x\| \leq \|z - x\|$ and $\|Tz - y\| \leq \|z - y\|$. Now $x - z = (1 - \lambda)(x - y)$ and $z - y = \lambda(x - y)$. Thus,

$$\|x - y\| \leq \|x - Tz\| + \|Tz - y\| \leq \|x - z\| + \|z - y\|$$

$$= (1 - \lambda)\|x - y\| + \lambda\|x - y\| = \|x - y\| .$$

Hence,

$$||x - Tz|| + ||Tz - y|| = ||x - Tz + Tz - y|| .$$

If $x - Tz = 0$, then $||Tz - y|| = ||x - y|| \leq ||z - y|| = \lambda||x - y||$, whence $1 \leq \lambda$, which is not true. Similarly, $Tz - y = 0$ implies $1 \leq 1 - \lambda$, whence $\lambda \leq 0$, which is not true. Since H is strictly convex, therefore, there exists $\alpha > 0$ such that $Tz - x = \alpha(y - Tz)$, whence $Tz = (1 - \beta)x + \beta y$, where $\beta = \frac{\alpha}{1 + \alpha}$. We have $Tz - x = \beta(y - x)$ and so $\beta||y - x|| = ||Tz - x|| \leq ||z - x|| = (1 - \lambda)||x - y||$, which gives $\beta \leq 1 - \lambda$. Using $Tz - y = (1 - \beta)(x - y)$, a similar argument gives $\beta \geq 1 - \lambda$. Thus, $\beta = 1 - \lambda$ and so $Tz = \lambda x + (1 - \lambda)y = z$, that is $z \in F$.

THEOREM 3.2. Let H be a Hilbert space and $T : H \to H$ be a mapping satisfying condition (α) with nonempty fixed points set. Let $S : H \to H$ be a mapping defined as in Theorem 1.1. Suppose the mapping S is continuous and $I - S$ demiclosed. Then for each $x_0 \in H$, the sequence of iterates $\{S^n x_0\}$ converges weakly to a fixed point.

Proof. Since F is nonempty we see that a ball B about some fixed point and containing x_0 is mapped into itself by S. Consequently, B contains the sequence of iterates $S^n x_0$. So, we may restrict ourselves to a mapping of a ball into itself. It follows from Lemma 3.1 that F is convex. The continuity of S implies F is closed. Thus, F being bounded, closed and convex is weakly compact. Observe that for $y \in F$, it follows from the definition of S that $\{||S^n x_0 - y||\}$ is nonincreasing. Define $g : F \to R^+$, (R^+ = nonnegative real numbers)

$$g(y) = \lim_{n\to\infty}||S^n x_0 - y|| .$$

Then g is continuous. Indeed, $g(z) = \lim_n||S^n x_0 - z|| = \lim_n||S^n x_0 - y + y - z|| \leq ||\lim_n S^n x_0 - y|| + ||y - z|| = g(y) + ||y - z||$. From this inequality it follows that $|g(y) - g(z)| \leq ||y - z||$, and g is continuous. Moreover, g is convex. In fact, $g(\lambda y + (1 - \lambda)z) = \lim_n||S^n x_0 -$

$(\lambda y + (1 - \lambda)z)\| \leq \lambda \lim_n \|S^n x_0 - y\| + (1 - \lambda) \lim_n \|S^n x_0 - z\| = \lambda g(y) + (1 - \lambda)g(z)$. So by proposition II.6 [6], we conclude that g is weakly lower semicontinuous. Applying proposition 1.4 [6], we obtain the existence of a point $u \in F$ such that $g(u) = d = \inf_{y \in F} g(y)$.

Note that u is unique. For suppose not, that is there exists another point $v \in F$ such that $g(v) = d$. Since g is convex, for $0 \leq \lambda \leq 1$, we have $g(\lambda u + (1 - \lambda)v) \leq \lambda g(u) + (1 - \lambda)g(v) = d$. Thus, $d = g(\lambda u + (1 - \lambda)v)$. Since u and v are in F, the sequences $\{\|S^n x_0 - u\|\}$ and $\{\|S^n x_0 - v\|\}$ are nonincreasing. Therefore, $\|x_n - u\| = \|S^n x_0 - u\| \to d$ and $\|x_n - v\| = \|S^n x_0 - v\| \to d$. Thus, from the uniform convexity of H, we conclude that $\|x_n - u - (x_n - v)\| \to 0$, i.e. $v = u$.

It remains to show that the sequence $\{S^n x_0\}$ converges weakly to u. Suppose not, then by the reflexivity of H and the boundedness of the sequence $\{S^n x_0\}$, there exists a subsequence $\{S^{n(j)} x_0\}$ of $\{S^n x_0\}$ whose limit z is different from u. Since S is asymptotically regular, the sequence $\{(I - S)^{n(j)} S x_0\}$ tends to zero as $n \to \infty$. Since by hypothesis $(I - S)$ is demiclosed, $(I - S)(z) = 0$, that is, z is a fixed point of S. We must show that $z = u$. Indeed we have

$$\|S^{n(j)} x_0 - u\|^2 = \|S^{n(j)} x_0 - z + z - u\|^2$$

$$= \|S^{n(j)} x_0 - z\|^2 + \|z - u\|^2$$

$$+ 2\mathrm{Re}(S^{n(j)} x_0 - z, z - u).$$

Taking limits we have

$$d^2 = g^2(u) = g^2(z) + \|z - u\|^2.$$

Since $g^2(z) \geq d^2$, we must have $\|z - u\|^2 \leq 0$. Hence, $z = u$.

REMARK 3.1. Special cases of Theorem 3.1 for mapping S are Theorem 3 [18] and Theorem 4 [27].

REMARK 3.2. For $\alpha_0 = \lambda$, $\alpha_1 = 1 - \lambda$, $\alpha_2 = \alpha_3 = \ldots = \alpha_m = 0$, special case of Theorem 3.1 is Theorem 2.11 [22].

REMARK 3.3. Special cases of Theorem 3.2 are Theorem 8 [4], Theorem 3.1 [32].

REFERENCES

1. U. Barbuti and S. Guerra, "On an extension of a theorem due to J. B. Diaz and F. T. Metcalf", Atti. Accad. Naz. Lincei Rend. Cl. Fix. Mat. Natur. 51 (1971), 29-31.
2. F. E. Browder, "Nonexpansive nonlinear operators in Banach spaces", Proc. Nat. Acad. Sci. U.S.A. 54 (1965), 1041-1044.
3. F. E. Browder and W. V. Petryshyn, "The solution by iteration of nonlinear functional equations in Banach spaces", Bull. Amer. Math. Soc. 72 (1966), 571-575.
4. F. E. Browder and W. V. Petryshyn, "Construction of fixed points of nonlinear mappings in Hilbert space", J. Math. Anal. Appl. 20 (1967).
5. K. M. Das, S. P. Singh and B. Watson, "A note on Mann iterates for quasi-nonexpansive mappings", Nonlinear Analysis, Theory, Methods and Applications, 6 (1981), 675-676.
6. D. G. de Figueiredo, "Topics in nonlinear functional analysis", Lecture Series No. 48, University of Maryland (1967).
7. J. B. Diaz and F. T. Metcalf, "On the set of subsequential limits", Trans. Amer. Math. Soc. 135 (1969), 459-485.
8. W. G. Doston Jr., "On the Mann iterative process", Trans. Amer. Math. Soc. 149 (1970), 65-73.
9. M. Edelstein, "A remark on a theorem of M. A. Krasnoselski", Amer. Math. Monthly 73 (1966), 509-510.
10. A. Genel and J. Lindenstrauss, "An example concerning fixed points", Israel J. Math. 22 (1975), 81-86.
11. G. E. Hardy and T. D. Rogers, "A generalization of a fixed point theorem of Reich", Canad. Math. Bull. 16 (1973), 201-206.
12. K. Iseki, "A fixed point theorem is striclty convex Banach space", Math. Sem. Notes 2 (1974), 23-25.
13. K. Iseki, "A generalization of Petryshyn-Riggio-Singh's theorem", Math. Sem. Notes 2 (1974), 138-140.
14. S. Ishikawa, "Fixed points and iteration of a nonexpansive mapping in a Banach space", Proc. Amer. Math. Soc. 59 (1976), 65-71.
15. S. Ishikawa, "Fixed points by new iteration method", Proc. Amer. Math. Soc. 44 (1974), 147-150.
16. R. Kannan, "Some results on fixed points", Bull. Calcutta Math. Soc. 60 (1968), 71-76.
17. R. Kannan, "Construction of fixed points of a class of nonlinear mappings", J. Math. Anal. Appl. 38 (1973), 430-438.

18. W. A. Kirk, "On successive approximations for nonexpansive mappings in Banach spaces", Glasgow Math. J. 12 (1971), 6-9.

19. M. A. Krasnoselski, "Two remarks about the method of successive approximations", Uspekhi Mat. Nauk 19 (1955), 123-127.

20. Ivar Massabo, "On the construction of fixed points for a class of nonlinear mappings", Boll. U.M.I. 10 (1974), 512-528.

21. R. D. Nussbaum, "The fixed point index and fixed point theorems for k-set contraction", Ph.D. Dissertation, University of Chicago (1969).

22. Z. Opial, "Nonexpansive and monotone mappings in Banach spaces", Lecture Note No. 1, Brown University (1967).

23. W. V. Petryshyn, "Structure of fixed point sets for k-set contractions", Arch. Rat. Mech. Anal. 40 (1971), 312-328.

24. W. V. Petryshyn and T. E. Williamson Jr., "Strong and weak convergence of the sequence of successive approximations for quasi-nonexpansive mappings", J. Math. Anal. Appl. 43 (1973), 459-497.

25. B. K. Ray, "Remarks on the fixed point theorem of Diaz and Metcalf", Bull. Acad. Polon. Sci. Ser. Sci. Math. Astr. Phys. 24 (1976), 899-900.

26. B. K. Ray and H. Chatterjee, "Some results on fixed points in metric and Banach spaces", Bull. Acad. Polon. Sci. Ser. Sci. Math. Astr. Phys. 25 (1977), 1243-1247.

27. B. K. Ray and B. E. Rhoades, "A class of fixed point theorems", Math. Sem. Notes 7 (1977), 477-489.

28. B. E. Rhoades, "Fixed point iteration using infinite matrices", Trans. Amer. Math. Soc. 196 (1974), 161-176.

29. B. E. Rhoades, "A comparison of various definitions of contractive mappings", Trans. Amer. Math. Soc. 226 (1977), 257-290.

30. H. Schaefer, "Über die Methode sukzessive Approximationen", Jbr. Deutch. Math. Verein, 59 (1957), 131-140.

31. K. L. Singh, "Convergence of sequence of iterates", Atti. Accad. Naz. Lincei Cl. Sci. Fis. Mat. Natur. 62 (1977), 154-159.

32. K. L. Singh, "Sequence of iterates of generalized contractions", Fund. Math. 105 (1980), 115-126.

33. S. P. Singh, "On the convergence of sequence of iterates", Atti. Accad. Naz. Lincei Cl. Sci. Fis. Mat. Natur. 57 (1974), 502-504.

34. S. P. Singh and M. I. Riggio, "On a theorem of J. B. Diaz and F. T. Metcalf", Atti. Accad. Naz. Lincei Cl. Sci. Fis. Mat. Natur. 53 (1972), 246-249.

35. C. S. Wong, "Approximation to fixed points of generalized nonexpansive mappings", Proc. Amer. Math. Soc. 54 (1974), 93-97.

36. R. K. Yadav, "Fixed point theorems in generalized metric spaces", Banaras Math. J. (1969), -.

A NOTE ON A THEOREM OF KY FAN*

S. P. Singh and B. Watson

Memorial University of Newfoundland
St. John's, Newfoundland
Canada

In [2], Ky Fan proved the following result.

THEOREM A. Let C be a nonempty compact convex subset of a normed linear space X. For any continuous function $f : C \to X$, there exists a $y \in C$ such that

$$\|y - f(y)\| = d(f(y), C).$$

This result was used to get the Schauder fixed point theorem as a corollary.

S. Reich [4] extended this result to noncompact subsets of a Banach space X. He proved the following:

THEOREM B. Let C be a closed, convex subset of a Banach space X with the Oshman property. If $f : C \to X$ is continuous and $f(C)$ is relatively compact, then there is a point $y \in C$ such that

$$\|y - f(y)\| = d(f(y), C).$$

Reich [4] used Theorem B to prove a few fixed point theorems.

Oshman [3] has characterized those reflexive Banach spaces in which the metric projection on every closed convex subset is upper semicontinuous.

*This work was supported in part by NSERC grant #A5154 and #A4646

We need the following definitions and results, see [1].

Assume that X is a Banach space and $f : X \to X$ is a continuous mapping.

(i) For every bounded set $A \subset X$, $\alpha(A)$, the measure of noncompactness, is defined as

$$\alpha(A) = \inf\{\varepsilon > 0 \;/\; A \text{ has a finite } \varepsilon\text{-net}\}.$$

(ii) $\overline{A} = \cap\{F \subset X \;/\; A \subset F,\ F \text{ closed}\}$, any $A \subset X$.

(iii) $\overline{\overline{A}} = \cap\{C \subset X \;/\; A \subset C.\ C \text{ convex}\}$, any $A \subset X$.

(iv) $Kf = \bigcup_{n \geq 0} \{K_n \;/\; K_o = \overline{\overline{K}},\ K_n = f(K_{n-1}) \subset X\}$

for a compact nonempty subset K of X.

DEFINITION. A continuous mapping $f : X \to X$ is said to be sequentially condensing or s-condensing if, for every bounded $Kf \subset X$ with $\alpha(Kf) > 0$ and $f(Kf)$ bounded, f satisfies the condition

$$\alpha(f(Kf)) < \alpha(Kf).$$

The retraction map on X is s-condensing but not condensing [5].

THEOREM C [1]. Let C be a nonempty closed bounded convex subset of a Banach space X and $f : C \to C$ be s-condensing. Then f has a fixed point in C.

THEOREM D [1]. If $f : C \to X$ is s-condensing and $r : X \to C$ is a retraction, then $rf : C \to C$ is s-condensing.

The aim of this note is to prove the following result.

THEOREM 1. Let B be an open ball about the origin of radius $r > 0$ in a Banach space X. If $f : \overline{B} \to X$ is s-condensing, then there exists a $y \in \overline{B}$ such that

$$\|y - f(y)\| = d(f(y), \overline{B}).$$

<u>Proof.</u> Let R be the retraction of X onto $\overline{B}$ given by

$$R(x) = \begin{cases} x & \text{if } \|x\| \leq r. \\ \dfrac{rx}{\|x\|} & \text{if } \|x\| \geq r. \end{cases}$$

Define $T(x) = R\ f(x)$. Then $T : \overline{B} \to \overline{B}$ is s-condensing by Theorem D. Therefore, by Theorem C, T has a fixed point $y \in \overline{B}$, i.e. $Ty = y$. Now, if $\|f(y)\| \leq r$, then

$$\begin{aligned} &\|y - f(y)\| = \|T(y) - f(y)\| \\ &= \|R\ f(y) - f(y)\| \\ &= 0. \end{aligned}$$

If $\|f(y)\| \geq r$, then

$$\begin{aligned} &\|y - f(y)\| = \|R\ f(y) - f(y)\| \\ &= \left\|\frac{r\ f(y)}{\|f(y)\|} - f(y)\right\| \\ &= \|f(y)\| - r. \end{aligned}$$

For every $x \in \overline{B}$,

$$\begin{aligned} &\|f(y)\| - r \leq \|f(y)\| - \|x\| \\ &\leq \|f(y) - x\| . \end{aligned}$$

Hence, $\|y - f(y)\| \leq d(f(y), \overline{B})$. But we always have $\|y - f(y)\| \geq d(f(y), \overline{B})$.

This completes the proof. (T. C. Lim, "A Note on a Theorem of Ky Fan", Canad. Math. Bull. 22 (1979), 513-515 is worth mentioning.)

REFERENCES

1. Daher, S. J., "Fixed point theorems for nonlinear operators with a sequential condition", Nonlinear Analysis, TMA 3, No. 1 (1979), 59-63.

2. Fan, Ky, "Extensions of two fixed point theorems of F. E. Browder", Math. Z. 112 (1969), 234-240.

3. Oshman, E. V., "On the continuity of metric projection in Banach space", Math. USSR Sb. 9 (1969), 171-182.

4. Reich, S., "Approximate selections, best approximations, fixed points and invariant sets", J. Math. Anal. Appl. 62, No. 1 (1978), 104-112.

5. Sadovskii, B. N., "A fixed point principle", Functional Anal. Appl., 1(2) (1967), 151-153.

ON ASYMPTOTICALLY QUASI NONEXPANSIVE FAMILIES OF MAPPINGS

P. Srivastava

Allahabad University
Allahabad, U.P.
India

S. C. Srivastava

K. A. Degree College
Allabahad, U.P.
India

Abstract

The concept of a quasi nonexpansive mapping was introduced by Dotson Jr. In the context of a family of mappings a weaker concept, viz. that of an asymptotically quasi nonexpansive family of mappings has been introduced in this paper. We have also obtained a common fixed point for a weakly continuous family of self mappings in a strictly convex normed linear space which improves upon and extends a result of DeMarr.

2. Introduction

While studying fixed point theorems, several authors have used the concept of an asymptotically nonexpansive family of mappings, e.g. Holmes and Narayanaswami [3], and Kiang [4]. In the present paper, we have introduced the concept of an asymptotically quasi nonexpansive family of mappings with respect to the set of all common fixed points of the family. Thus the latter is a weaker form of the former. Moreover, the concept of an asymptotically quasi nonexpansive family of mappings weakens the concept of a family of quasi nonexpansive mappings given by Dotson Jr. (see Example 2.3). Dotson Jr. [2] proved a number of fixed point results employing the concept of quasi nonexpansive mappings. The object of this paper is to show that in the results of Dotson Jr. [2] quasi nonexpansive mappings may be replaced by an asymptotically quasi nonexpansive family of mappings.

Common fixed point of a nonexpansive commutative family of self mappings of a Banach space was obtained by DeMarr [1]. In section 3.3, we have obtained a common fixed point of a weakly continuous commutative family of self mappings of a strictly normed linear space generalizing the result of DeMarr.

2. Definitions And An Example

2.1 DEFINITION [2]. A self mapping T of a subset C of a normed linear space is said to be quasi nonexpansive provided T has at least one fixed point in C, and if $p \in C$ is any fixed point of T, then $\|T(x) - p\| \leq \|x - p\|$ holds for all $x \in C$.

If each member of a family G of self mappings of C is quasi nonexpansive, then G is said to be quasi nonexpansive.

2.2 DEFINITION. A family G of mappings $f : C \to C$, (C is a subset of a normed linear space) is to be termed asymptotically quasi non-expansive if G has at least one common fixed point in C and there is a $g \in G$ such that $\|fg(x) - p\| \leq \|x - p\|$, for all $f \in G$, every $x \in C$ and each common fixed point p of G.

2.3 EXAMPLE. Let $I = [0, 1]$. Define a mapping $f : I \to I$ by

$$f(x) = \frac{1}{2}, \quad 0 \leq x \leq \frac{1}{2},$$

$$= \frac{1 - x}{2}, \quad \text{otherwise.}$$

The point $\frac{1}{2}$ is a fixed point of f. For $x = \frac{3}{4}$, $p = \frac{1}{2}$, $\|f(x) - p\| = \frac{3}{8}$ and $\|x - p\| = \frac{1}{4}$. Thus f is not quasi nonexpansive and hence, the semigroup G generated by f is not a quasi nonexpansive family. It can be easily seen that G is an asymptotically quasi nonexpansive family of mappings.

3. LEMMAS. Given below are two lemmas required in the sequel. The method of proof of the second lemma is a generalized technique of [2].

3.1 LEMMA. Let $G : X \to X$ be a commutative semigroup of mappings on X. If $x \notin F(G)$ (the set of all common fixed points of G), then for any $g \in G$, there is a $T \in G$ such that $Tg(x) \neq x$. Hence $F(G) = F(Gg)$.

Proof. Let $x \notin F(G)$. Then there is a $T_o \in G$ such that $T_o(x) \neq x$. Let $g \in G$ and suppose that for every $T \in G$, $Tg(x) = x$. In particular, $T_o^2 g(x) = T_o(x)$ and hence $T_o(x) = x$ which is a contradiction.

3.2 LEMMA. If C is a closed and convex subset of a strictly convex normed linear space, and $G : C \to C$ is a commutative semigroup of asymptotically quasi nonexpansive mappings, then $F(G)$ is a nonempty, closed and convex subset of C.

Proof. Under the hypothesis, $F(G) \neq \phi$. Suppose $F(G)$ is not closed. Then there is a point $x_o \in \overline{F(G)}$ such that $x_o \notin F(G)$. By Lemma 3.1, for any $g \in G$, there is a $T_o \in G$ such that $T_o g(x_o) \neq x_0$. Set

$$r = \frac{1}{3}\|T_o g(x_o) - x_o\| > 0. \tag{i}$$

Since $x_o \in \overline{F(G)}$, there is a $y \in F(G)$ such that $\|y - x_o\| < r$. As G is asymptotically quasi nonexpansive, there exists a $g_o \in G$ such that $\|Tg_o(x) - y\| \leq \|x - y\|$, for any $T \in G$ and $x \in C$. Choose T_o in (i) corresponding to g_o. Then

$$\begin{aligned} 3r = \|T_o g_o(x_o) - x_o\| &\leq \|T_o g_o(x_o) - y\| + \|y - x_o\| \\ &\leq \|x_o - y\| + \|y - x_o\| \\ &< r + r = 2r. \end{aligned}$$

This contradiction establishes that $F(G)$ is closed. We proceed to show that $F(G)$ is convex. Suppose $a, b \in F(G)$, $a \neq b$ and $0 < t < 1$. Then $c = (1 - t)\, a + tb \in C$. Since G is asymptotically quasi nonexpansive, there is a $g \in G$ such that $\|Tg(c) - a\| \leq \|c - a\|$ and $\|Tg(c) - b\| \leq \|c - b\|$ for all $T \in G$. Noting that $c - a = t(b - a)$ and $c - b = (1 - t)(a - b)$, we have

$$\begin{aligned} \|b - a\| &\leq \|b - Tg(c)\| + \|Tg(c) - a\| \\ &\leq \|c - b\| + \|c - a\| \\ &= \|b - a\| . \end{aligned}$$

Hence

$$\|b - Tg(c) + Tg(c) - a\| = \|b - Tg(c)\| + \|Tg(c) - a\| .$$

If $b - Tg(c) = 0$, then

$$\begin{aligned} \|b - a\| &= \|Tg(c) - a\| \\ &\leq \|c - a\| \\ &= t\|b - a\| . \end{aligned}$$

Hence, $1 \leq t$, which is not true. Similarly $Tg(c) - a = 0$ implies that $1 \leq 1 - t$, i.e. $t \leq 0$ which is again false. Since the space is strictly convex, there exists $r > 0$ such that $Tg(c) - a = r(b - Tg(c))$; and, therefore, $Tg(c) = (1 - s)a + sb$, where $s = \frac{r}{r + 1}$. We have

$$\begin{aligned} s\|b - a\| &= \|Tg(c) - a\| \\ &\leq \|c - a\| \\ &= t\|b - a\| , \end{aligned}$$

which gives $s \leq t$. Using $Tg(c) - b = (1 - s)(a - b)$, a similar argument gives that $s \geq t$. Thus $s = t$ and $Tg(c) = (1 - t)a + tb = c$. Hence $c \in F(Gg) = F(G)$, by Lemma 1, and $F(G)$ is convex.

We prove the following.

4.1 THEOREM. Let C be a compact convex subset of a strictly convex normed linear space and let G be a commutative semigroup of asymptotically quasi nonexpansive self mappings on C. If f is any continuous mapping on C into C such that $fg = gf$ for all $g \in G$, then $F(G) \cap F(f) \neq \phi$.

Proof. By Lemma 3.2, $F(G)$ is nonempty, closed and convex. Consequently $F(G)$ is compact and convex. Since $gf = fg$, for every $g \in G$, $f(F(G)) \subseteq F(G)$. Hence by Tychonoff theorem f has a fixed point in $F(G)$.

4.2 THEOREM. Let C be a closed, bounded and convex subset of a uniformly convex Banach space and let G be a commutative semigroup of asymptotically quasi nonexpansive self mappings of C. If f is a self mapping of C which is either nonexpansive or weakly continuous and $fg = gf$ for all $g \in G$, then $F(G) \cap F(f) \neq \phi$.

Proof of this theorem runs parallel to corresponding theorem of [2].

4.3 THEOREM. Suppose C is a weakly compact convex subset of a strictly convex normed linear space and G is a commutative semigroup of asymptotically quasi nonexpansive self mappings. Let $\{T_\alpha\}$ be a commutative family of weakly continuous mappings of C into itself such that $T_\alpha g = gT_\alpha$ for each α and $g \in G$. Then $\bigcap_\alpha F(T_\alpha) \neq \phi$.

Proof. By Lemma 3.2, $F(G)$ is a nonempty, closed and convex subset of C and hence $F(G)$ is a weakly compact and convex subset of C. Since $T_\alpha g = gT_\alpha$ for each α and $g \in G$, we have $T_\alpha(F(G)) \subseteq F(G)$ for each α. Also since weak topology is locally convex, by Tychonoff theorem, $F(T_\alpha) \neq \phi$. Also $F(T_\alpha)$ is weakly closed for each α. If $\{F(T_\alpha)\}$ has the finite intersection property, the required result follows from the weak compactness of C.

Suppose $F(T_{\alpha_i})$, $i = 1, 2, \ldots, n$, are n of these which have nonempty intersection. Then $D = \bigcap_{i=1}^{n} F(T_{\alpha_i})$ is a nonempty weakly closed subset of C and is therefore weakly compact. As $\{T_\alpha\}$ is a commutative family, $T_{\alpha_{n+1}} F(T_{\alpha_i}) \subseteq F(T_{\alpha_i})$, $i = 1, 2, \ldots, n$, and $T_{\alpha_{n+1}}(D) \subseteq D$. Since $T_{\alpha_{n+1}}$ is weakly continuous and D is weakly compact and convex, by Tychonoff theorem, $T_{\alpha_{n+1}}$ has a fixed point in D and $\bigcap_{i=1}^{n+1} F(T_{\alpha_i}) \neq \phi$. This proves that $\{F(T_\alpha)\}$ has the finite intersection property and the proof is complete.

REFERENCES

1. R. DeMarr, "Common fixed points for commuting contraction mappings", Pac. J. Math. 13 (1963), 1139-1141.

2. W. G. Dotson, Jr., "Fixed points of quasi-nonexpansive mappings", J. Austral. Math. Soc. 13 (1972), 167-170.

3. R. D. Holmes and P. P. Narayanaswami, "On asymptotically nonexpansive semigroup of mappings", Canad. Math. Bull. 13 (1970), 209-214.

4. M. T. Kiang, "Fixed point theorems for certain classes of semigroups of mappings", Trans. Amer. Math. Soc. 189 (1974), 63-76.

SOME REMARKS AND EXAMPLES CONCERNING THE FIXED POINT PROPERTY OF POLYHEDRA

S. Thomeier

Memorial University of Newfoundland
St. John's, Newfoundland
Canada

ABSTRACT

Even for simply connected polyhedra the fixed point property is known not to be invariant under topological products and several other geometric constructions. Mapping cones, mapping cylinders, suspensions and join products are used to study the behaviour of the fixed point property under such constructions within more restrictive categories. A result, involving join products, related to a conjecture of Fadell in S_0, is discussed, and some relatively low-dimensional examples for the non-invariance of the fixed point property under products and suspensions are given.

1. Introduction

It is known that the fixed point property (f.p.p.) behaves badly with respect to many geometric constructions. It is, for example, not invariant under topological products, not even if one restricts consideration to certain classes of very nice spaces, such as (finite) polyhedra, or even simply connected polyhedra. If X has the f.p.p., even the product $X \times I$ with the closed unit interval $I = [0,1]$ need not have the f.p.p., as one sees from the famous examples of Fadell - Lopez and Bredon ([3],[7],[8]). Even the product of manifolds with f.p.p. need not have f.p.p. as Husseini [9] has shown. (See [1],[3],[4],[7] and the forthcoming paper [5] for the history of this old problem, and for a general exposition.) Essentially the same examples can also be used to show that also certain other constructions (suspensions, smash products, join products) do not preserve the f.p.p., again not even for simply connected polyhedra.

All these examples involve polyhedra of fairly high dimensions. In the Fadell-Lopez example, for instance, a factor space is of dimension 17, whereas in the Husseini example, each factor space has dimension 56, so the product is 112 - dimensional. In this context, Fadell [8] has asked the following question over ten years ago: if $SX \times Y$ has f.p.p. (where SX denotes the suspension of X), does this imply that $X \times Y$ has the f.p.p.? He also stated the

CONJECTURE (Fadell). Suppose X and Y belong to S_0 (i.e. are simply connected polyhedra satisfying the Shi condition), and that X and all its suspensions SX, S^2X, ... have the f.p.p. Then, if Y has the f.p.p., so does $X \times Y$.

Although this is certainly not always true outside the class S_0, nothing general seems to be known about it. Nevertheless, the following result is somewhat related.

THEOREM 1.1. Suppose X and Y belong to class S (i.e. are polyhedra satisfying the Shi condition), and that the suspensions SX and SY and also the join product $X * Y$ all have the f.p.p. Then also $X \times Y$ and its suspension $S(X \times Y)$ have the f.p.p.

A proof of this result and a discussion of join products constitute the content of section 2.

Finally, in section 3, some examples for the non-invariance of the f.p.p. under products and suspensions are given which involve polyhedra of dimensions considerably lower than those in the examples mentioned above.

2. Join Products

The join product $X * Y$ of two topological spaces X and Y is obtained as the quotient space of $X \times Y \times I$ under the following identifications: $(x, y, 0) \sim (x, y', 0)$ and $(x, y, 1) \sim (x', y, 1)$, for all $x, x' \in X$ and all $y, y' \in Y$. For any specific $x \in X$, $y \in Y$, the "line segment" from x to y in $X * Y$ is the subset $[x, y] = \{(x, y, t) / 0 \leq t \leq 1\}$, obviously each point of $X * Y$ with $t \neq 0, 1$ lies on a unique such line segment.

EXAMPLE 1. For Y a single point y_o, the join $X * \{y_o\}$ is the cone CX over X, which clearly is always contractible.

EXAMPLE 2. The join $E^m * E^n$ of two closed balls of dimensions m and n is homeomorphic to E^{m+n+1}.

EXAMPLE 3. The join $S^m * S^n$ of two spheres of dimensions m and n is homeomorphic to a sphere S^{m+n+1}.

EXAMPLE 4. For the special case $Y = S^o$ = 0-sphere $\{S_o, S_1\}$, the join product $X * S^o = Sx$ is the suspension of X; obviously $E^m * S^n \equiv E^m * \underbrace{S^o * S^o * \ldots * S^o}_{n+1}$ is homeomorphic to E^{m+n+1}.

If $f : X \to X'$ and $g : Y \to Y'$ are (continuous) maps, the join product $f * g : X * Y \to X' * Y'$ is defined by $(f * g)(x, y, t)$ $(f(x), g(x), t)$, for $x \in X$, $y \in Y$, $t \in I = [0, 1]$. If, in particular, $Y = Y' = S^0$ and $g = 1_{S^0}$ = identity map on S^0, then $f * 1_{S^0} = S(f)$: $SX \to SX^1$ is the suspension of f. The mapping cyclinder $M(f)$ of a map $f : X \to Y$ is the subspace of $X * Y$ that includes all line segments $[x, f(x)]$ for $x \in X$, together with the points of Y. When the top X in $M(f)$ is identified to a point, the resulting quotient space $C(f)$ is the mapping cone of f.

Another construction which, however, is known to behave exceptionally well with respect to the f.p.p., is the one-point-union ("wedge") $Z = X \vee Y$ of two spaces, which is obtained from the disjoint union $X \cup Y$ by identifying a point $x_0 \in X$ with a point $y_0 \in Y$, so that $Z = X \cup Y$, with $X \cap Y$ being a single point. Therefore, $X \vee Y$ can also be viewed as the subspace $(X \times y_0) \cup (x_0 \times Y)$ of $X \times Y$.

The importance of the wedge for fixed point theory stems from the easily established fact that, if X and Y each have the f.p.p., then so does $X \vee Y$, and vice versa. This is actually true for the wedge of any number of spaces:

PROPOSITION 2.1. Let $Z = \vee X_\lambda$ be the one-point union of the topological spaces X_λ, $\lambda \in \Lambda$(= arbitrary index set); i.e. $X_\lambda \cap X_\mu = \{p\}$, for $\lambda \neq \mu$. Then Z has the f.p.p. if and only if each X_λ has the f.p.p.

Proof. First, it is clear that one has retraction maps $r_\lambda : Z \to X_\lambda$, for each λ defined by $r_\lambda(z) = z$, if $z \in X_\lambda$, and $r_\lambda(z) = p$, otherwise. Therefore, each X_λ is a retract of Z and hence inherits the f.p.p.. Suppose now, conversely, that all X_λ have the f.p.p., and let $f : Z \to Z$ be a map with $f(p)$ in X_μ. Then $r_\mu \cdot f|X_\mu : X_\mu \to X_\mu$ is continuous and has a fixed point $x \in X_\mu$. There are two possible cases:

(i) if $x = p$, then, by $f(p) \in X_\mu$ and $r_\mu \mid X_\mu = 1_{X_\mu}$ (= identity map on X_μ), it follows that $f(p) = p$;

(ii) if $x \neq p$, then $f(x)$ must be in X_μ, because all points outside X_μ are sent to p by r_μ. Since $r_\mu \mid X_\mu = 1_{X_\mu}$ again, one has $x = (r_\mu \cdot f \mid X_\mu)(x)$; i.e. x is a fixed point of f.

Despite the non-preservation of the f.p.p. under topological products and suspensions, it is also known and easily seen that (a) if $X \times Y$ has the f.p.p., then both X and Y have the f.p.p., and (b) if SX has the f.p.p., then so does X. Also, the join product does not preserve the f.p.p. ([7]), but one has the following.

PROPOSITION 2.2. If the join product $X * Y$ possesses the f.p.p., then at least one of X, Y possesses the f.p.p.

Proof. If, on the contrary, both X and Y were without the f.p.p., there would be fixed point free maps $f : X \to X$ and $g : Y \to Y$. Consider their join product $f * g : X * Y \to X * Y$, defined by $(f * g)(x, y, t) = (f(x), g(x), t)$, for all $x \in X$, $y \in Y$, $t \in I = [0, 1]$. Since $f(x) \neq x$ and $g(y) \neq y$ always, clearly $f * g$ would be without fixed points, contradicting the assumption.

A more thorough discussion of the behaviour of the f.p.p. with respect to the join product and other geometric constructions is contained in [15].

Proof of Theorem 1.1. The following facts and observations will be needed:

(i) Let X be a compact ANR, and let Y be a polyhedron of type S (i.e., satisfying the SHI condition) which has the f.p.p. Then, for any map $f : X \to Y$, the mapping cylinder $M(f)$ has the f.p.p. ([3], page 147). Since Y is a retract (actually, a deformation retract) of $M(f)$, and retracts of spaces with f.p.p. again have f.p.p. it follows therefore that for polyhedra of type S the f.p.p. is a homotopy property (i.e. if X and Y are polyhedra of type S and are of the same homotopy type, then X has the f.p.p. if and only if Y has it); whereas the f.p.p. is not a homotopy property in general (not even for simply connected polyhedra, as can be seen from the Fadell-Lopez example).

(ii) For polyhedra X and Y the suspension $S(X \times Y)$ of the product $X \times Y$ is of the same homotopy type as the wedge $(X * Y) \vee SX \vee SY$ of the join $X * Y$ with the suspensions SX and SY (see e.g. [6], [11]). Therefore, if both X and Y are polyhedra of type S, with SX, SY and $X * Y$ all having the f.p.p. it follows that $(X * Y) \vee SX \vee SY$ has the f.p.p. and, by (i) and (ii) above, that also $S(X \times Y)$, and hence $X \times Y$, all possess the f.p.p., too. This establishes Theorem 1.1. for the class S of all polyhedra of type S, and therefore also for the subclass S_0 of all simply connected such polyhedra.

On the other hand, it is obvious from (ii) that, for X and Y being polyhedra of type S, if $S(X \times Y)$ has the f.p.p., then also $X * Y$, SX and SY (and hence also X and Y) all must have the f.p.p.

3. Some Examples

As already mentioned in the introduction, the well-known examples for the non-preservation of the f.p.p. under topological products and suspensions are relatively high-dimensional. The Fadell-Lopez example, for instance, involves the factor space $Z = S\mathbf{C}P^8 \vee X$, the wedge of the suspension of complex projective 8-space with X, a quotient space of $\mathbf{C}P^2 \cup S^2 \times S^2 \cup \mathbf{C}P^4$. This Z is a polyhedron of dimension 17 with the f.p.p.. Similarly, $Z' = S\mathbf{C}P^6 \vee X$ also has the f.p.p., but both $Z \times I$ and SZ' fail to have the f.p.p.

For manifolds, the Husseini example is a 112-dimensional product of two 56-dimensional factors. The main reason for this high-dimensionality is the fact that a suitable stable element in the homotopy groups of spheres is used in the underlying construction according to the method of Bredon (see [9] and the forthcoming paper [5] for a detailed description of this construction).

By making use of the same results and methods, but using instead unstable or metastable elements in the homotopy of spheres with suitable properties, one can construct lower-dimensional example spaces. In the following, a few samples are given.

EXAMPLE 3.1. Let $\phi = \nu_6 \circ \eta_5$ in $\pi_9(S^5)$, where $\eta_5 = S^3(\eta_2)$ is the triple suspension of the generator η_2 in $\pi_3(S^2)$, and $\nu_6 = S^2(\nu_4)$ is the double suspension of the generator ν_4 in $\pi_7(S^4)$. The element ϕ is then the generator of $\pi_9(S^5)$; it is itself the suspension of an element from $\pi_8(S^4)$ and is known to have order 2 (see [16]). Therefore, by [2], the mapping cone $X = C(\phi) = S^5 \cup_\phi e^{10}$, obtained by attaching a 10-dimensional cell to S^5 by means of the attaching map ϕ, has the f.p.p. (actually, has no self-maps of zero Lefschetz number), but its suspension $SX = S^6 \cup_{S\phi} e^{11}$ fails to have the f.p.p. because $S(\phi) = 0$; in fact, SX is the wedge $S^6 \vee S^{11}$ of spheres, clearly without the f.p.p. Obviously, the dimension of X is 10 in this example. This is, however, not the lowest-dimensional example that can be constructed. There is a 4-dimensional example of a mapping cone with f.p.p. (with an attaching map that is more complicated, namely a Whitehead product), whose suspension fails to have the f.p.p.

On the other hand, not every mapping cone with the f.p.p. constructed in this way has the property that by suspension it looses the f.p.p. Consider, for instance, $\psi = \eta_3 \cdot \eta_4$, the generator of $\pi_5(S^3)$ of order 2. Its mapping cone $Y = C(\psi) = S^3 \cup_\psi e^6$ also has, for the same reasons, the f.p.p., but its suspension $SY = S^4 \cup_{S\psi} e^7$ still has the f.p.p. as well, because $S(\psi)$ is non-zero.

EXAMPLE 3.2. Now consider elements α and β in the group $\pi_{15}(S^5) \cong \mathbb{Z}_9 \oplus \mathbb{Z}_8 \oplus \mathbb{Z}_2$, which are of order $p = 3$ and $q = 2$, respectively; e.g. take $\beta = \eta_5 \circ \mu_6$ of order 2 (in the notation of [16]), and $\alpha = 3g$, where g generates the 3-primary component $\mathbb{Z}_9$ of $\pi_{15}(S^5)$. One easily verifies that these are suspension elements (either by the method of [13] or by an EHP-sequence argument as in [14]). Then, again by [2], the mapping cones $X = C(\alpha) = S^5 \cup_\alpha e^{16}$ and $Y = C(\beta) = S^5 \cup_\beta e^{16}$ both have the (homotopy invariant) f.p.p. but one can show that their product $X \times Y$ possesses fixed point free self-maps (in fact, fixed point free self-maps none of whose iterates has fixed points either). The technical details will be contained in [15].

Since the dimension of both factors is only 16, it would be interesting to modify the construction of [9] in such a way that this lower-dimensional example of polyhedra would yield a lower-dimensional example of manifolds.

The dimensions in this example can, in fact, be reduced even further by replacing the element β in this construction by the generating element $\gamma = \eta_5 \circ \eta_6$ in $\pi_7(S^5)$ of order $q = 2$. Its mapping cone $Z = C(\gamma) = S^5 \cup_\gamma e^8$ has again the f.p.p. and is of dimension 8 only; the product space $X \times Z = C(\alpha) \times C(\gamma)$ again fails to have the f.p.p. and is only 24-dimensional.

REFERENCES

1. R. H. Bing, "The elusive fixed point property", Amer. Math. Monthly 76, (1969), 119-132.

2. G. Bredon, "Some examples for the fixed point property", Pacif. J. Math. 38, (1971), 571-575.

3. R. F. Brown, "The Lefschetz fixed point theorem", Scott-Foreman and Co., Chicago, 1971.

4. R. F. Brown, "On some old problems of fixed-point theory", Rocky Mount. J. Math. 4, (1974), 3-14.

5. R. F. Brown, "The fixed point property and cartesian products", (to appear).

6. R. Brown, "Elements of modern topology", McGraw-Hill, London, 1968.

7. E. Fadell, "Recent results in the fixed point theory of continuous maps", Bull. Amer. Math. Soc. 76, (1970), 10-29.

8. E. Fadell, "Some examples in fixed point theory", Pacif. J. Math. 33, (1970), 89-100.

9. S. Y. Husseini, "The products of manifolds with the f.p.p. need not have the f.p.p.", Amer. J. Math. 99, (1977), 919-931.

10. W. Lopez, "An example in the fixed point theory of polyhedra", Bull. Amer. Math. Soc. 73, (1967), 922-924.

11. C. R. F. Maunder, "Algebraic topology", Van Nostrand, London, 1970.

12. E. Spanier, "Algebraic topology", McGraw-Hill, New York, 1966.

13. S. Thomeier, "Über eine Beziehung zwischen unstabilen und stabilen Homotopiegruppen von Sphären", Arch. Math. 15, (1964), 351-353.

14. S. Thomeier, "Einige Ergebnisse über Homotopiegruppen von Sphären", Math. Ann. 164, (1966), 225-250.

15. S. Thomeier, "Geometric constructions and the fixed point property", (to appear).

16. H. Toda, "Composition methods in homotopy groups of spheres", Ann. of Math. Stud. no. 49, Princeton University Press, Princeton, 1962.

A DUAL VIEW OF A THEOREM OF BAILLON

Barry Turett

Oakland University
Rochester, Michigan
U. S. A.

In [1], Baillon has shown that if X is a reflexive Banach space whose modulus of smoothness ρ_X satisfies $\lim_{\tau \downarrow 0} \frac{\rho_X(\tau)}{\tau} < \frac{1}{2}$, then X has the fixed point property for nonexpansive mappings; in particular, uniformly smooth Banach spaces have the fixed point property for nonexpansive mappings. Since the modulus of smoothness of a Banach space X is, in a certain sense, dual to the modulus of rotundity of the dual space X^* [11, p. 61] and since the modulus of rotundity is frequently used to obtain fixed points for nonexpansive and uniformly lipschitzian mappings ([2], [7]), it is tempting to prove Baillon's result using the modulus of rotundity of X^*. Yielding to this temptation results in a proof of Baillon's theorem that shows that if X is a Banach space satisfying the above condition on ρ_X, then both X and X^* are superreflexive Banach spaces with normal structure and thus both have the fixed point property for nonexpansive mappings.

Recall that the modulus of smoothness of a Banach space X is defined, for $\tau > 0$, by

$$\rho_X(\tau) = \sup \{\frac{\|x + \tau y\| + \|x - \tau y\| - 2}{2} : x, y \in S_X\}$$

where S_X denotes the unit sphere of X.

A Banach space X is said to be uniformly smooth if $\lim_{\tau \downarrow 0} \frac{\rho_X(\tau)}{\tau} = 0$. The modulus of rotundity of a Banach space X is defined by

$$\delta_X(\varepsilon) = \inf\{1 - \frac{\|x + y\|}{2} : x, y \in S_X, \|x - y\| \geq \varepsilon\}$$

where $0 \leq \varepsilon \leq 2$. A Banach space X is said to be uniformly rotund if $\delta_X(\varepsilon) > 0$ for each $\varepsilon > 0$. The characteristic of rotundity of a Banach space X, $\varepsilon_0(X)$, is then defined to be $\sup\{\varepsilon \in [0, 2] : \delta_X(\varepsilon) = 0\}$. It is clear that a Banach space X is uniformly rotund if and only if $\varepsilon_0(X) = 0$.

The first proposition, connecting the smoothness of the unit ball B_X of X with the rotundity of the unit ball B_{X^*} of X^*, is essentially classical. Šmulian [12] connected Fréchet differentiability of the norm of X with slices of the unit ball of X^*. More recently, Giles, Gregory and Sims [6] proved a refinement of Šmulian's result. The equivalence of conditions (a), (b) and (c) in the first proposition is merely a uniform version of their Lemma 2.1. Giles, Gregory and Sims noted that if $x \in S_X$ and $\varepsilon > 0$, then x determines a slice of B_{X^*} of diameter less than ε if and only if, for some $\delta(\varepsilon, x) > 0$,
$\sup\{\frac{\|x + \tau y\| + \|x - \tau y\| - 2}{\tau} : \|y\| = 1, 0 < \tau < \delta\} < \varepsilon$.

Before stating proposition 1, a few more definitions are needed. If x is a norm-one element in X and δ is a positive real number, then the x,δ-slice of B_{X^*}, denoted by $S(B_{X^*}, x, \delta)$, is the set $\{x^* \in B_{X^*} : x^*(x) > 1 - \delta\}$. The diameter of a set C is denoted by diam C. Finally, the coefficient $\kappa_0(X^*)$ which appears in condition (d) of proposition 1, is connected with the intersection balls in X^*. This coefficient was initially defined by Lifschitz [10] in his study of fixed points of uniformly lipschitzian mappings. Since this coefficient is not central to this paper, the reader is referred to Downing and Turett [4] for a proof that conditions (c) and (d) in proposition 1 are equivalent.

PROPOSITION 1. Let X be a Banach space. The following assertions are equivalent:

(a) $\lim_{\tau \downarrow 0} \frac{\rho_X(\tau)}{\tau} < \frac{1}{2}$.

(b) There exists $\varepsilon_0 > 0$ and $\delta_0 > 0$ such that diam $S(B_{X^*}, x, \delta_0)$

$\leq 1 - \varepsilon_0$ for every $x \in S_X$.

(c) $\varepsilon_0(X^*) < 1$.

(d) $\kappa_0(X^*) > 1$.

Proof. Suppose $\lim\limits_{\tau \downarrow 0} \frac{\rho_X(\tau)}{\tau} < \frac{1}{2}$. There exists $\varepsilon_0 > 0$ and $\tilde{\delta} > 0$ such that if $0 < \tau < \tilde{\delta}$ and x and y are norm-one elements in X, then

$$\frac{\|x + \tau y\| + \|x - \tau y\| - 2}{\tau} < 1 - \varepsilon_0. \qquad (*)$$

By Lemma 2.1 in [6], for each x in S_X, there exists $\tilde{\delta}(x) > 0$ with $\operatorname{diam} S(B_{X^*}, x, \tilde{\delta}(x)) < 1 - \varepsilon_0$. Let $\delta(x) = \sup\{\delta > 0 : \operatorname{diam} S(B_{X^*}, x, \delta) \leq 1 - \frac{1}{2}\varepsilon_0\}$. Note $\delta(x) \geq \tilde{\delta}(x) > 0$. Define $\delta_0 = \inf\{\delta(x) : x \in S_X\}$.

Suppose $\delta_0 = 0$. Then there exists $x_n \in S_X$ such that $\operatorname{diam} S(B_{X^*}, x_n, \frac{1}{n^2}) > 1 - \frac{1}{2}\varepsilon_0$. Choose x_n^* and y_n^* in $S(B_{X^*}, x_n, \frac{1}{n^2})$ satisfying $\|x_n^* - y_n^*\| > 1 - \frac{1}{2}\varepsilon_0$. Then choose y_n in S_X with $(x_n^* - y_n^*)(y_n) > 1 - \frac{1}{2}\varepsilon_0$. Then

$$\begin{aligned} &\|x_n + \tfrac{1}{n}y_n\| + \|x_n - \tfrac{1}{n}y_n\| \\ &\quad \geq x_n^*(x_n + \tfrac{1}{2}y_n) + y_n^*(x_n - \tfrac{1}{n}y_n) \\ &\quad > 2 - \tfrac{2}{n^2} + \tfrac{1}{n}(x_n^* - y_n^*)(y_n) \\ &\quad > 2 - \tfrac{2}{n^2} + \tfrac{1}{n}(1 - \tfrac{1}{2}\varepsilon_0). \end{aligned}$$

Therefore, for n large enough so that $\frac{1}{n} < \min\{\frac{\varepsilon_0}{4}, \tilde{\delta}\}$,

$$\frac{\|x_n + \frac{1}{n}y_n\| + \|x_n - \frac{1}{n}y_n\| - 2}{1/n} > 1 - \varepsilon_0.$$

Since this contradicts $(*)$, $\delta_0 > 0$. Thus, $\operatorname{diam} S(B_{X^*}, x, \delta_0) \leq \operatorname{diam} S(B_{X^*}, x, \delta(x)) < 1 - \varepsilon_0$ for each x in S_X. This completes the proof of (a) implies (b).

Suppose that (b) is satisfed and $\varepsilon_0(X^*) \geq 1$. Then, for each natural number n, $\delta_{X^*}(1 - \frac{1}{n}) = 0$. Therefore, for each n, there exists norm-one x_n^* and y_n^* in X^* with $\|x_n^* - y_n^*\| \geq 1 - \frac{1}{n}$ and

$\|\frac{1}{2}(x_n^* + y_n^*)\| \geq 1 - \frac{1}{2n}$. Choose x_n in S_X satisfying $x_n^*(x_n) + y_n^*(x_n) > 2 - \frac{2}{n}$ and note that $x_n^*(x_n) > 1 - \frac{2}{n}$ and $y_n^*(x_n) > 1 - \frac{2}{n}$. Therefore, $\operatorname{diam} S(B_{X^*}, x_n, \frac{2}{n}) \geq \|x_n^* - y_n^*\| \geq 1 - \frac{1}{n}$. But then $\sup_n \operatorname{diam} S(B_{X^*}, x_n, \frac{2}{n}) \geq 1$, a contradiction to condition (b). Therefore, $\varepsilon_0(X^*) < 1$ and the proof of (b) implies (c) is complete.

Suppose $\varepsilon_0(X^*) < 1$. Then there exists $\varepsilon > 0$ with $\delta = \delta_{X^*}(1 - \varepsilon) > 0$. Let x and y be norm-one elements in X and let (λ_n) be a sequence of positive real numbers converging to 0. Pick norm-one x_n^* and y_n^* in X^* such that $x_n^*\left(\frac{x + \lambda_n y}{\|x + \lambda_n y\|}\right) = 1$ and $y_n^*\left(\frac{x - \lambda_n y}{\|x - \lambda_n y\|}\right) = 1$. Then

$$\begin{aligned} \frac{\|x + \lambda_n y\| + \|x - \lambda_n y\| - 2}{\lambda_n} &\leq (x_n^* - y_n^*)(y) \\ &\leq \|x_n^* - y_n^*\|. \end{aligned} \qquad (**)$$

Choosing a natural number N so that $\left\|x - \frac{x + \lambda_n y}{\|x + \lambda_n y\|}\right\| < \delta$ and $\left\|x - \frac{x - \lambda_n y}{\|x - \lambda_n y\|}\right\| < \delta$ for $n \geq N$ yields

$$\begin{aligned} x_n^*(x) &= x_n^*\left(\frac{x + \lambda_n y}{\|x + \lambda_n y\|}\right) + x_n^*\left(x - \frac{x + \lambda_n y}{\|x + \lambda_n y\|}\right) \\ &\geq 1 - \left\|x - \frac{x + \lambda_n y}{\|x + \lambda_n y\|}\right\| \\ &> 1 - \delta. \end{aligned}$$

Similarly, for $n \geq N$, $y_n^*(x) > 1 - \delta$. Thus, both x_n^* and y_n^*, and hence $\frac{x_n^* + y_n^*}{2}$, are in $S(B_{X^*}, x, \delta)$ for $n \geq N$. Using the definition of $\delta_X(1 - \varepsilon)$, $\|x_n^* - y_n^*\| < 1 - \varepsilon$. It follows from (**) that $\lim_{\tau \downarrow 0} \frac{\rho_X(\tau)}{\tau} < \frac{1}{2}$. This completes the proof of (c) implies (a) and the proof of proposition 1.

One final comment concerning the proof of this proposition seems in order. The proof shows that if $\alpha > 0$, $\lim_{\tau\downarrow 0} \frac{\rho_X(\tau)}{\tau} < \alpha$ if and only if $\varepsilon_0(X^*) < 2\alpha$. So, if $\lim_{\tau\downarrow 0} \frac{\rho_X(\tau)}{\tau} = 0$, the above proof is a proof that X is uniformly smooth if and only if X^* is uniformly rotund.

These observations combine with several well-known facts about superreflexive Banach spaces to yield the following corollary.

COROLLARY 2. If X is a Banach space whose modulus of smoothness satisfies $\lim_{\tau\downarrow 0} \frac{\rho_X(\tau)}{\tau} < 1$, then both X and X^* are superreflexive.

<u>Proof</u>. Since $\lim_{\tau\downarrow 0} \frac{\rho_X(\tau)}{\tau} < 1$ is equivalent to $\varepsilon_0(X^*) < 2$, X^* is uniformly non-square [5, p. 188]. Thus X^* and X are superreflexive [5, p. 193].

We are now ready to prove the main result.

PROPOSITION 3. If X is a Banach space with $\varepsilon_0(X^*) < 1$, then both X and X^* are superreflexive Banach spaces with normal structure.

<u>Proof</u>. The superreflexivity of X and X^* is the conclusion of corollary 2 and the normal structure of X^* is known [3, p. 430]. Thus, we need only show that X has normal structure. So assume $\varepsilon_0(X^*) < 1$ but X fails to have normal structure.

Step 1. There exists a sequence (x_n) in B_X such that (x_n) converges weakly to 0, $(\|x_n\|)$ converges to 1, diam $\{x_1, x_2, \ldots\} = 1$, $x_1 = 0$, and dist $(x_n, \text{co}\{x_1,\ldots,x_{n-1}\}) > 1 - \frac{1}{n}$.

By a method of Brodskii-Milman (see [8, p. 676]) there exists a sequence $(x_n)_{n=2}^{\infty}$ satisfying $(x_n)_{n=2}^{\infty}$ converges weakly to 0, $(\|x_n\|)_{n=2}^{\infty}$ converges to 1, diam $\{x_2, x_3, \ldots\} = 1$, and dist $(x_n, \text{co}\{x_2,\ldots,x_{n-1}\}) > 1 - \frac{1}{n}$. Defining $x_1 = 0$ yields the desired sequence. The only condition which is not obvious is the last one. Since (x_n) tends weakly to 0, $\lim_{n\to\infty}$ dist $(0, \text{co}\{x_2,\ldots,x_{n-1}\}) = 0$. Choose $y_n \in \text{co}\{x_2,\ldots,x_{n-1}\}$ such that $(\|y_n\|)$ tends to 0. Then, for $\lambda_k \geq 0$ with $\sum_{k=1}^{n-1} \lambda_k = 1$,

$$\|x_n - \sum_{k=1}^{n-1} \lambda_k x_k\|$$

$$\geq \|x_n - (\lambda_1 y_n + \sum_{k=2}^{n-1} \lambda_k x_k)\| - \lambda_1 \|y_n\|$$

$$\geq \text{dist}\,(x_n, \text{co}\{x_2,\ldots,x_{n-1}\}) - \|y_n\|$$

$$\geq 1 - \frac{1}{n} - \|y_n\| .$$

Since $(\|y_n\|)$ converges to 0, $\lim_{n\to\infty} d(x_n, \text{co}\{x_1,\ldots,x_{n-1}\}) = 1$. Taking a subsequence of (x_n) if necessary completes the proof of step 1.

Step 2. For each $n \in N$, there exists $x_n^* \in S_{X^*}$ such that $1 \geq x_n^*(x_n) - x_n^*(x_j) > 1 - \frac{1}{n}$ for $j = 1,\ldots,n-1$.

Consider x_n and $\text{co}\{x_1,\ldots,x_{n-1}\}$. Let $z_n \in \text{co}\{x_1,\ldots,x_{n-1}\}$ such that $\|x_n - z_n\| = \text{dist}\,(x_n, \text{co}\{x_1,\ldots,x_{n-1}\})$. Applying the basic separation theorem to $\text{co}\{x_1,\ldots,x_{n-1}\}$ and $B(x_n, \|x_n - z_n\|)$ yields a norm-one x_n^* such that $x_n^*(x_n - z_n) = \|x_n - z_n\|$ and $\min x_n^*(B(x_n, \|x_n - z_n\|)) = x_n^*(z_n) = \max x_n^*(\text{co}\{x_1,\ldots,x_{n-1}\})$. Thus, for $j = 1,\ldots,n-1$,

$$1 - \frac{1}{n} < d(x_n, \text{co}\{x_1,\ldots,x_{n-1}\})$$

$$= \|x_n - z_n\|$$

$$= x_n^*(x_n) - x_n^*(z_n)$$

$$\leq x_n^*(x_n) - x_n^*(x_j)$$

$$\leq \|x_n - x_j\|$$

$$\leq 1.$$

Step 3. Given $\varepsilon > 0$, there exist $n_0, j_0 \in N$ with $n_0 > j_0$ such that $-x_{j_0}^*, x_{n_0}^* \in S(B_{X^*}, \frac{x_{n_0} - x_{j_0}}{\|x_{n_0} - x_{j_0}\|}, \varepsilon)$

and $\|x_{n_0}^* - (-x_{j_0}^*)\| > 1 - \varepsilon$.

Note that setting $j = 1$ in step 2 yields $x_n^*(x_n) > 1 - \frac{1}{n}$ for $n \geq 2$. Choose $j_0 \in N$ with $j_0 > 1$ and $\frac{1}{j_0} < \frac{\varepsilon}{2}$. Since $x_n \overset{w}{\to} 0$, there exists $n_0 \in N$ with $n_0 > j_0$ and $|x_{j_0}^*(x_{n_0})| < \frac{\varepsilon}{2}$. Then

$$
\begin{aligned}
-x_{j_0}^*\left(\frac{x_{n_0} - x_{j_0}}{\|x_{n_0} - x_{j_0}\|}\right) &\geq -x_{j_0}^*(x_{n_0} - x_{j_0}) \\
&= x_{j_0}^*(x_{j_0}) - x_{j_0}^*(x_{n_0}) \\
&> 1 - \frac{1}{j_0} - \frac{\varepsilon}{2} \\
&> 1 - \varepsilon
\end{aligned}
$$

and

$$
\begin{aligned}
x_{n_0}^*\left(\frac{x_{n_0} - x_{j_0}}{\|x_{n_0} - x_{j_0}\|}\right) &\geq x_{n_0}^*(x_{n_0} - x_{j_0}) \\
&> 1 - \frac{1}{n_0} \\
&> 1 - \varepsilon.
\end{aligned}
$$

Thus $-x_{j_0}^*$ and $x_{n_0}^*$ are in the desired slice of B_{X^*}. Now

$$
\begin{aligned}
\|x_{n_0}^* - (-x_{j_0}^*)\| &= \|x_{n_0}^* + x_{j_0}^*\| \\
&\geq (x_{n_0}^* + x_{j_0}^*)(x_{n_0}) \\
&> 1 - \frac{1}{n_0} - \frac{\varepsilon}{2} \\
&> 1 - \varepsilon.
\end{aligned}
$$

This completes the proof of step 3.

Step 3 implies that $\varepsilon_0(X^*) \geq 1$, a contradiction. Thus, if $\varepsilon_0(X^*) < 1$, X has normal structure. This completes the proof of proposition 3.

Baillon's theorem now follows immediately from propositions 1 and 3 and a well-known theorem of Kirk's [9].

COROLLARY 4. (Baillon). If X is a Banach space whose modulus of smoothness satisfies $\lim_{\tau \downarrow 0} \frac{\rho_X(\tau)}{\tau} < \frac{1}{2}$, then X has the fixed point property for nonexpansive mappings. In particular, uniformly smooth Banach spaces have the fixed point property for nonexpansive mappings.

REFERENCES

1. J. B. Baillon, "Quelques aspects de la theorie des points fixes dans les espaces de Banach I", Seminaire d'analyse fonctionnelle, École Polytechnique, 1978-1979.
2. F. E. Browder, "Nonexpansive nonlinear operators in a Banach space", Proc. Nat. Acad. Sci. U.S.A. 54 (1965), 1041-1044.
3. W. L. Bynum, "Normal structure coefficients for Banach spaces", Pacific J. Math. 86 (1980), 427-436.
4. D. J. Downing and B. Turett, "Some properties of the characteristic of convexity relating to fixed point theory", Pac. J. Math., to appear.
5. D. van Dulst, "Reflexive and superreflexive Banach spaces", Math. Centre Tracts 102, Amsterdam, 1978.
6. J. R. Giles, D. A. Gregory and B. Sims, "Characterization of normed linear spaced with Mazur's intersection property", Bull. Austral. Math. Soc. 18 (1978), 105-123.
7. K. Goebel, W. A. Kirk, and R. L. Thele, "Uniformly lipschitzian families of transformations in Banach spaces", Can. J. Math. 26 (1974), 1245-1256.
8. J. P. Gossez and E. Lami Dozo, "Structure normale et base de Schauder", Bull. Acad. Royale Belgique (5e ser.) 55 (1969), 673-681.
9. W. A. Kirk, "A fixed point theorem for mappings which do not increase distance", Amer. Math. Monthly 72 (1965), 1004-1006.
10. E. A. Lifschitz, "Fixed point theorems for operators in strongly convex spaces", Voronež Gos. Univ. Trudy Math. Fak. 16 (1975), 23-28, (Russian).
11. J. Lindenstrauss and L. Tzafrifi, "Classical Banach Spaces II", Ergebnisse der Math. und ihrer Grenzgebiete, Band 97, New York, 1979.
12. V. L. Šmulian, "Sur la derivabilite de la norme dans l'espace de Banach", Doklady (C. R. Acad. Sci. URRS) 27 (1940), 643-648.

SCHAUDER'S FIXED POINT THEOREM AND THE INVARIANT SUBSPACE PROBLEM*

B. S. Yadav

University of Delhi
Delhi
India

Introduction

One of the most outstanding, difficult and long open problem in operator theory is the so-called 'Invariant Subspace Problem' which asks the question: Does every bounded linear operator on an infinite-dimensional separable complex Hilbert space have a nontrivial invariant subspace? The problem which is so easy to state is, in fact, very complex in nature. Although it has drawn the attention of many mighty mathematicians and a lot of research work has been done over the decades on its ramifications, no solution of the problem so far seems to be in sight [48], [51].

The interest of Functional-Analysts in the study of invariant subspaces of linear operators was aroused by the celebrated results of Beurling [6] in which he characterized the lattice of invariant subspaces of the unilateral shift of multiplicity one on Hilbert space. Since then a vast amount of work has appeared regarding existence and determination of invariant subspaces of various types of operators. A beautiful account of the literature that centres around Beurling's theorem can be found in Helson [19] and Radjavi and Rosenthal [41].

An interesting aspect of the invariant subspace problem has been to establish existence of nontrivial invariant subspaces of compact operators.

*To appear in the 'Proceedings of International Conference on Non-linear Analysis and Applications' held at the Memorial University of Newfoundland, St. John's, Canada, June 1-3, 1981.

We shall discuss here the brilliant technique developed by a young Russian mathematician, Victor Lomonosov, to show that every non-zero compact operator has a nontrivial hyperinvariant subspace. Lomonosov's technique makes an ingenious use of Schauder's fixed point theorem which essentially belongs to the realm of linear analysis. His result has far reaching consequences and gives, in particular, a new insight into the general invariant subspace problem, the implications of which are being investigated currently by many operator-theorists.

To put Lomonosov's achievement in a proper perspective and thus highlight the role of Schauder's fixed point theorem in this development, we shall briefly survey in this article those parts of the invariant subspace theory which mainly concern the existence and characterization of invariant subspaces of compact operators. We work in the context of a Hilbert space, although many of the results that we discuss remain true for general Banach spaces.

1. Preliminaries

We denote by H an infinite-dimensional separable complex Hilbert space and by $B(H)$ the Banach algebra of all bounded linear operators on H. A closed subspace M of H is invariant under an operator T in $B(H)$ if $T(M) \subset M$, and hyperinvariant if M is invariant under every operator in $B(H)$ that commutes with T. M is nontrivial if $M \neq \{0\}, H$. If A is a subalgebra of $B(H)$, a subspace is invariant under A if it is invariant under every operator in A and A is called transitive if its only invariant subspaces are the trivial ones. The collection of all invariant subspaces of an operator T (algebra A) forms a complete lattice under the set inclusion as the partial order. We shall denote this lattice by Lat T (Lat A).

An operator T is called compact if it maps every bounded set on a relatively compact set, and polynomially compact if there is a polynomial p such that $p(T)$ is compact. T is quasi-nilpotent if its spectrum consists of zero only.

Let $\{e_n\}_{n=0}^{\infty}$ be an orthonormal basis of H and let $w = \{w_n\}_{n=0}^{\infty}$ be a bounded sequence of complex numbers. The operator S_w in $B(H)$ defined by $S_w e_o = 0$, and $S_w e_n = w_n e_{n-1}$ for $n > 0$ is called a weighted backward unilateral shift with the weight sequence w. If $w_n = 1$ for all n, S_w is simply the backward unilateral shift and is denoted by S. If $T_w e_n = w_{n+1} e_{n+1}$ for all $n \geq 0$, then T_w is called a unilateral forward weighted shift with the weight sequence w and the corresponding T for $w_n = 1$ for all n, is the forward unilateral shift. Weighted shifts were first introduced by Kelley [24] and have been extensively studied thereafter by many mathematicians. (See Shields [47].) It is easy to see that the adjoint of a backward unilateral shift (weighted or unweighted) is a forward unilateral shift and vice versa. Since for any operator T, the lattice of invariant subspace of its adjoint T^*, $\text{Lat } T^* = \{M^T : M \in \text{Lat } T\}$, one needs study only the invariant subspace lattices of shifts of any one type.

§2. Characterization Of Invariant Subspace Lattices

While the solution of the general invariant subspace problem remains elusive, there are operators whose invariant subspaces have been completely characterized. As already remarked Beurling [6] characterized the invariant subspaces of simple shifts. Another operator whose invariant subspaces have been fully known is the Volterra integral operator V on the Hilbert space $L^2(0, 1)$ defined by

$$V(f)(x) = \int_0^x f(t)dt, \quad 0 \leq x \leq 1.$$

V is a quasinilpotent compact operator [11]. Let us consider the subspaces

$$M_\alpha = \{f \in L^2(0, 1) : f = 0 \text{ a.e. on } [0, \alpha], 0 \leq \alpha \leq 1\}.$$

of $L^2(0, 1)$. Obviously each M_α is invariant under V. What is important is that these are the only invariant subspaces of V.

THEOREM. If V is the Volterra operator on $L^2(0, 1)$, then

$$\text{Lat } V = \{M_\alpha : \alpha \in [0, 1]\}.$$

The invariant subspaces of the Volterra operator were first obtained by

Dixmier [9] on the real space $L^2(0, 1)$. The invariant subspaces of V on the complex Hilbert space $L^2(0, 1)$ were characterized by Donoghue [10] and Brodskii [7] independently of each other. See also Kalisch [22], [23], Waterman [50], Sarason [46] and Ong [38].

There is yet another quasinilpotent compact operator whose invariant subspaces have been determined. A weighted backward unilateral shift S_w is called a Donoghue operator if the sequence $\{|w_n|\}_{n=1}^{\infty}$ is monotonically decreasing and square-summable.

THEOREM. If S_w is a Donoghue operator, then

$$\text{Lat } S_w = \{\{0\}, M_0', M_1' \ldots, M_n', \ldots, H\},$$

where

$$M_n' = \bigvee_{k=0}^{n} \{e_k\},$$

the closed subspace generated by $\{e_0, e_1, \ldots, e_n\}$. This theorem is due to Nikilskii [32]; see also Nikolskii [33] and Korenbljum [27]. The special case for which $w_n = 2^{-n}$ was earlier obtained by Donoghue [10]. Recently Yadav and Chatterjee [57] have proved a more general result. We shall sketch the proof of this result which is based on the techniques of the authors' earlier work [54], [55], [56]; see also Yadav and Chatterjee [58].

Consider the Hilbert space ℓ^2 of all square-summable complex sequences $x = \{x_0, x_1, x_2, \ldots\}$ with the norm defined by

$$\|x\| = (\sum_{m=0}^{\infty} |x_m|^2)^{1/2}.$$

If $w = \{w_m\}_{m=0}^{\infty}$ is a bounded sequence of complex numbers, then the weighted forward shift T_w with the weight sequence w on ℓ^2 appears as

$$T_w x = \{0, w_0 x_0, w_1 x_1, \ldots\}.$$

We shall assume, without any loss of generality, that the weights w_m

are positive real numbers; see [18 - Prob. 2] or [17]. We say that a sequence $\{w_m\}_{m=0}^{\infty}$ is of bounded variation if

$$\sum_{m=0}^{\infty} |w_m - w_{m+1}| < \infty.$$

It is easy to see that every monotonically decreasing sequence is a sequence of bounded variation but the converse is not true.

THEOREM. If $\{w_m\}_{m=0}^{\infty}$ is of bounded variation and

$$\delta = \sup_{m\geq 2,n} \sum_{k=0}^{\infty} \left(\frac{w_{k+m} \cdots w_{k+n}}{w_m w_{m+1} \cdots w_n}\right)^2 < \infty,$$

then

$$\text{Lat } T_w = \{\ell^2, M_o, M_1, \ldots, \{0\}\},$$

where $M_k = \{x \in \ell^2 : x_m = 0, m \leq k\}$.

Proof. Let $\{e_m\}_{m=0}^{\infty}$ be the standard orthonormal basis of ℓ^2 and let M be an invariant subspace of T_w. If a vector $x = \{x_m\}_{m=0}^{\infty}$ with $x_o \neq 0$ is in M, we claim that $M = \ell^2$. In fact, we have

$$\left\|\frac{T^n x}{x_o w_o w_1 \cdots w_{n-1}} - e_n\right\|^2$$

$$= \sum_{m=0}^{\infty} \left(\frac{w_{m+1} \cdots w_{m+n}}{w_o w_1 \cdots w_{n-1}}\right)^2 \left|\frac{x_{m+1}}{x_o}\right|^2$$

$$\leq \frac{w_n^2 \|x\|^2}{w_o^2 w_1^2 |x_o|^2} \sum_{m=0}^{\infty} \left(\frac{w_{m+2} \cdots w_{m+n}}{w_2 \cdots w_n}\right)^2 w_{m+1}^2$$

$$= \frac{w_n^2 \|x\|^2}{w_o^2 w_1^2 |x_o|^2} \sum_{m=0}^{\infty} \sum_{k=0}^{m} \left(\frac{w_{k+2} \cdots w_{k+n}}{w_2 \cdots w_n}\right)^2 (w_{m+1}^2 - w_{m+2}^2) \quad \text{(by Abel's transformation [59])}$$

$$\leq \frac{\delta\, w_n^2 \|x\|^2}{w_o^2 w_1^2 |x_o|^2} \sum_{m=0}^{\infty} (w_{m+1} + w_{m+2})(w_{m+1} - w_{m+2})$$

$$\leq \frac{2\mu\delta\, w_n^2 \|x\|^2}{w_o^2 w_1^2 |x_o|^2} \sum_{m=0}^{\infty} |w_{m+1} - w_{m+2}|$$

$$\leq C w_n^2,$$

where $\mu = \sup_n \{w_n\}$ and C is a constant. Since $\sum_{n=0}^{\infty} w_n^2 < \infty$, it follows by Paley-Wiener theorem [43; page 208] that the system

$$\left\{\frac{T^n x}{x_o w_o w_1 \ldots w_{n-1}}\right\}_{n=0}^{\infty}$$

is a Riesz basis in ℓ^2, and hence we have $M = \ell^2$. Now if $x_o = 0$ and k is the least positive integer such that $x_k \neq 0$, then we can similarly show that

$$\bigvee_{n=0}^{\infty} \{T^n x\} = M_k.$$

This shows that every cyclic subspace of T is an M_k. Since the span of any number of M_k's is again M_k, our theorem follows:

§3. Existence Of Invariant Subspaces

The pleasing situation that we have discussed in §2 regarding the structure of LatT holds only for a handful of operators. For a general operator T, the structure of LatT is too difficult to be described, particularly when it is not known if it contains at all elements other than the trivial ones. It follows via Fuglede's theorem that every non-scalar normal operator has a nontrivial hyperinvariant subspace. Even results on the existence of nontrivial invariant subspaces of various normal-like non-normal operators are known [41; Chapter 6]. Perhaps the most striking recent result in this direction is due to Scott Brown that every subnormal operator has a nontrivial invariant subspace [8]. The study of existence of invariant subspaces of compact operators has been since long an interesting area of research in operator theory. It all started with von Neumann (unpublished) and Aronszajn and Smith [1] who proved that every non-zero compact operator has a nontrivial invariant subspace. This result was extended to polynomially compact operators by Bernstein and Robinson [5] by using the techniques of non-standard analysis due to Robinson [44]. Halmos [16] translated their proof into standard analysis. By essentially chiselling the method of proof by Halmos, Arveson and Feldman [3] obtained the following more general result.

THEOREM. If T is a quasinilpotent operator such that the uniformally closed algebra generated by T contains a non-zero compact operator, then T has a nontrivial invariant subspace.

The path leading to this much progress has been all through rather difficult. However, Lomonosov [30] stunned operator-theorists by proving a much stronger result, and that too by using an entirely different technique. It follows as an easy corollary of his result that every non-zero compact operator has a non-trivial hyperinvariant subspace. Lomonosov makes an ingenious use of Schauder's fixed point theorem in proving his result. Not only that his result has superceded all previous significant results based on the von Neumann-Aronszajn-Smith-techniques, it has left a thought-provoking influence even on the general invariant subspace problem. We discuss here his result and some of its applications; see Pearcy [39] and Pearcy and Shields [40].

4. Lomonosov's Technique

Schauder's fixed point theorem states that every compact convex subset C of a Banach space is a fixed point space, i.e., if f is a continuous mapping of C into itself, then there exists a point x_0 in C such that $f(x_0) = x_0$ [13; p. 456]. By using Mazur's theorem that the closed convex hull of a compact set is also compact [3; p. 416], it can be shown that the conclusion of Schauder's fixed point theorem holds even if C is a closed (not necessarily compact) convex subset such that $f(C) \subset K$, where K is a compact subset of C. In fact, it is in this form that Lomonosov makes use of Schauder's theorem in the proof his main result:

<u>LOMONOSOV'S THEOREM</u>. If a non-scalar operator T in $B(H)$ commutes with a non-zero compact operator K, then T has a nontrivial hyperinvariant subspace. In particular, every non-zero compact operator has a nontrivial hyperinvariant subspace.

The part of Lomonosov's proof which makes a decisive use of Schauder's theorem is carved out in [40] in the form of the following lemma:

LOMONOSOV'S LEMMA. Let $\mathcal{A}$ be a transitive subslagebra of $B(H)$ and let K be any non-zero compact operator. Then there exists an operator T in $\mathcal{A}$ and a non-zero vector x in H such that $TKx = x$.

For a proof, we may and shall assume $\|K\| = 1$. Choose a vector x_o such that $\|Kx_o\| \geq 1$ which, in particular, means that $\|x_o\| > 1$. Let $S = \{x \in H : \|x - x_o\| \leq 1\}$ and $D = \overline{KS}$. Obviously, $0 \notin S$. Also, as $\|K\| = 1$ and $\|Kx_o\| > 1$, we see that $0 \notin D$. Now define

$$\mathcal{U}(T) = \{y : \|Ty - x_o\| < 1\},\ T \in \mathcal{A}.$$

For any $y \neq 0$, we observe that $\overline{[\mathcal{A}y]}$ is an invariant subspace of $\mathcal{A}$. Since $\mathcal{A}$ is transitive, $\overline{[\mathcal{A}y]} = H$. Thus for each $y \neq 0$, there is a $T \in \mathcal{A}$ such that $\|Ty - x_o\| < 1$ which means $y \in \mathcal{U}(T)$. Hence

$$D \subset H - \{0\} = \bigcup_{T \in \mathcal{A}} \mathcal{U}(T).$$

Since D is compact and each $\mathcal{U}(T)$ is open, there exists $T_1, T_2 \ldots T_n$ in $\mathcal{A}$ such that

$$D \subset \bigcup_{n=1}^{\infty} \mathcal{U}(T).$$

Now for $y \in D$ and $i = 1, 2\ldots, n$, define

$$\alpha_i(y) = \max\{0,\ 1 - \|T_i y - x_o\|\}.$$

For each i, α_i is continuous and $0 \leq \alpha_i(y) \leq 1$. Also for each $y \in D$, there is an i such that $y \in \mathcal{U}(T_i)$, and hence

$$\sum_{i=1}^{n} \alpha_i(y) > 0 \quad \text{for all} \quad y \in D.$$

Thus we can define, for $y \in D$ and $i = 1, 2, \ldots, n$,

$$\beta_i(y) = \alpha_i(y) / \sum_{i=1}^{n} \alpha_i(y).$$

Again, for each i, β_i is continuous, $0 \leq \beta_i(y) \leq 1$, and

$$\sum_{i=1}^{n} \beta_i(y) = 1 \quad \text{for all} \quad y \in D.$$

Finally, we define a function $\psi : D \to H$ by

$$\psi(y) = \sum_{i=1}^{n} \beta_i(y) T_i y, \; y \in D.$$

It is easy to verify that ψ is continuous. Also, as for every $x \in S$,

$$\sum_{i=1}^{n} \beta_i(Kx) x_0 = x_0,$$

we have

$$\|\psi(kx) - x_0\| \leq \sum_{i=1}^{n} \beta_i(Kx) \|T_i Kx - x_0\| .$$

Since $\alpha_i(Kx) = 0$ whenever $\|T_i Kx - x_0\| \geq 1$, it follows that

$$\|(Kx) - x_0\| \leq \sum_{i=1}^{n} \beta_i(Kx) = 1.$$

Thus $\psi(KS) \subset S$, and hence $\psi(D) \subset S$. Now putting $\Phi(x) = \psi(Kx)$, $x \in S$, we see that Φ is a continuous function of S into itself with $\overline{\Phi(S)}$

as a compact subset of S. Therefore, by the extended version of Schauder's fixed point theorem stated above, there exists a point $x_1 \in S$ such that $\Phi(x_1) = x_1$. Now define an operator T by

$$T = \sum_{i=1}^{n} \beta_i(Kx_1)T_i.$$

Then $T \in \mathcal{A}$ and $TKx_1 = \psi(Kx_1) = \Phi(x_1) = x_1$. This completes the proof of the lemma.

To prove the theorem, let $\mathcal{A}$ denote the commutant of T. We must show that $\mathcal{A}$ is not transitive. Suppose $\mathcal{A}$ is transitive. Then by Lomonosov's lemma, there exists an operator $A \in \mathcal{A}$ and a vector x such that $AKx = x$. Thus the compact operator AK has 1 as an eigenvalue. Let E be the associated finite-dimensional eigenspace. Since T commutes with AK, we have $AK(Tx) = T(AKx) = Tx$ for all $x \in E$, and hence $Tx \in E$. Thus T maps E into itself. It follows that T has an eigenvalue. If M is the associated eigenspace, then $M \neq H$ as T is non-scalar. This shows that M is a nontrivial invariant subspace of $\mathcal{A}$, a contradiction.

A number of remarks are in order about Lomonosov's theorem. Firstly, the theorem is based on a brilliant new technique absolutely unrelated with the line of action hitherto followed by operator-theorists working in this area. Secondly, it gives a result much stronger than the earlier best known result that every polynomially-compact operator has a nontrivial invariant subspace. Thirdly, although it is natural to think of replacing the use of Schauder's fixed point theorem in the proof of Lomonosov's lemma, say by the Banach's contraction principle, or to prove the lemma without using any fixed point theorem at all, nobody has so far suceeded in giving an alternative proof of the theorem. However, a simple proof of the special case that every non-zero compact operator has a nontrivial hyperinvariant subspace has been given by M. Hildon without using any fixed point theorem; see Michaels [31]. He starts exactly as in the proof of Lomonosov's lemma, but avoids the use of Schauder's theorem by assuming the compact operator to be quasinilpotent which can be done without any loss of generality. As a matter of fact, if a non-zero compact operator is not quasinilpotent, then it must have at least one non-zero eigenvalue, and the corresponding eigenspace as a nontrivial hyperinvariant subspace.

Lastly we remark that a number of extensions and applications of Lomonosov's work have been obtained in various directions; see, for example, [15], [25], [26], [35], [37], [42] and [45]. However, to what extent these efforts lead to a solution of the general invariant subspace problem for which Lomonosov's theorem seems to have given a great impetus remains anybody's guess.

5. The Transitive Algebra Problem

One of the important applications of Lomonosov's lemma is to obtain a partial solution of another long open problem in operator theory, the so-called 'transitive algebra problem': Is every transitive subslgebra of B(H) weakly dense in B(H)? The problem was first raised by Kadison [21] and continues to be still unsolved. However, partial solutions of the problem have been obtained by many mathematicians in the sense that if a transitive algebra satisfies certain additional condition, then it is weakly dense in B(H).

The theory of transitive operator algebras was actually initiated by Arveson [2] who obtained a number of partial solutions of the transitive algebra problem. In case H is finite-dimensional, the fact that every transitive algebra of B(H) coincides with B(H) is the classical theorem of Burnside [20, p. 276]. In the general case, Arveson [2] showed that every transitive algebra containing a maximal abelian self-adjoint algebra or a unilateral shift is weakly dense in B(H). For generalizations of these results, we refer to Douglas and Pearcy [12], Nordgren [34] and Nordgren, Radjavi and Rosenthal [36]. In particular, it is shown in [36] that every transitive algebra containing either a non-zero operator of finite rank or a Donoghue operator - see every such operator is compact - is weakly dense in B(H). Recently, Yadav and Chatterjee [53] have generalized the last result by showing that a transitive algebra containing a certain type of weighted shifts with weight sequence more general than that of a Donoghue operator is weakly dense in B(H). In yet another paper [55], they have shown that if a transitive algebra $\mathcal{A}$ contains a weighted backward shift with weights $w_m = \|A^m x\| / \|A^{m-1}x\|$, where x is a non-zero vector and A is a contraction of class C_1. [49, p. 27], then $\mathcal{A}$ is weakly dense in B(H). See also Yadav and Bansal [52]. These results assume significance in the light of the conjecture that a transitive algebra containing any weighted shift is weakly dense in $B(H)$. All these partial solutions are based on the work of Arveson [2]. However, there are partial solutions independent of Arveson's work. One such solution was given by Lambert [28] who proved that a transitive algebra containing a strictly cyclic algebra is weakly dense in $B(H)$; see also Embry [14]. We shall discuss here the partial solution given by Lomonosov [30]; see also [4].

THEOREM. If a transitive algebra A contains a non-zero compact operator K, then A is weakly dense in $B(H)$.

Proof. Since the closure of A in the weak topology coincides with the closure of A in the strong topology [41; p. 118], we may assume that A is strongly, and hence uniformly, closed. Now, by Lomonosov's lemma, there is an operator $T \in A$ such that the compact operator $K_1 = TK$ has 1 as an eigenvalue. Since $\sigma(K_1)$ is at most denumerable with zero as the only possible limit point, let G_1 and G_2 be two disjoint open sets such that $1 \in G_1$ and $\sigma(K_1) - \{1\} \subset G_2$ and let f be an analytic function defined on $G_1 \cup G_2$ such that it equals 1 on G_1 and vanishes on G_2. Then the operator $f(K_1)$ defined by the Riesz functional calculus [13; Chapter VIII] is an idempotent operator with its range as the finite-dimensional eigenspace of K_1 corresponding to the eigenvalue 1. Now we invoke Lavrentiev's theorem [29] to approximate $f(z)$ by polynomials $p(z)$ subnormally on $G_1 \cup G_2$. Since A is uniformly closed, we conclude that $f(K_1) \in A$. Again as $f(K_1)$ is finite-dimensional, it follows from [36; Cor. 2] that A is weakly dense in $B(H)$.

The transitive algebra problem is intimately related to the invariant subspace problem. In fact, it can be quickly seen [41; p. 138] that an affirmative answer to the transitive algebra problem implies that every operator which is not a multiple of the identity has a nontrivial hyperinvariant subspace. It is not surprising, therefore, that Lomonosov's theorem leaves via compact operators a great impact even on the genral invariant subspace problem.

§6. The Invariant Subspace Problem

We now discuss the insight provided by Pearcy [39] and Pearcy and Shields

[40] into the status in which the invariant subspace problem has been left by Lomonosov's theorem. If two operators A and B in B(H) commute, we write $A \leftrightarrow B$. Denote by $\mathcal{K}$ the two-sided ideal of compact operators, and by $\mathcal{A}$ the class of all operators $A \neq \lambda I$ in B(H). Again let $\mathcal{B}$ denote the class of all those operators B in B(H) which satisfy the relation: $K \leftrightarrow A \leftrightarrow B$ for some $K \neq 0$ in $\mathcal{K}$ and some A in $\mathcal{A}$. The question is: Does $\mathcal{B}$ coincide with B(H)? The affirmative answer to this question would imply an affirmative answer to the invariant subspace problem. In fact, if $\mathcal{B}$ = B(H), then for every operator B in B(H), there exists a $K \neq 0$ in $\mathcal{K}$ and an A in $\mathcal{A}$ such that $K \leftrightarrow A \leftrightarrow B$. By Lomonosov's theorem, A will have a nontrivial hyperinvariant subspace which in turn will serve as a nontrivial invariant subspace of B. If $\mathcal{B} \neq$ B(H), the problem reduces to finding out an example of an operator B which does not commute with any operator $A \neq \lambda I$ which commutes with a non-zero compact operator.

Lomonosov's theorem can be restated as follows: For an operator $A \neq \lambda I$, let $K \neq 0$ be in K such that the commutator AK-KA is of rank zero. Then A has a nontrivial hyperinvariant subspace. Kim, Pearcy and Shields [25] have shown that if A is any operator in B(H) such that for a compact operator K, the commutator AK - KA is of rank one, then A has a nontrivial hyperinvariant subspace. See also Kim, Pearcy and Shields [26]. Thus, in order to show that the invariant subspace problem does not have an affirmative answer, one must show that there exists an operator A such that for no compact operator K, the commutators AK - KA is of rank one.

Lastly, we remark that most of the mathematicians working in the invariant subspace theory have a strong feeling that there should exist an operator in B(H) with no invariant subspaces other than the two trivial ones. In the light of Lomonosov's work, it seems that either this conjecture turns out to be eventually true, or else every operator which is not a multiple of the identity will have not only a nontrivial invariant subspace but a nontrivial hyperinvariant subspace.

REFERENCES

1. N. Aronszajn and K. T. Smith, "Invariant subspaces of completely continuous operators", Ann. of Math., 60 (1954), 345-350.
2. W. B. Arveson, "A density theorem for operator algebras", Duke Math. J., 34 (1967), 635-647.

3. W. B. Arveson and J. Felman, "A note on invariant subspaces", Michigan Math. J., 15 (1968), 60-64.

4. E. Azoff, "Compact operators in reductive algebras", Canadian J. Math. 27 (1975), 152-154.

5. A. R. Bernstein and A. Robinson, "Solution of an invariant subspace problem of K. T. Smith and P. R. Halmos", Pacific J. Math., 16 (1966), 421-431.

6. A. Beurling, "On two problems concerning linear transformations in Hilbert space", Acta Math., 81 (1949), 239-255.

7. M. S. Brodskii, "On a problem of I. M. Gelfand", Uspehi Mat. Nauk., 12 (1957), 129-132.

8. Scott W. Brown, "Some invariant subspaces for subnormal operators", Integ. Equa. Operator Theory, 1/3 (1978), 310-333.

9. J. Dixmier, "Les opêrateurs permutables à l'operatéur integral", Fas. 2, Portugal Math., 8 (1949), 73-84.

10. W. F. Donoghue, "The lattice of invariant subspaces of a completely continuous quasinilpotent transformation", Pacific J. Math., 7 (1957), 1031-1035.

11. R. G. Douglas, "Banach algebra techniques in operator theory", Academic Press, New York, 1972.

12. R. G. Douglas and C. Pearcy, "Hyperinvariant subspaces and transitive algebras", Michigan Math., J., 19 (1972), 1-12.

13. N. Dunford and J. T. Schwartz, "Linear operators, Part I", General Theory, Interscience, New York, 1957.

14. Mary R. Embry, "Strictly cyclic operator algebras on a Banach space", Pacific J. Math., 45 (1973), 443-452.

15. C. K. Fong, E. A. Nordgren, M. Radjabalipour, H. Radjavi and P. Rosenthal, "Extensions of Lomonosov's invariant subspace theorem", Acta Sci. Math. (Szeged), 41 (1979), 55-62.

16. P. R. Halmos, "Invariant subspaces of polynomially compact operators", Pacific J. Math., 16 (1966), 433-437.

17. P. R. Halmos, "A Hilbert space problem book", Van Nostrand, Princeton, 1967.

18. P. R. Halmos, "Ten problems in Hilbert space", Bull. Amer. Math. Soc., 76 (1970), 887-933.

19. H. Helson, "Lectures on invariant subspaces", Academic Press, New York, 1964.

20. N. Jacobson, "Lectures in abstract algebra II", Linear Algebra, Van Nostrand, Princeton, 1953.

21. R. V. Kadison, "On the orghogonalization of operator representations", Amer. J. Math., 78 (1955), 600-621.

22. G. K. Kalisch, "On similarity, reducing maniforlds, and unitary equivalence of certain Volterra operators", Ann. of Math., 66 (1957), 481-494.

23. G. K. Kalisch, "A functional analytic proof of Titchmarsh's convolution theorem", J. Math. Anal. Appl., 5 (1962), 176-183.

24. R. L. Kelley, "Weighted shifts on Hilbert space", Ph.D. Dissertation, University of Michigan, Ann Arbor, 1966.

25. H. W. Kim, C. Pearcy and A. L. Shields, "Rank-one commutators and hyperinvariant subspaces", Michigan Math. J., 22 (1975), 193-194.

26. H. W. Kim, C. Pearcy and A. L. Shields, "Sufficient conditions for rank-one commutators and hyperinvariant subspaces", Michigan Math. J., 22 (1976), 235-243.

27. B. I. Korenbljum, "Invariant subspaces of the shift operator in certain weighted Hilbert spaces of sequences", Mat. Sbornik, 39 (131) (1972), 110-137 = Math. USSR Sbornik, 18 (1972), 111-131 = Soviet Math. Dokl., 13 (1972), 272-275.

28. A. Lambert, "Strictly cyclic operator algebras", Pacific J. Math., 39 (1971), 717-726.

29. M. Lavrentiev, "Sur les fonctions d'une variable complexe representables par des séries de polynomes", Actualités, Sci., Indust., no. 441, Hermann, Paris, 1936.

30. V. J. Lomonosov, "On invariant subspaces of families of operators commuting with a completely continuous operator", Funkcional. Anal.i Prilozen, 7 (1973), 55-56 (Russian).

31. A. J. Michaels, "Hildon's simple proof of Lomonosov's invariant subspace theorem", Advances in Mathematics, 25 (1977), 56-58.

32. N. K. Nikolskii, "Invariant subspaces of certain completely continuous operators", Vestnik Leningrad Univ. (Math. 1), 7 (1965), 68-77, (Russian).

33. N. K. Nikolskii, "Invariant subspaces of weighted shift operators", Math. USSR-Sb., 3 (1967), 159-176.

34. E. A. Nordgren, "Transitive operator algebras", J. Math. Anal. Appl., 32 (1970), 639-643.

35. E. A. Nordgren, M. Radjabalipour, H. Radjavi and P. Rosenthal, "Algebras intertwining compact operators", Acta Sci. Math. (Szeged), 39 (1977), 115-119.

36. E. A. Nordgren, H. Radjavi and P. Rosenthal, "On density of transitive algebras", Acta Sci. Math. (Szeged), 30 (1969), 175-179.

37. E. A. Nordgren and P. Rosenthal, "Algebras containing unilateral shifts of finite-rank operators", Duke Math. J., 40 (1973), 419-424.

38. Boon-Hua Ong, "Invariant subspace lattices for a class of operators", Pacific J. Math. 94 (1981), 385-405.

39. C. Pearcy, "Some recent developments in operator theory", CBMS, Amer. Math. Soc., Providence, Rhode Island, 1977.

40. C. Pearcy and A. L. Shields, "A survey of Lomonosov technique in the theory of invariant subspaces in: Topics in Operator Theory", Amer. Math. Soc. Surveys, No. 13 (1974), 220-229.

41. H. Radjavi and P. Rosenthal, "Invariant subspaces", Springer-Verlag, New York, 1973.

42. M. Radjabalipour and P. Rosenthal, "Compact operator ranges and transitive algebras", J. London Math. Soc., 17 (1978), 522-524.

43. F. Riesz and B. Sz. Nagy, "Functional Analysis", Ungar. New York, 1955.

44. A. Robinson, "Non-standard analysis", North Holland, Amsterdam, 1966.

45. P. Rosenthal, "Applications of Lomonosov's lemma to non-selfadjoint operator algebras", Spectral Theory Symposium, Proc. Royal Irish Acad. Sect. A., 74 (1974), 271-281.

46. D. Sarason, "Invariant subspaces, In: Topics in Operator Theory", Amer. Math. Soc. Surveys, No. 13 (1974), 1-47.

47. A. L. Shields, "Weighted shift operators and analytic function theory, In: Topics in Operator Theory", Amer. Math. Soc. Surveys, No. 13 (1974), 49-128.

48. U. N. Singh, "The invariant subspace problem and related questions", Presidential Address, 43rd Conf. On Indian Math. Soc., Aligarh, December 1977.

49. B. Sz. Nagy and C. Foias, "Harmonic analysis of operators on Hilbert space", Akademiai Kiado, Budapest, 1970.

50. R. E. Waterman, "Invariant subspaces, similarity and isometric equivalence of certain commuting operators in L_p", Pacific J. Math., 48 (1973), 593-613.

51. B. S. Yadav, "Recent developments in Hilbert space operator theory", Nieuw Arch. Voor Wisk. (3), XXVII (1979), 398-406.

52. B. S. Yadav and R. Bansal, "On systems of N-variable weighted shifts", (to appear).

53. B. S. Yadav and S. Chatterjee, "On a partial solution of the transitive algebra problem", Acta Sci. Math. (Szeged), 42 (1980), 211-215.

54. B. S. Yadav and S. Chatterjee, "Cyclic vectors of weighted shifts on ℓ^p spaces", Proc. Amer. Math. Soc., 80 (1980), 95-99.

55. B. S. Yadav and S. Chatterjee, "Cyclic vectors of Lambert's weighted shifts", Proc. Amer. Math. Soc., 80 (1980), 100-104.

56. B. S. Yadav and S. Chatterjee, "Invariant subspaces of weighted shifts", Glasnik Math., 15 (35) (1980), 321-325.

57. B. S. Yadav and S. Chatterjee, "On a characterization of invariant subspace lattices of weighted shifts", Proc. Amer. Math. Soc. (to appear).

58. B. S. Yadav and S. Chatterjee, "Invariant subspace lattices of Lambert's weighted shifts", J. Austral. Math. Soc. (to appear).

59. A. Zygmund, "Trigonometric series", 2nd Ed., Vol. I, Cambridge Univ. Press, New York, 1959.

APPLICATIONS

ON FRACTIONAL CALCULUS AND ITS APPLICATIONS TO THE THEORY OF ORDINARY DIFFERENTIAL EQUATIONS OF GENERALIZED ORDER

M. A. Al-Bassam

University of Kuwait
Kuwait

INTRODUCTION

Historical Resume

The concept of fractional calculus is as old as that of the integral on the classical calculus. At the time Leibnitz adopted the notation $\frac{d^n y}{dx^n}$ for the nth order derivative, L'Hospital asked "what if $n = \frac{1}{2}$?" Leibnitz [1] in a letter dated September 30, 1965, from Hannover to L'Hospital wrote: "$d^{\frac{1}{2}}x$ will be equal to $x\sqrt{dx:x}$, an apparent paradox from which one day useful consequences will be drawn". In 1730 Euler [2] suggested that the fractional derivative "may be expedited with the help of the interpolation of series" as he explained earlier in his work. Laplace [3], in 1812, used some expressions containing fractional derivatives in his work on the analytic theory of probability. In 1819 Lacrois [4] generalized the integral n^{th} derivative of $D_x^n x^m$ by induction and found that $D_x^{\frac{1}{2}} x = 2\sqrt{\frac{x}{\pi}}$.

The first major study of fractional calculus was published in a memoire by Liouville in 1832 [(5], [6]) in which he assumed the series representation $f(x) = \sum_{n=0}^{\infty} C_n e^{a_n x}$ and defined the derivative of an arbitrary order $D^{\nu} f(x) = \sum_{n=0}^{\infty} C_n a_n^{\nu} e^{a_n x}$. However, this definition has the disadvantage that it is valid only for values of ν for which the series converges.

In 1847, Riemann [7] used a generalization of Taylor's series expansion and derived a definition of fractional derivative in an integral form. It may be pointed out that the work of Riemann was done in 1847 but was not published until 1876. Prior to the publication of Riemann's work, two important papers appeared. Holmgren [8] in 1865 gave a more general and useful definition to the concept of generalized derivative in four different forms, each of which is obtained by applying a linear fractional transformation to the form

$$D_x^{\mu} f(x) = \frac{1}{\Gamma(m-\mu)} D_x^m (x-x_o)^{m-\mu} \int_o^1 (1-\mu)^{m-\mu-1} f[x_o + (x-x_o)u]\,du.$$

Grünwald [9], (1967), made use of a definition of generalized derivative as the limit of a difference quotient and obtained an integral form similar to that of Holmgren except that Holmgren introduced the derivative in his definition.

On the basis of Grünwald definition, E. L. Post [13] developed a definition for a functional operator of generalized differentiation and obtained some of its properties.

Since then, the definite integral form of Riemann-Holmgren-Grünwald has become a subject of extensive study by many authors (for details see [32], [33]), just to mention but a few, Hardy and Littlewood [11] (1917-1928) obtained some properties of the integral of fractional order by imposing different assumptions on the integrand function such as L^p and Lipschitzian functions. However, H. Weyl [10] was the first who studied the theory of the integral under the last condition. M. Riesz [15] (1937) showed some properties of the integral of fractional order which is a generalization of the Riemann form to more than one dimension. He has given the definition

$$I^{\alpha} f(P) = \frac{1}{H_m(\alpha)} \int_{D_s^P} f(Q)\, r_{PQ}^{\alpha-m}\, dQ,$$

where α is real and $> m-2$, $H_m(\alpha) = \Pi^{\frac{m-2}{2}}\, 2^{\alpha-1}\, \Gamma(\frac{\alpha}{2})\Gamma(\frac{\alpha+2-m}{2})$ and

$r_{PQ} = \sqrt{(x_1 - t_1)^2 - (x_2 - t_2)^2 - \ldots - (x_m - t_m)^2}$, and D_s^P is the domain bounded by the aft cone $D^P : r_{PQ} = 0$ with $x_1 - t_1 > 0$ and the surface S is the hyperspace of $m - 1$ dimensions such that the generators of the cone intersect this surface in one and only one point. He also introduced the form

$$I^\alpha f(P) = \frac{1}{H_m(\alpha)} \int_\Omega f(q)\, r_{PQ}^{\alpha - m}\, dQ,$$

where

$$H_m(\alpha) = \frac{\pi^{\frac{m}{2}} 2^\alpha \Gamma(\frac{\alpha}{2})}{\Gamma(\frac{m-\alpha}{2})},$$

and r_{PQ} is the Euclidean distance between the two points $P(x_1, x_2, \ldots, x_m)$ and $Q(t_1, t_2, \ldots, t_m)$ and Ω is the hyperspace of m dimensions. He used the properties of these integrals to solve the Cauchy problem for the hyperbolic linear partial differential equations and to study problems in potential theory. Also, Copson [17] (1943) and Fremberg [18], (1946), used this integral to mark further developments with regard to the solution of a certain type of partial differential equations.

It is to be noted that many authors have extended the definition of Cauchy integral to fractional complex integral and obtained interesting properties (for details and references see [32], [33], [34].

The author has shown the equivalence between the forms of Holmgren and Riesz [19], and consequently a modified and extended definition has been adopted. Based on this definition many interesting properties have been obtained. Also, he has applied these properties of the integral, which he called "H-R transform" as to its relation to Holmgren and Riesz, to some existence theorems of ordinary differential equations and to the development of methods of solving differential equations through H-R transform equations, [see research papers listed in the References]. Also, the author has extended the definition of the integral into a two dimensional form and applied it to the study of solutions of certain types of partial differential equations and boundary value problems [22], ([32], pp. 91-105). Mambriani (1940) [16], depending on Holmgren work, applied the generalized derivative operators to solve certain types of differential equations. He transformed the hypergeometric differential equations into some forms of differential equations which, in a similar work, this writer called them "Transform Equations". H. T. Davis [14] (1936) applied the fractional calculus to differential equations. In fact, there are so many authors who, either have contributed directly to fractional calculus or through its application to other fields [32], [33], [34].

Contents And Notation

This paper is mainly based on the work of the author and consists of three parts. The first part deals with the definition of the operator of generalized order and some properties of fractional calculus. The second part contains some existence theorems of ordinary differential equations of generalized order. The last part includes some existence theorems of series solutions of certain types of ordinary differential equations of generalized order.

Throughout the work m and n will be used to represent positive integers or zero. "Class $C^{(n)}$" will be used to mean the class of functions with continuous n^{th} derivative. "$R\alpha$" and "$\mathcal{F}\alpha$" will denote the real and imaginary parts of the complex number α, respectively. Other notations such as "$D^n_x f$" or "$f^{(n)}(x)$" and $\Gamma(x)$ represent the nth derivative of f and the Gamma function, respectively. (C, Lip) is the set of continuous functions which satisfy Lipschitz condition.

I. SOME PROPERTIES OF THE FRACTIONAL CALCULUS

DEFINITION 1. If $f(x)$ is a real valued function of class $C^{(n)}$, i.e. $f \in C^{(n)}$ on the interval $a \leqq {x \atop t} \leqq b$ and $R\alpha + n > 0$, then

$$\text{(H)}\quad {}_a^x I^{\alpha} f = \frac{1}{\Gamma(\alpha+n)} D^n_x \int_a^x (x-t)^{\alpha+n-1} f(t)\,dt, \quad (n = 0, 1, 2, \ldots)$$

or

$$(R) = \sum_{i=0}^{n-1} \frac{f^{(i)}(a)(x-a)^{\alpha+i}}{\Gamma(\alpha+i+1)} + \frac{1}{\Gamma(\alpha+n)} \int_a^x (x-t)^{\alpha+n-1} f^{(n)}(t)\,dt,$$

$(n = 1, 2, \ldots)$.

It is clear that (H) can be written as

$$(H_1) \quad {}_a^x I^{\alpha} f = D_x^m \, {}_a^x I^{\alpha+m} f.$$

1. The Identity and Derivative Properties of ${}_a^x I^{\alpha}$

THEOREM 1.1. (i) If $f \in C^{(m+n)}$ on $[a, b]$ and $R\alpha + m > 0$, then

$$D_x^n \, {}_a^x I^{\alpha} f = {}_a^x I^{\alpha-n} f$$

(ii) If $f(x)$ is of class $C^{(0)}$ on $[a, b]$, then

$$\lim_{\alpha \to 0^+} {}_a^x I^{\alpha} f = f(x),$$

for every $x \in [a, b]$.

COROLLARY. If $f(x)$ is a function of class $C^{(n)}$ on $[a, b]$, then

$${}_a^x I^{-n} f = D_x^n f = f^{(n)}(x).$$

THEOREM 1.2. (i) If $f(x)$ is of class $C^{(0)}$ on $c \leqq x \lessgtr a \leqq b$ and $R\alpha > 0$, then

$$\lim_{x \to a} {}_a^x I^{\alpha} f = 0$$

(ii) If $f \in C^{(n)}$ on (a, b) and $R\alpha + n > 0$, then the above equality holds if $f^{(i)}(a) = 0$ $(i = 0, 1, 2, \ldots, n-1)$, otherwise the limit does not exist.

2. Other Properties

THEOREM 1.3. (i) If $f \in C^{(n+1)}$ and $g \in C^{(0)}$ on (a, b) and if $R\alpha > 0$, then

$$\underset{a}{\overset{x}{I}}{}^{\alpha} fg = \sum_{i=0}^{n} \binom{-\alpha}{i} f^{(i)}(x) \underset{a}{\overset{x}{I}}{}^{\alpha+i} g$$

$$+ \frac{(-1)^{n+1}}{\Gamma(\alpha)} \int_a^x f^{(n+1)}(t) [\underset{a}{\overset{t}{I}}{}^{n+1} g(t)(x - t)^{\alpha-1}] \, dt.$$

(ii) If $f(x) \in C^{(m+n-1)}$ and $g(x) \in C^{(m)}$ on (a, b) and $m - R\beta > 0$, then

$$\underset{a}{\overset{x}{I}}{}^{-\beta} fg = \sum_{p=0}^{n} \binom{\beta}{p} f^{(p)}(x) \underset{a}{\overset{x}{I}}{}^{-\beta+p} g$$

$$+ \sum_{p=n+1}^{m+n} = \sum_{i=0}^{n} \binom{m}{p-i} \binom{\beta-m}{i} f^{(p)}(x) \underset{a}{\overset{x}{I}}{}^{p-\beta} g$$

$$+ \frac{(-1)^{n+1}}{\Gamma(m-\beta)} D_x^m \int_a^x f^{(n+1)}(t) [\underset{a}{\overset{t}{I}}{}^{n+1} (x - t)^{m-\beta-1} g(t)] \, dt.$$

COROLLARY. If $f(x)$ is a polynomial of degree n, then for $R\alpha > 0$, $R\beta > 0$,

$$\underset{a}{\overset{x}{I}}{}^{\alpha} fg = \sum_{i=0}^{n} \binom{-\alpha}{i} f^{(i)}(x) \underset{a}{\overset{x}{I}}{}^{\alpha+i} g$$

$$\underset{a}{\overset{x}{I}}{}^{-\beta} fg = \sum_{p=0}^{n} \binom{\beta}{p} f^{(p)}(x) \underset{a}{\overset{x}{I}}{}^{-\beta+p} g,$$

where $\binom{\alpha}{m} = \dfrac{\Gamma(\alpha+1)}{\Gamma(m+1)\Gamma(\alpha-m+1)}$.

THEOREM 1.4. If $f \in C^{(n)}$ on $[a, b]$ and $0 < q < 1$, then

$$\overline{\left|\int_a^x \frac{f(t)\,dt}{(b-t)^{q+n}}\right.} = \Gamma(-q-n)\, {}_a^xI^{-q-n} f$$

where "$\overline{|\ \ }$" denotes "Hadamard finite part of the infinite integral" ([12], pp. 133-158) and ([19], pp. 9-11).

THEOREM 1.5. The following relation

$${}_a^xI^{\alpha}\, {}_a^xI^{\beta} f = {}_a^xI^{\alpha+\beta} f$$

holds if

(i) $R\alpha > 0$, $R\beta > 0$ and $f \in C^{(0)}$ on $[a, b]$

(ii) $R\alpha > 0$, $R\beta \leqq 0$ $(R\beta + m > 0)$, such that $\beta \neq -m$ and $f \in C^{(m)}$ on $[a, b]$. When $\beta = -m$, then

$${}_a^xI^{\alpha}\, {}_a^xI^{-m} f = {}_a^xI^{\alpha} f^{(m)}$$

$$= {}_a^xI^{\alpha-m} f - \sum_{i=0}^{m-1} \frac{f^{(i)}(a)(x-a)^{\alpha-m+i}}{\Gamma(\alpha+i-m+1)} .$$

(iii) $R\alpha \leqq 0$ $(R\alpha + n > 0)$, $R\beta > 0$ and $f \in C^{(n)}$ on $[a, b]$.

(iv) $R\alpha \leqq 0$, $R\beta \geqq 0$, $\beta \neq -m$ and $f \in C^{(m+n)}$ on $[a, b]$. When $\beta = -m$ the same relation in (ii) for this value also holds.

REMARK 1. For details on these properties and theorems may be found in [19].

REMARK 2. (H) can be shown to have the convolution ring property as defined by A. Erdeyli in his book "Operational Calculus and Generalized

Functions", pp. 14-25.

REMARK 3. A study of the operator of generalized order $\underset{-\infty}{\overset{x}{I}}{}^{\alpha}$ has been made by the author ([20], pp. 2-6).

II. SOME EXISTENCE THEOREMS

Theorem 2.1. If $f(x, y)$ is a real-valued continuous and Lipschitzian function, i.e. $f(x, y) \in (C, \text{Lip})$, on a domain D of the (xy)-plane and $\sup_{(x,y)\in D} |f(x, y)| = M < \infty$, then there exists a unique and continuous solution $y = y(x)$ which satisfies the differential equation

$$y^{(\alpha)} = f(x, y), \quad (0 < \alpha \leqq 1),$$

together with the condition $y^{(\alpha-1)}(a) = \eta$ in the region $R \subset D$:

$$a < x \leqq a + h, \quad |(x - a)^{1-\alpha}y(x) - b| \leqq k,$$

where

$$b = \frac{\eta}{\Gamma(\alpha)}, \quad k > \frac{Mh}{\Gamma(\alpha + 1)}, \quad \text{and} \quad y^{(\alpha)} = \underset{a}{\overset{x}{I}}{}^{-\alpha}y.$$

THEOREM 2.2. If $f(x, y) \varepsilon$ (c. Lip) on a domain D of the (xy)-plane and

$$\sup_{(x,y)\varepsilon D} |f(x, y)| = M < \infty,$$

then there exists a unique and continuous solution $y = y(x)$ which satisfies the differential equation

$$y^{(\alpha)} = f(x, y), \quad n - 1 < \alpha \leqq n, \quad (n = 1, 2, \ldots),$$

together with the initial conditions $y^{(\alpha-j)}(a) = n_j$, $(j = 1, 2, \ldots, n)$ in the region $R_n \subset D$ defined by

$$a < x \leqq a + h, \; |(x - a)^{n-\alpha} y(x) - b_n| \leqq k,$$

where

$$b_n = \frac{\eta_n}{\Gamma(\alpha - n + 1)}, \; \sum_{i=0}^{n-1} \frac{h^{n-i}\eta_i}{\Gamma(\alpha - i + 1)} < k \quad \text{and} \quad \eta_0 = M.$$

REMARK. When $\alpha = n$ (a positive integer), Theorem 2.2 represents the well known existence and uniqueness theorem of the differential system:

$$y^{(n)} = f(x, y)$$

$$y^{(n-j)}(a) = \eta_j, \; (j = 1, 2, \ldots, n), \; y^{(0)}(a) = y(a),$$

in the region defined by: $|x - a| \leqq h$, $|y(x) - b_n| \leqq k$; where

$$b_n = \eta_n, \; \{\sum_{r=1}^{n} \eta_{n-r} \frac{h^r}{\Gamma(r + 1)}\} < k \quad \text{and} \quad \eta_0 = M \geqq |f(x, y)|.$$

THEOREM 2.3. With the hypothesis of Theorem 2.1, suppose that $y = Y(x)$ is the solution for which $Y^{(\alpha-1)}(a) = \eta + \delta$. Then for $a < x \leqq a + h$, $\delta > 0$,

$$|Y(x) - y(x)| \leqq \delta \, E_\alpha(A; |x - a|),$$

where $E_\alpha(A; x)$ is the function defined in ([21], pp. 359-360).

This implies that a small change in the initial conditions may cause only a small change in the solution throughout the interval.

THEOREM 2.4. Let $f_i(x, y_1, y_2, \ldots, y_m)$, $(i = 1, 2, \ldots, m)$, be real-valued and continuous functions on a domain D of the real space $(x, y_1, y_2, \ldots, y_m)$. Let $f_i \in$ (Lip) with respect to $y_1, y_2, \ldots, y_m$, i.e., there exist a positive constant A such that if $(x, Y_1, Y_2, \ldots, Y_m)$ and $(x, Z_1, Z_2, \ldots, Z_m)$ are contained in D then

$$|f_i(x, Y_1, Y_2, \ldots, Y_m) - f_i(x, Z_1, Z_2, \ldots, Z_m)| \leqq A \sum_{k=1}^{m} |Y_k - Z_k| ,$$

$(i = 1, 2, \ldots, m)$.

Let

$$\sup_{\substack{(x,y_1,y_2,\ldots,y_m)\varepsilon D \\ (i=1,2,\ldots,m)}} |f_i(x, y_1, \ldots, y_m)| = M < \infty.$$

Then there exists a unique set of continuous solutions $y_1(x)$, $y_2(x)$, ..., $y_m(x)$ which satisfy the differential equations

$$y_i^{(\alpha)} = f_i(x, y_1, y_2, \ldots, y_m), \ 0 < \alpha \leqq 1,$$

together with the initial conditions $y_i^{(\alpha-1)}(a) = \eta_i$, $(i = 1, 2, \ldots, m)$, on the set $R \subset D$ defined by

$$a < x \leqq a + h, \ |(x - a)^{1-\alpha} y_i(x) - b_i| \leqq k,$$

where $k > \frac{Mh}{\Gamma(\alpha + 1)}$, $b_i = \frac{\eta_i}{\Gamma(\alpha)}$, $(i = 1, 2, \ldots, m)$.

REMARK. Theorem 2.4 can be extended to include the case $n - 1 < \alpha \leqq n$, in the same way Theorem 2.1 was generalized by Theorem 2.2.

THEOREM 2.5. If $f(x, y, y^{(\alpha)}, y^{(2\alpha)}, \ldots, y^{(\overline{n-1}\alpha)}) \varepsilon C$ and is Lipschitzian with respect to $y, y^{(\alpha)}, \ldots, y^{(\overline{n-1}\alpha)}$, in some domain D of the space $(x, y, y^{(\alpha)}, \ldots, y^{(\overline{n-1}\alpha)})$, then there exists an interval of x in which the equation

$$y^{(n\alpha)} = f(x, y, y^{(\alpha)}, y^{(2\alpha)}, \ldots, y^{(\overline{n-1}\alpha)}), \ 0 < a \leqq 1,$$

has a unique and continuous solution for which $y^{(k\alpha-1)}(a) = c_k$, $(k = 1, 2, \ldots, n)$.

REMARK. The theorem may be extended to include the cases when $m - 1 < \alpha \leqq m$, $(m = 1, 2, \ldots)$, in a way similar to that described by Theorem 2.2 as an extension of Theorem 2.1.

Linear Equations

The definition of linear differential equations of generalized order is used here in the same sense of that adopted for the differential equations of integer-order.

According to Theorem 2.4, the system

$$y_i^{(\alpha)}(x) = \sum_{j=1}^{m} p_{ij} y_j + q_i, \quad 0 < \alpha \leqq 1,$$

$$y_i^{(\alpha-1)}(a) = \eta_i,$$

$(i = 1, 2, \ldots, m)$, where p_{ij} and q_i are continuous functions of x in an interval, (a, b) has a unique set of continuous solutions $y_1(x)$, $y_2(x), \ldots, y_m(x)$. Also we have the following:

THEOREM 2.6. If $p_i(x)$, $(i = 1, 2, \ldots, n)$ and $F(x)$ are continuous functions of x in an interval (a, b), $p_0(x) \neq 0$, then the differential equation

$$\sum_{i=0}^{n} p_i(x) y^{(\overline{n-i\alpha})}(x) = F(x), \quad 0 < \alpha \leqq 1,$$

where $y^{(0)}(x) = y(x)$, admits of a unique and continuous solution in (a, b) which satisfies the initial conditions $y^{(k\alpha-1)}(a) = \eta_k$, $(k = 1, 2, \ldots, n)$.

REMARK. It should be noted that Theorem 2.6 may be extended to include the cases $m - 1 < \alpha \leqq m$, $(m = 2, 3, \ldots)$. In such cases the solution satisfies initial conditions of the forms $y^{(j\alpha-k)}(a) = \eta_{jk}$, $(j = 1, 2, \ldots, n)$, $k = 1, 2, \ldots, m)$. This may be illustrated by the following example:

Let us consider the equation

$$y^{(2\alpha)} = v(x), \quad n - 1 < \alpha \leqq n.$$

We find that

(1) when $0 < \alpha \leqq 1$, we have

$$y(x) = {}_a^x I^{2\alpha} v + \eta_2 \frac{(x-a)^{2\alpha-1}}{\Gamma(2\alpha)} + \eta_1 \frac{(x-a)^{\alpha-1}}{\Gamma(\alpha)},$$

where

$$y^{(\alpha-1)}(a) = \eta_1, \quad y^{(2\alpha-1)}(a) = \eta_2;$$

and

(2) when $1 < \alpha \leqq 2$, we have

$$y(x) = {}_a^x I^{2\alpha} v + \eta_1 \frac{(x-a)^{2\alpha-1}}{\Gamma(2\alpha)} + \eta_2 \frac{(x-a)^{2\alpha-2}}{\Gamma(2\alpha-1)}$$

$$+ \xi_1 \frac{(x-a)^{\alpha-1}}{\Gamma(\alpha)} + \xi_2 \frac{(x-a)^{\alpha-2}}{\Gamma(\alpha-1)},$$

where $y^{(2\alpha-k)}(a) = \eta_k$ and $y^{(\alpha-k)}(a) = \xi_k$, $(k = 1, 2)$.

In general if $n - 1 < \alpha \leqq n$ then we have

$$y(x) = {}_a^x I^{2\alpha} v + \sum_{i=1}^{n} \eta_i \frac{(x-a)^{2\alpha-i}}{\Gamma(2\alpha-i+1)} + \sum_{j=1}^{n} \xi_j \frac{(x-a)^{\alpha-j}}{\Gamma(\alpha-j+1)},$$

where $y^{(2\alpha-k)}(a) = \eta_k$, and $y^{(\alpha-k)}(a) = \xi_k$, $(k = 1, 2, \ldots, n)$.

REMARK 1. The last two terms of the right hand side of the solution in case (1) of the Example represent the finite parts of divergent integrals in the Hadamard sense as indicated by Theorem 1.4 (also see [19], p. 12), and so the solution is not reduced to that of $y' = v(x)$ when $\alpha = \frac{1}{2}$.

But, certainly it represents the solution of $y'' = v(x)$ for $\alpha = 1$. However, if we follow the steps for solving the given equation we find that:

$$y^{(2\alpha)} = {}_a^x I^{-2\alpha} y = D_x^2 \, {}_a^x I^{2-2\alpha} y = v(x),$$

from which we may have

$${}_a^x I^{2-2\alpha} y = {}_a^x I^2 v + c_1(x - a) + c_2,$$

c_1 and c_2 are arbitrary constants, and

$${}_a^x I^{1-2\alpha} y = {}_a^x I v + c_1 \quad \text{implies that} \quad c_1 = \eta_2 = y^{)2\alpha-1)}(a),$$

$${}_a^x I^{1-\alpha} y = {}_a^x I^{\alpha+1} v + \eta_2 \frac{(x - a)^\alpha}{\Gamma(\alpha + 1)} + c_2, \quad \text{so that}$$

$$c_2 = \eta_1 = y^{(\alpha-1)}(a).$$

It may be noticed that the third equation is reduced to the solution of $y' = v$ for $\alpha = \frac{1}{2}$, and the last equation represents the solution of $y'' = v(x)$ for $\alpha = 1$. Thus the solution in (1), after performing the required operations, is obtained from the following:

$$y(x) = {}_a^x I^{2\alpha} v + \eta_2 \, {}_a^x I^{\alpha-1} \frac{(x - a)^\alpha}{\Gamma(\alpha + 1)} + \eta_1 \, {}_a^x I^{\alpha-1} 1,$$

where

$${}_a^x I^{\alpha-1} = Dx \, {}_a^x I^\alpha.$$

REMARK 2. Proofs, examples and further details may be found in [24].

III. SOME EXISTENCE THEOREMS OF SERIES SOLUTIONS

In a previous work the author has defined and studied the generalized operations on y, i.e., $y^{(\alpha)} = \overset{x}{\underset{a}{I}}{}^{-\alpha} y$, together with its operational properties (see [19])*. In another paper some of the properties of T^{α}-functions have been investigated (see [21]).

Based on the above mentioned research we have the following:

THEOREM 3.1. If $P(x) \varepsilon T^{\alpha}$ on an interval $L : 0 < A \leq x \leq B < R$; then there exists a series solution $y = Y(x)$ which satisfies the differential equation of order α, $n < \alpha < n+1$, $(n = 1, 2, \ldots)$

$$y^{(\alpha)} + P(x)y = 0,$$

and is such that $y = Y(x; a_0)$.

Proof. Since $P(x) \varepsilon T^{\alpha}$ we may write

$$P(x) = \sum_{n=0}^{\infty} b_n (x - a)^{n\alpha}.$$

Let us assume that

$$y(x) = \sum_{n=0}^{\infty} a_n (x - a)^{n\alpha + c}, \quad c > 0,$$

where $a \varepsilon L$. Then we find that

$$\sum_{n=0}^{\infty} \frac{\Gamma(n\alpha + c + 1)}{\Gamma(\overline{n-1}\alpha + c + 1)} a_n (x - a)^{\overline{n-1}\alpha + c}$$

$$+ \sum_{n=0}^{\infty} b_n (x - a)^{n\alpha} \sum_{n=0}^{\infty} a_n (x - a)^{n\alpha + c} = 0.$$

---

*Some of these properties are included in Part I.

By performing the indicated multiplication in the second term we obtain

$$\sum_{n=0}^{\infty} \frac{\Gamma(n\alpha + c + 1)}{\Gamma(\overline{n-1}\alpha + c + 1)} a_n (x - a)^{\overline{n-1}\alpha+c}$$

$$+ \sum_{n=1}^{\infty} \sum_{k=0}^{n-1} a_k b_{n-k-1} (x - a)^{\overline{n-1}\alpha+c} = 0,$$

from which we have for $n = 0$,

$$\frac{\Gamma(c + 1)}{\Gamma(c - \alpha + 1)} a_o = 0.$$

Assuming that $a_o \neq 0$, this implies that $c = \alpha - j - 1$, $(j = 0, 1, \ldots, n - 1)$. For $n = 1$, we have

$$\frac{\Gamma(\alpha + c + 1)}{\Gamma(c + 1)} a_1 + a_o b_o = 0.$$

If we take $j = 0$, then for $n \geq 2$ we have

$$\frac{\Gamma(\overline{n+1}\alpha)}{\Gamma(n\alpha)} a_n + \sum_{k=0}^{n-1} a_k b_{n-k-1} = 0,$$

or

$$a_n = \frac{\Gamma(n\alpha)}{\Gamma(\overline{n+1}\alpha)} \sum_{k=0}^{n-1} a_k b_{n-k-1}.$$

Since $P(x) \varepsilon T^{\alpha}$, the series $\sum_{n=0}^{\infty} b_n (x - a)^{n\alpha}$ is absolutely and uniformly convergent in an interval $[A_1, B_1]$, $A_1 > 0$. Therefore there exists $M_1 > 0$, $r_1 > 0$ such that $|b_n| < \frac{M_1}{r_1^n}$ for every n. Suppose that

$M_2 > 0$, $r_2 > 0$ such that $|a_1| < \frac{M_2}{r_2}$. Choose $r = \min(r_1, r_2)$ and $M > \max(M_1, M_2, |a_1| r_2, r, 1)$, then we have $|a_1| < \frac{M}{r}$ and $|b_n| < \frac{M}{r^n}$.

Assume that $|a_k| < \frac{M^k}{r^k}$ for $1 \leq k < n$. Then we have

$$|a_n| \leq \frac{\Gamma(n\alpha)}{\Gamma(\overline{n+1}\alpha)} \sum_{k=0}^{n-1} |a_k| \, |b_{n-k-1}|$$

$$< \frac{\Gamma(n\alpha)}{\Gamma(\overline{n+1}\alpha)} \sum_{k=0}^{n-1} \frac{M^{k+1}}{r^{n-1}}$$

$$< \frac{\Gamma(n\alpha)}{\Gamma(\overline{n+1}\alpha)} \sum_{k=0}^{n-1} \frac{M^n}{r^n} = \frac{n\Gamma(n\alpha)}{\Gamma(\overline{n+1}\alpha)} \frac{M^n}{r^n} < \frac{M^n}{r^n}.$$

Thus we have $|a_n| < \frac{M^n}{r^n}$ for all n. This shows that the series converges for $0 < x - a < (\frac{r}{M})^{\frac{1}{\alpha}}$, and the solution depends upon the choice of the arbitrary constant a_o, i.e. $y = Y(x; a_o)$.

EXAMPLE. Consider the equation

$$y^{(\alpha)} + \lambda y = 0, \tag{1}$$

$m < \alpha < m + 1$, $(m = 1, 2, 3, \ldots)$ and λ is a number.

To find the solution of (1), let us assume that

$$y = \sum_{n=0}^{\infty} a_n (x - a)^{n\alpha+c}, \quad c > 0.$$

Then we have

$$\sum_{n=0}^{\infty} \frac{\Gamma(n\alpha + c + 1)}{\Gamma(\overline{n-1}\alpha + c + 1)} a_n (x - a)^{\overline{n-1}\alpha+c}$$

$$+ \lambda \sum_{n=0}^{\infty} a_n (x - a)^{n\alpha+c} = 0.$$

For $n = 0$, $\frac{\Gamma(c + 1)}{\Gamma(c - \alpha + 1)} a_o = 0$.

Assuming that $a_o \neq 0$, then $\frac{\Gamma(c + 1)}{\Gamma(c - \alpha + 1)} = 0$ if $c = \alpha - k - 1$, $(k = 0, 1, 2, \ldots, m-1)$. Thus for $n > 1$ we have

$$\frac{\Gamma(\overline{n+1}\alpha - k)}{\Gamma(n\alpha - k)} a_n + \lambda a_{n-1} = 0,$$

from which we find that

$$a_n = (-\lambda)^n \frac{\Gamma(\alpha - k)}{\Gamma(\overline{n+1}\alpha - k)} a_o.$$

Consequently we have the solutions

$$y_{k,\alpha} = C_k \sum_{n=1}^{\infty} (-\lambda)^{n-1} \frac{(x - a)^{n\alpha-k-1}}{\Gamma(n\alpha - k)},$$

where $C_k = \Gamma(\alpha - k)a_o$, $(k = 0, 1, 2, \ldots, m-1)$. In general, the solution may be written as

$$y_\alpha = \sum_{k=0}^{m-1} y_{k,\alpha}. \tag{3}$$

Some Particular Cases of (1):

(i) If $1 < \alpha < 2$, then from (3) we have

(i - 1) $$y_\alpha = y_{o,\alpha} = a_o \Gamma(\alpha) \sum_{n=1}^{\infty} (-\lambda)^{n-1} \frac{(x - a)^{n\alpha-1}}{\Gamma(n\alpha)}$$

We notice that as $\alpha \to 1^-$, equation (1) together with its solution (i - 1) take the forms:

$$y' + \lambda y = 0$$

and its solution

$$\lim_{\alpha \to 1^-} y_{o,\alpha} = y_{o,1} = e^{-\lambda(x-a)}.$$

(ii) When $2 < \alpha < 3$, we have

$$y_\alpha = y_{o,\alpha} + y_{1,\alpha},$$

where $y_{o,\alpha}$ is given by (i - 1) and

$$y_{1,\alpha} = a_o \Gamma(\alpha - 1) \sum_{n=1}^{\infty} (-\lambda)^{n-1} \frac{(x - a)^{n\alpha - 2}}{\Gamma(n\alpha - 1)}.$$

As $\alpha \to 2^-$ we observe that

$$y_{o,\alpha} \to y_{o,2} = a_o \Gamma(2) \sum_{n=1}^{\infty} (-\lambda)^{n-1} \frac{(x - a)^{2n-1}}{\Gamma(2n)}$$

$$= \frac{a_o}{\lambda} \sin \lambda(x - a),$$

and

$$y_{1,\alpha} \to y_{1,2} = a_o \cos \lambda(x - a),$$

which are the solutions of

$$y'' + \lambda y = 0.$$

REMARK. This example was discussed by J. H. Barrett (see references in

[24] and [26]). He obtained similar results by solving the equivalent integral equation by the method of successive substitution.

THEOREM 3.2. If $P(x)$, $Q(x) \varepsilon T^{\alpha}$ on the intervals

$$I_1 : 0 < x - a < R_1$$

$$I_2 : 0 < x - a < R_2, \; a \geq 0,$$

respectively, then there exists a series solution $y = Y(x)$ which satisfies the differential equation of order α, $m < \alpha < m + 1$, $(m = 1, 2, \ldots)$:

$$y^{(2\alpha)} + P(x)y^{(\alpha)} + Q(x)y = 0.$$

To show this we assume that the equation has a solution of the form:

$$y(x) = \sum_{n=0}^{\infty} a_n (x - a)^{n\alpha+c},$$

$a_0 \neq 0$ and c is a real number which may be determined later.

Since $P(x) \varepsilon T^{\alpha}$ on I_1 and $Q(x) \varepsilon T^{\alpha}$ on I_2 we may write

$$P(x) = \sum_{n=0}^{\infty} b_n (x - a)^{n\alpha}$$

$$Q(x) = \sum_{n=0}^{\infty} c_n (x - a)^{n\alpha}.$$

Then we have

$$y^{(\alpha)} = \sum_{n=0}^{\infty} a_n \frac{\Gamma(n + c + 1)}{\Gamma(\overline{n-1}\alpha + c + 1)} (x - a)^{\overline{n-1}\alpha+c}$$

$$y^{(2\alpha)} = \sum_{n=0}^{\infty} a_n \frac{\Gamma(n\alpha + c + 1)}{\Gamma(\overline{n-2}\alpha + c + 1)} (x - a)^{\overline{n-2}\alpha+c}.$$

Substituting these values in the differential equation we find that

$$\sum_{n=0}^{\infty} a_n \frac{\Gamma(n\alpha + c + 1)}{\Gamma(\overline{n-2}\alpha + c + 1)} (x - a)^{\overline{n-2}\alpha+c}$$

$$+ \sum_{n=0}^{\infty} \sum_{k=0}^{n} a_k b_{n-k} \frac{\Gamma(k\alpha + c + 1)}{\Gamma(\overline{n-2}\alpha + c + 1)} (x - a)^{\overline{n-1}\alpha+c}$$

$$+ \sum_{n=0}^{\infty} \sum_{k=0}^{n} a_k c_{n-k} (x - a)^{n\alpha+c} = 0.$$

This may be written in the form:

$$\sum_{n=0}^{\infty} a_n \frac{\Gamma(n\alpha + c + 1)}{\Gamma(\overline{n-2}\alpha + c + 1)} (x - a)^{\overline{n-2}\alpha+c}$$

$$+ \sum_{n=1}^{\infty} \sum_{k=0}^{n-1} a_k b_{n-k-1} \frac{\Gamma(k\alpha + c + 1)}{\Gamma(\overline{k-1}\alpha + c + 1)} (x - a)^{\overline{n-2}\alpha+c}$$

$$+ \sum_{n=2}^{\infty} \sum_{k=0}^{n-2} a_k c_{n-k-2} (x - a)^{\overline{n-2}\alpha+c} = 0.$$

Thus we have

$$a_0 \frac{\Gamma(c + 1)}{\Gamma(c - 2\alpha + 1)} = 0, \qquad (1)$$

$$a_0 \frac{\Gamma(\alpha + c + 1)}{\Gamma(c - \alpha + 1)} + a_0 b_0 \frac{\Gamma(c + 1)}{\Gamma(c - \alpha + 1)} = 0,$$

or

$$\frac{1}{\Gamma(c - \alpha + 1)} [a_1\, \Gamma(\alpha + c + 1) + a_0 b_0 \Gamma(c + 1)] = 0$$

and for $n \geqq 2$,

$$a_n \frac{\Gamma(n\alpha + c + 1)}{\Gamma(\overline{n-2}\alpha + c + 1)} + \sum_{k=0}^{n-1} a_k b_{n-k-1} \frac{\Gamma(k\alpha + c + 1)}{\Gamma(\overline{k-1}\alpha + c + 1)} \tag{3}$$

$$+ \sum_{k=0}^{n-2} a_k c_{n-k-2} = 0.$$

From (1) we observe that $a_0 \neq 0$ implies that $c = 2\alpha - j - 1$, $(j = 0, 1, 2, \ldots)$. Also, we notice that (2) is satisfied when $c = \alpha - i - 1$ $(i = 0, 1, 2, \ldots)$ and this implies that $a_0 = 0$*. Therefore we may choose $c = 2\alpha - j - 1$ with $j = 0$. In this case all values of a_n, $n > 0$ depend upon the value of a_0 and consequently the solution, if it exists, depends upon a_0.

Thus when $c = 2\alpha - 1$ and for $n \geq 2$ we have

$$a_n \frac{\Gamma(\overline{n+2}\alpha)}{\Gamma(n\alpha)} + \sum_{k=0}^{n-1} a_k b_{n-k-1} \frac{\Gamma(\overline{k+2}\alpha)}{\Gamma(\overline{k+1}\alpha)} + \sum_{k=0}^{n-2} a_k c_{n-k-2} = 0$$

or

$$a_n = - \frac{\Gamma(n\alpha)}{\Gamma(\overline{n+2}\alpha)} \left[\sum_{k=0}^{n-1} a_k b_{n-k-1} \frac{\Gamma(\overline{k+2}\alpha)}{\Gamma(\overline{k+1}\alpha)} + \sum_{k=0}^{n-2} a_k c_{n-k-2} \right].$$

Since $P(x)$ and $Q(x)$ belong to T^α, the series $\sum b_n (x - a)^{n\alpha}$ and $\sum c_n (x - a)^{n\alpha}$ are absolutely and uniformly convergent in I_1 and I_2 respectively. Therefore there exist positive numbers M_1, M_2, r_1, r_2 such that

$$|b_n| < \frac{M_1}{2r_1^n} \quad \text{and} \quad |c_n| < \frac{M_2}{2r_2^n} \quad \text{for every } n.$$

*If the condition $c > 0$ is assumed then, $(j = 1, 2, \ldots, 2m - 1)$ and $(i = 0, 1, \ldots, m - 1)$.

Suppose that $M_3 > 0$, $r_3 > 0$ such that $|a_1| < \frac{M_3}{r_3}$. Choose $r = \min(r_1, r_2, r_3)$ and $M > \max(M_1, M_2, M_3, |a_1| r_3, r, 1)$. Then we have

$$|a_1| < \frac{M}{r}, \ |b_n| < \frac{M}{2r^n} \text{ and } |c_n| < \frac{M}{2r^n} .$$

Now assume that $|a_k| < \frac{M^k}{r^k}$ for $1 \leq k < n$. Then we have from (3.1):

$$|a_n| \leq \frac{\Gamma(n\alpha)}{\Gamma(\overline{n+2}\alpha)} \frac{\Gamma(\overline{n+1}\alpha)}{\Gamma(n\alpha)} [\sum_{k=0}^{n-1} |a_k| \, |b_{n-k-1}|$$

$$+ \sum_{k=1}^{n-1} |a_{k-1}| \, |c_{n-k-1}|]$$

$$< \frac{\Gamma(\overline{n+1}\alpha)}{\Gamma(\overline{n+2}\alpha)} [\sum_{k=0}^{n-1} \frac{M^k}{r^k} \cdot \frac{M}{2r^{n-k-1}} + \sum_{k=1}^{n-1} \frac{M^{k-1}}{r^{k-1}} \frac{}{2r^{n-k-1}}]$$

$$< \frac{\Gamma(\overline{n+1}\alpha)}{\Gamma(\overline{n+2}\alpha)} [\sum_{k=0}^{n-1} \frac{M^{k+1}}{2r^{n-1}} \frac{M^{n-k-1}}{r} + \sum_{k=1}^{n-1} \frac{M^k}{2r^{n-1}} \frac{M^{n-k}}{r}]$$

$$< \frac{\Gamma(\overline{n+1}\alpha)}{\Gamma(\overline{n+2}\alpha)} [\sum_{k=0}^{n-1} \frac{M^n}{2r^n} + \sum_{k=1}^{n-1} \frac{M^n}{2r^n}]$$

$$< \frac{\Gamma(\overline{n+1}\alpha)}{\Gamma(\overline{n+2}\alpha)} \sum_{k=0}^{n-1} \frac{M^n}{r^n}$$

$$< \frac{\Gamma(\overline{n+1}\alpha)}{\Gamma(\overline{n+2}\alpha)} \frac{(n+1)M^n}{r^n} < \frac{M^n}{r^n} .$$

Thus we have $|a_n| < \frac{M^n}{r^n}$ for all n.

This shows that the series converges for $0 < x - a < (\frac{r}{M})^{\frac{1}{\alpha}}$ and the solution depends upon the choice of a_0.

EXAMPLE. Consider the equation

$$y^{(2\alpha)} + x^{2\alpha} y = 0, \quad m < \alpha < m + 1, \quad (m = 1, 2, \ldots). \tag{1}$$

To find the solution of (1) we may assume that

$$y = \sum_{n=0}^{\infty} a_n x^{n\alpha + c},$$

where c is to be determined. Then we have

$$\sum_{n=0}^{\infty} \frac{\Gamma(n\alpha + c + 1)}{\Gamma(\overline{n-2}\alpha + c + 1)} a_n x^{\overline{n-2}\alpha + c} + \sum_{n=0}^{\infty} a_n x^{\overline{n+2}\alpha + c} = 0.$$

This may be written as

$$\frac{\Gamma(c + 1)}{\Gamma(c - 2\alpha + 1)} a_0 x^{c-2\alpha} + \frac{\Gamma(\alpha + c + 1)}{\Gamma(c - \alpha + 1)} a_1 x^{c-\alpha} + \frac{\Gamma(2\alpha + c + 1)}{\Gamma(c + 1)} a_2 x^{c}$$
$$+ \frac{\Gamma(3\alpha + c + 1)}{\Gamma(c + \alpha + 1)} a_3 x^{\alpha + c}$$

$$+ \sum_{n=4}^{\infty} \frac{\Gamma(n\alpha + c + 1)}{\Gamma(\overline{n-2}\alpha + c + 1)} a_n + a_{n-4}) \, x^{\overline{n-2}\alpha + c} = 0,$$

which implies that

$$\frac{\Gamma(c + 1)}{\Gamma(c - 2\alpha + 1)} a_0 = 0, \quad \frac{\Gamma(\alpha + c + 1)}{\Gamma(c - \alpha + 1)} = a_1 = 0,$$

$$\frac{\Gamma(2\alpha + c + 1)}{\Gamma(c + 1)} a_2 = 0, \quad \frac{\Gamma(3\alpha + c + 1)}{\Gamma(c + \alpha + 1)} a_3 = 0.$$

We may consider the following cases:

CASE 1. If $a_0 \neq 0$, then

$c = 2\alpha - j - 1, \ (j = 0, 1, 2, \ldots).$

$a_1 = a_2 = a_3 = 0.$

CASE 2. If $a_1 \neq 0$, then $c = \alpha - i - 1, \ (i = 0, 1, 2, \ldots)$. Also

$a_0 = a_2 = a_3 = 0.$

CASE 1. If $c = 2\alpha - k - 1$, then we have for $n \geq 4$,

$$a_n = - \frac{\Gamma(n\alpha - j)}{\Gamma(\overline{n+2}\alpha - j)} a_{n-4}.$$

Thus we have

$$a_{4k} = (-1^k \frac{\Gamma(4\alpha - j)\Gamma(8\alpha - j)\ldots\Gamma(4k - j)}{\Gamma(6\alpha - j)\Gamma(10\alpha - j)\ldots(4\overline{k+2}\alpha - j)} a_0,$$

and a set of solutions is given by

$$y_j(\alpha; a_0; x)$$

$$= a_0 \sum_{k=0}^{\infty} (-1)^k \frac{\Gamma(4\alpha - j)\Gamma(8\alpha - j)\ldots\Gamma(4k - j)}{(6\alpha - j)\Gamma(10\alpha - j)\ldots\Gamma(4\overline{k+2}\alpha - j)} x^{\overline{4k+2}-j-1}.$$

CASE 2. Let $c = \alpha - i - 1$. Then we have for $n \geq 5$,

$$a_n = - \frac{\Gamma(\overline{n-1}\alpha - i)}{\Gamma(\overline{n+1}\alpha - i)} a_{n-4}.$$

Thus we have

$$a_{4k+1} = (-1)^k \frac{\Gamma(4\alpha - i)(8\alpha - i)\ldots\Gamma(4k\alpha - i)}{\Gamma(6\alpha - i)\Gamma(10\alpha - i)\ldots\Gamma(\overline{4k+2}\alpha - i)} a_1,$$

and a set of solutions may be given by

$$y_i(\alpha;\, a_1;\, x) = a_1 \sum_{k=0}^{\infty} (-1)^k \frac{\Gamma(4\alpha - i)\Gamma(8\alpha - i)\ldots\Gamma(4k\alpha - i)x^{4k+2\alpha-i-1}}{\Gamma(6\alpha - i)\Gamma(10\alpha - i)\ldots\Gamma(4k + 2\alpha - i)}$$

REMARKS. It may be of interest to note that the series solution of the equation

$$y'' + x^2 y = 0$$

can be obtained from the cases above by assuming that $1 < \alpha < 2$ and $\alpha \to 1^+$. Thus we find

(i) $$y_0(1^+;a_0;x) = a_0 \left[x + \sum_{k=1}^{\infty} \frac{(-1)^k x^{4k+1}}{2^{2k}\Gamma(k+1)5.9\ldots(4k+1)}\right]$$

$$y_1(1^+;a_0;x) = a_0 \left[1 + \sum_{k=1}^{\infty} \frac{(-1)^k x^{4k}}{2^{2k}\Gamma(k+1)3.7.11\ldots(4k-1)}\right]$$

(ii) $$y_0(1^+;a_1;x) = a_1 \left[x + \sum_{k=1}^{\infty} \frac{(-1)^k x^{4k+1}}{2^{2k}\Gamma(k+1)5.9\ldots(4k+1)}\right]$$

$$y_1(1^+;a_1;x) = a_1 \left[1 + \sum_{k=1}^{\infty} \frac{(-1)^k x^{4k}}{2^{2k}\Gamma(k+1)3.7\ldots(4k-1)}\right].$$

So it may be concluded that the series solution of the equation $y(x)$ can be expressed by either

$$y(x) \equiv y_0(1^+;\, a_0;\, x) + y_1(1^+;\, a_1;\, x)$$

or

$$y(x) \equiv y_1(1^+;\, a_0;\, x) + y_0(1^+;\, a_1;\, x).$$

REFERENCES

1. G. W. Leibnitz, "Leibnitzen Mathematische Schriften", Volume 2, pp. 301-302 and Volume 4, p. 25, Olms Verlag., Hildesheim, Germany (1962).

2. L. Euler, "De Progressionibus Transcentibus", Sev Quarium Termini Algebraice Dari Nequeunt", Comment. Acad. Sci. Imperialis Pelropotitanae 5, 38-57, (1738).

3. P. S. Laplace, "Théorie Analytie des Probabilités", Courcier, Paris (1820).

4. S. F. Lacrois, "Traité du Calcul Differentiel et du Calcul Integral", Vol. 3, (1819).

5. J. L. Liouville, "Mémoire sur le calcul des differentielles à indices quelconques", J. Ecole Polytech., 13, Section 12, (1832).

6. J. L. Liouville, "Mémoire sur le changement de la variable dans le calcul des differentielle à indices quelconques", J. Ecole Polytech., 15, Section 24, (1835).

7. B. Riemann, "Versuch einer allgemeinen Auffassung der Integration und Differentiation", Gesammelte Mathematische Werke, Leipzig, (1876).

8. H. J. Holmgren, "Om differentialkalkylen med indices of hvilken natur som helst", Kungl. Svenska Vetenskaps-Akademins Handlingar, Bd. 5, n.11, Stockholm (1865-66), p.1-83.

9. K. A. Grünwald, "Uber begrenzte Derivationen und deren anwendung", Zeitschrift fur Math. und Physik, Vol. 12, (1867), pp.441-480.

10. H. Weyl, "Bermerkungen Zum Bergriff des differential - quotienten gebrochner ordnung", Vierteljahrsschriff der naturf. Ges. in Zurich, Vol. 62, (1917), pp. 296-302.

11. G. H. Hardy, "Some properties of integrals of fractional order", Messenger of Mathematics, Vol. 47, (1918), pp. 145-150.

G. H. Hardy, and J. Littlewood, "Some properties of fractional integrals", Proc. London Math. Soc. (2), Vol. 24, (1925). Also Mathematische Zeitschrift, Vol. 27, (1927-1928), pp. 565-606.

12. J. Hadamard, "Lectures on Cauchy's problem", Yale University Press (1923).

13. E. L. Post, "Generalized differentiation", Transactions of the American Mathematical Society, Vol. 32, (1930), pp. 723-781.

14. H. T. Davis, "The theory of linear operators", Principia Press, Bloomington, Indiana, (1936).

15. M. Riesz, "L'integral de Riemann-Liouville et le problem de Cauchy", Acta Mathematica, Vol. 81, (1949), pp. 1-223.

16. A. Mambriani, "The derivative of any order and the solution of hypergeometrie equation", (translated title), Boll. Un. Math., Ital. (2), 9-10, (1940). Also,
A. Mambriani, "On the partial derivative of any order and the solution of Euler and Poisson Equation", (translated title), Ann. Scoula Norm. Sup. Pisa (2), 11, (1942), pp. 79-97.

17. E. T. Copson, "Some applications of Marcel Riesz's integral of fractional order", Proc. Roy. Soc. Edinburgh, A 61, (1943), pp. 260-272.

18. N. E. Fremberg, "A study of generalized hyperbolic potentials", Dissertation Meddelanden Fran Lunds Universitets Matematiska Seminarium, Band 7, (1946).

19. M. A. Al-Bassam, "Some properties of Holmgren-Riesz transform", Annali della Scoula Normale superiore de Pisa, Scienze Fisiche e Mathematiche, Serie III, Vol. XV, Fasc. 1-11, (1961), pp. 1-24.

20. M. A. Al-Bassam, "Concerning Holmgren-Riesz transform equations of Causs-Riemann type", Rendiconti del Circolo Matematico di Palermo, Serie II, Vol. XI, (1962), pp. 1-20.

21. M. A. Al-Bassam, "On functional representation by a certain type of generalized power series", Annali della Scoula Normale Superiore di Pisa, Serie III, Vol. XVI, Fasc. IV, (1962), pp. 351-366.

22. M. A. Al-Bassam, "Some properties of Holmgren-Riez transform in two dimensions", Annali della Scoula Normale Superiore di Pisa, Serie IIII, Vol. XVI, Fasc. 1, (1962), pp. 75-90.

23. M. A. Al-Bassam, "On certain types of H-R transform equations and their equivalent differential equations", Journal fur die reine und angewandte Mathematik, Band 216, (1964), pp. 91-100.

24. M. A. Al-Bassam, "Some existence theorems on differential equations of generalized order", Journal fur die reine und angewandte Mathematik, Band 218, (1965), pp. 70-78.

25. M. A. Al-Bassam, "On an integro-differential equation of Lagendre-Velterra type", Portugaliae Math., Vol. 25, Fasc. 1, (1966), pp. 53-61.

26. M. A. Al-Bassam, "Existence of series solution of a type of differential equation of generalized order", Bulletin, College of Science, Univ. of Baghdad, Vol. 9, (1966), pp. 175-180.

27. M. A. Al-Bassam, "H-R transform equations of laguerre type", Bulletin, College of Science, University of Baghdad, Vol. 9, (1966), pp. 181-184.

28. M. A. Al-Bassam, "On Laplace's second order linear differential equations and their equivalent H-R transform equations", Journal fur die reine und angewandte Mathematik, Band 225, (1967), pp. 76-84.

29. M. A. Al-Bassam, "On the existence of series solution of differential equation of generalized order", Portugaliae Mathematica, Vol. 29, Fasc. 1, (1970), pp. 5-11.

30. M. A. Al-Bassam, "On some differential and integro-differential equations associated with Jacobi's differential equation", Journal fur die reine und angewandte Mathematik, Band 288, (1976), pp. 211-217.

31. M. A. Al-Bassam, "Some operational properties of the H-R operator", Appeared in the Annali della Scoula Normale Superiore di Pisa, classe di Scienze, Vol. XXVII, III, (1973), pp. 491-506.

32. "Fractional Calculus and its Applications", Proceedings of the International Conference, Univ. of New Haven, June 1974, edited by B. Ross, Lecture Notes in Mathematics (457), Springer-Verlag, New York, (1975).

33. K. B. Oldham and J. Spanier, "The fractional calculus", Academic Press, New York, (1974).

34. Katsuynki Nishimoto, "On the fractional calculus", Dissertations in celebration of the 30th anniversary of College of Engineering, Nihon University, Japan, October, (1977).

ITERATED STIELTJES TRANSFORM OF GENERALIZED FUNCTIONS

L. S. Dube

Vanier College
Montreal, Quebec
Canada

1. Summary

Recently, extensions of classical integral transformations to generalized functions have comprised an active and interesting area of research. In this paper an iterated Stieltjes transform (S_2-transform) is extended to generalized functions by so called direct approach, and an inversion formula is established. Note that, for simplicity, throughout this paper, the transform of a function or generalized function is denoted by $F(x)$, and the actual transform is made evident by the context.

The Stieltjes transform of $\phi(t) \in L\,(o, \infty)$ is defined as

$$F(x) = \int_o^\infty \frac{\phi(t)}{(x + t)}\, dt, \quad x > o \tag{1}$$

Widder [8, p. 325] studied various properties of the transform (1) and proved the inversion formula:

$$\lim_{k\to\infty} L_{k,\,x}\,[F(x)] = \phi(x)$$

for almost all $x > o$, where the differential operator $L_{k,\,x}$ is defined as:

$$L_{k,\,x}\,[F(x)] = \frac{(-1)^{k-1}}{k!(k-2)!}\,[x^{2k-1}\,F(x)^{(k-1)}]^{(k)}$$

Benedetto [1], Zemanian [9, p. 244] and Pandey [6] have given the distributional extensions of the transform (1) following different approaches.

The transform (1) is generalized as:

$$F(x) = \int_o^\infty \frac{\phi(t)}{(x+t)^\rho}\,dt \quad \text{for } \rho > o \ (x > o) \tag{2}$$

Pollard [8] defined the operator

$$L_{k,\,x;\,\rho}\,(F(x)) = \frac{(-1)^{k-1}\,2^{\rho-1}\,(2k-1)!\Gamma\rho}{k!(k-2)!\Gamma(2k+\rho-1)}\,[x^{2k+\rho-2}F(x)^{(k-1)}]^{(k)} \tag{3}$$

and proved the inversion formula:

$$\lim_{k\to\infty} L_{k,\,x;\,\rho}\,(F(x)) = \phi(x) \quad \text{for almost all } x > o.$$

Pathak [7] and Erdélyi [5] extended the transform (2) independently and differently, to generalized functions.

The generalized iterated Stieltjes transform of $\phi(t) \in L\,(o,\infty)$ is defined as

$$F(x) = \int_o^\infty \frac{dy}{(x+y)^\rho} \int_o^\infty \frac{\phi(t)}{(y+t)^\rho}\,dt \quad \text{for } \rho > \frac{1}{2} \ (x > o) \tag{4}$$

The change in order of integration in (4) leads to so called generalized S_2-transform as:

$$F(x) = \int_o^\infty \phi(t)\,dt \int_o^\infty \frac{1}{(x+y)^\rho(y+t)^\rho}\,dy$$

$$\equiv \int_o^\infty K(x,\,t;\,\rho)\,\phi(t)\,dt;\ \rho > \frac{1}{2}\,. \tag{4$'$}$$

Boas and Widder [2] have studied transforms (4) and (4)' in great detail for the case $\rho = 1$ and have given corresponding inversion formulas. Recently the author [4] considered the classical transforms (4) and (4)' for the general $\rho > \frac{1}{2}$ and proved the inversion formula:

$$\lim_{k\to\infty} H_{k,\ x;\ \rho}(F(x)) = \phi(x) \quad \text{for almost all } x > o, \quad \text{where}$$

$$H_{k,\ x;\ \rho} \equiv L_{k,\ x;\ \rho}\,(L_{k,\ x;\ \rho}). \tag{5}$$

The distributional extension of (4)' for $\rho = 1$ was given by the author [3].

Our main objective in this paper is to extend the transform (4)' to generalized functions, and to prove the inversion formula (5) in the distributional sense.

The notation and the terminology of this work will follow that of [3] and [9]. I denotes the open interval (o, ∞) and all testing functions herein are defined on I. Throughout the work x, t and y are variables over I. $D(I)$ represents the space of infinitely differentiable functions defined on I having compact support. The topology of $D(I)$ is that which makes its dual the space $D'(I)$ of the Schwartz distributions.

2. The Testing Function Space S (I) And Its Dual S'(I)

We define $S_\rho(I)$ as the collection of all infinitely differentiable functions $\phi(t)$ defined on I such that

$$\gamma_k(\phi) = \sup_{o<t<\infty} \left| \xi(t)\left(t\frac{d}{dt}\right)^k \phi(t)\right| < \infty;$$

for each $k = o, 1, 2, \ldots$, when

$$\xi(t) = \begin{cases} t^\rho, & o < t \leqq 1 \\ 1, & t > 1 \end{cases} \tag{6}$$

ρ being a fixed real number greater than $\frac{1}{2}$.

We assign to $S_\rho(I)$ the topology generated by semi-norms $\{\gamma_k\}$, thereby making it a countably multi-normed space. A sequence $\{\phi_n(x)\}$ where each $\phi_n(x) \in S_\rho(I)$ converges in $S_\rho(I)$ to $\phi(x)$ if $\gamma_k(\phi_n - \phi)$ tends to zero as $n \to \infty$ for each $k = o, 1, 2, \ldots$. A sequence $\{\phi_n(x)\}$ is said to be a Cauchy sequence if $\gamma_k(\phi_m - \phi_n)$ tends to zero as m and n approach to ∞ independently of each other for each $k = o, 1, 2, \ldots$. It can be seen that $S_\rho(I)$ is a Frechet space, i.e., a complete countably multi-normed space. The dual space $S_\rho'(I)$ consists of all continuous linear functionals on $S_\rho(I)$ and is equipped with the usual weak topology.

It can be checked that the space $D(I)$ is contained in $S_\rho(I)$, and the topology of $D(I)$ is stronger than that induced on it by $S_\rho(I)$. Hence the restriction of any $F \in S_\rho'(I)$ to $D(I)$ is in $D'(I)$.

It can also be checked that if $f(t)$ is a function defined on I such that

$$\int_o^\infty \frac{|f(t)|}{\xi(t)}\, dt < \infty,$$

then $f(t)$ generates a regular generalized function on $S_\rho(I)$ by

$$\langle f, \phi\rangle = \int_o^\infty f(t)\, \phi(t)\, dt; \ \phi \in S_\rho(I).$$

3. The Generalized S_2 Transform Of Generalized Functions

Let $\displaystyle\int_o^\infty \frac{1}{(x + y)^\rho (y + t)^\rho}\, dy \equiv K(x, t; \rho)$, for $x, t > o$ $(\rho > \frac{1}{2})$.

The generalized S_2-transform of $f \in S_\rho'(I)$ is defined as a function $F(x)$ obtained by applying $f(t)$ on the kernel $K(x, t; \rho)$, i.e.,

$$F(x) = \langle f(t), K(x, t; \rho)\rangle, \ x > o.$$

We state following results without proof. The details will be published elsewhere.

THEOREM 1. For fixed $x > o$ and $\rho > \frac{1}{2}$, $K(x, t; \rho)$ belongs to $S_\rho(I)$.

THEOREM 2. Let $f \in S'_\rho(I)$ and let $F(x)$ denote the generalized S_2-transform of f as defined by (7). Then $F(x)$ is infinitely differentiable function for $x > o$, and that

$$F^{(k)}_{(x)} = \langle f(t), \frac{\partial^k}{\partial x^k} K(x, t; \rho)\rangle, k = o, 1, 2, \ldots .$$

THEOREM 3. (Inversion) Let $f \in S'_\rho(I)$ and let $F(x)$ be the generalized S_2 transform of f as defined by (7). Then for any $\phi \in D(I)$, $\lim_{k\to\infty} H_{k,\, x;\, \rho} \langle F(x), \phi(x)\rangle = \langle f, \phi\rangle$ where

$$H_{k,\, x;\, \rho}(F(x)) = L_{k,\, x;\, \rho}(L_{k,\, x;\, \rho}(F(x))).$$

$$\equiv [\frac{2^{\rho-1}(2k-1)!\Gamma\rho}{k!(k-2)!\Gamma(2k+-1)}]2[x^{2k+\rho-2}\{x^{2k+\rho-2}F^{(k-1)}_{(x)}\}^{(2k-1)}]^{(k)};$$

the differentiation herein referring to distributional differentiation.

REFERENCES

1. Benedetto, J. J., "Analytic representation of generalized functions", Math. Zeitschrift, 97 (1967), 303-319.

2. Boas, R. P., Jr. and Widder, D. V., "The iterated Stieltjes transform", Trans. Amer. Math. Soc., 45 (1939), 1-72.

3. Dube, L. S., "An inversion of S_2-transform of generalized functions", Pac. J. Math., 61 No. 2 (1975), 383-390.

4. Dube, L. S., "On classical generalized iterated Stieltjes Transform", (submitted).

5. Erdélyi, A., "Stieltjes transform of generalized function", Proc. Royal Soc. Edin. 76(A), (1977), 231-249.

6. Pandey, J. N., "On the Stieltjes Transform of generalized functions", Proc. Camb. Phil. Soc. 71 (1972), 85-96.

7. Pathak, R. S., "A distributional generalized Stieltjes transformation", Proc. Edin. Math. Soc. 20 (1976), 15-22.

8. Widder, D. V., "The Laplace Transform", Princeton (1946).

9. Zemanian, A. H., "Generalized integral transformations", Inter Science Publishers, (1968).

DEGREE THEORETIC METHODS IN OPTIMAL CONTROL

R. E. Gaines

Colorado State University
Fort Collins, Colorado
U. S. A.

J. K. Peterson

Michigan Technological University
Houghton, Michigan
U. S. A.

We consider optimal control problems of the type

$$\text{minimize } \{\ell(x(0), x(t)) + \int_0^T L(t, x(t), \dot{x}(t))dt \mid x \in A_n^1[0, T]\} \tag{1.1}$$

where $A_n^1[0, T] = \{x: [0, T] \to R^n \mid x$ is absolutely continuous$\}$. We assume

$\ell: R^n \times R^n \to R \cup \{+\infty\}$ is a proper, lower semicontinuous and convex on $R^n \times R^n$ (1.2)

$L: [0, T] \times R^n \times R^n \to R \cup \{+\infty\}$ is a proper convex normal integrand on $R^n \times R^n$ (1.3)

Under mild assumptions on L, (1.1) possesses a solution if and only if the associated system of differential inclusions

$$\begin{aligned} &\dot{x}(t) \in \partial_p H(t, x(t), p(t)) \text{ a.e.} \\ &-\dot{p}(t) \in \partial_x H(t, x(t), p(t)) \text{ a.e.} \\ &(p(0), -p(T)) \in \partial\ell(x(0), x(t)) \end{aligned} \tag{1.4}$$

possesses a solution (x, p). The "∂" symbols here refer to subgradients and the function $H: [0, T] \times R^n \times R^n \to R \cup \{\pm\infty\}$ is defined by

$$H(t, x, p) = \sup \{p.v - L(t, x, v) \mid v \in R^n\} \tag{1.5}$$

The system (1.4) can be put into operator form as follows. Let

$$L(x, p) = (\dot{x}, -\dot{p}, p(0), -p(T))$$

and

$$G(x, p) = \partial_p H(t, x, p) \times \partial_x H(t, x, p) \times \partial \ell(x(0), x(T))$$

where the domain of L and G is $A_n^1[0, T] \times A_n^1[0, T]$.

A standard technique for proving the existence of a solution to (1.4) involves the use of set-valued topological degree. However, all topological degrees to date insist that L and G be nonempty valued on some open bounded subset of their domain. Many boundary functionals $\ell(a, b)$ have subdifferentials that are empty valued at "most" points $(a, b) \in R^n \times R^n$. For example, if

$$\ell(a, b) = \begin{cases} 0, & \text{if } a = b \\ + \infty, & \text{otherwise} \end{cases}$$

then

$$\partial\ell(a, b) = \begin{cases} \{(r, s) \mid r + s = 0\}, & \text{if } a = b \\ \phi, & \text{otherwise.} \end{cases}$$

For this ℓ, the periodic boundary functional, the operator G is empty-valued unless $x(0) = x(T)$ and $p(0) = p(T)$. Standard topological degrees do not apply to operators such as these.

To handle this situation, we develop a notion of topological degree that allows the operator G to be empty-valued at points in its domain. The construction is carried out by approximating the functional ℓ by a sequence of functions $\{\ell_n\}_{n=1}^{\infty}$ for which $\partial\ell_n$ is nonempty on all of $R^n \times R^n$. We then define the degree of the pair (L, G) relative to an open bounded set Ω and the point 0 to be

$$d[(L, G), \Omega, 0] = \lim_n d\,[(L, G_n), \Omega, 0]$$

where $G_n(x, p) = \partial_p H(t, x, p) \times \partial_x H(t, x, p) \times \partial \ell_n(x(0), x(T))$ and $d[(L, G_n), \Omega, 0]$ is the standard coincidence degree of the pair (L, G_n) using the set valued topological degree of Webb.

We use this new degree to develop a comprehensive existence theory for optimal control problems of type (1.1). The details of our arguments and all technical assumptions will appear in the Journal of Mathematical Analysis and Applications.

OSCILLATION THEOREMS FOR nth ORDER DELAY DIFFERENTIAL EQUATIONS

S. R. Grace and B. S. Lalli

University of Saskatchewan
Saskatoon, Saskatchewan
Canada

Abstract

Using weighted integrals some sufficient conditions are obtained under which an even order delay differential equation

$$x^{(n)} + q(t)f(x[g(t)]) = 0$$

is oscillatory. The results established include some known oscillation criteria.

Oscillation criteria are obtained for a class of functional differential equations of the form

$$(*) \quad x^{(n)} + q(t)f(x[g(t)]) = 0, \ n \ \text{even}.$$

1. Introduction

Our goal is to discuss, using weighted integrals, the oscillatory behaviour of solutions of the nonlinear delay equation

$$x^{(n)} + q(t)f(x[g(t)]) = 0, \ n \ \text{even}. \tag{1}$$

and to extend some of the known results for equation (1) when $n = 2$ to equation (1) for $n \geq 2$.

In what follows we consider only the nontrivial solutions of equation (1) which are indefinitely continuable to the right. A solution of equation (1) is said to be oscillatory if it has arbitrarily large zeros, and non-oscillatory if it is eventually of constant sign. The equation (1) is said to be oscillatory if every solution of equation (1) is oscillatory.

Oscillation theory for equation (1) has been developed by many authors. We mention in particular the papers by Mahfoud [9, 10], Kartsatos [4], Dahiya and Singh [1], Kusano and Onose [7].

Our main results in the following sections generalize some earlier standard oscillation results for retarded differential equations.

We will have an occasion to use the following Lemmas respectively by Kiguradze and Grammatikopoulis, Sficas and Staikos.

LEMMA 1. Let u be a positive an n times differentiable function on $[t_0, \infty)$. If $u^{(n)}(t)$ is of constant sign and is not identically zero in any interval of the form $[t_1, \infty)$, then there exist a $t_u \geq t_0$ and an integer ℓ, $0 \leq \ell \leq n$ with $n + \ell$ even for $u^{(n)} \geq 0$ or $n + \ell$ odd for $u^{(n)} \leq 0$ and such that $\ell > 0$ implies that $u^{(K)}(t) > 0$, for $t \geq t_u$, $(K = 0, 1, \ldots, \ell-1)$, and $\ell \leq n - 1$ implies that

$$(-)^{\ell+K}u^{(K)}(t) > 0, \quad \text{for } t \geq t_u, \ (K = \ell, \ldots, n - 1).$$

LEMMA 2. If the function u is as in Lemma 1 and

$$u^{(n-1)}(t)u^{(n)}(t) \leq 0 \quad \text{for } t \geq t_u$$

then for every λ, $0 < \lambda < 1$, there exists a constant $M > 0$ such that

$$u(\lambda t) \geq Mt^{n-1}|u^{(n-1)}(t)|, \quad \text{for all large } t.$$

In the sequel it is assumed that:

(i) $q : [t_0, \infty) \to [0, \infty)$, is a continuous function and $q(t)$ is not eventually identically equal to zero on any interval (t_1, ∞).

(ii) $g : [t_0, \infty) \to R$, is a continuous function, $g(t) \leq t$, and $g(t) \to \infty$ as $t \to \infty$.

(iii) $f : R \to R$, is a continuous function and $xf(x) > 0$ for $x \neq 0$.

Let $T \geq 0$. Suppose there exist a twice continuously differentiable function $\sigma : [T, \infty) \to (0, \infty)$ and a continuously differentiable function $F : R \to R$ such that:

(iv) $\sigma(t) \leq g(t)$, $\sigma^{\cdot}(t) > 0$ for $t \geq T$, $(^{\cdot} = \frac{d}{dt})$ and $\delta(t) \to \infty$ as $t \to \infty$,

(v) $|f(x)| \geq |F(x)|$ and $xF(x) > 0$ for $x \neq 0$,

(vi) $F'(x) \geq 0$ for $x \neq 0$, $(' = \frac{d}{dx})$,

(vii) $\int_{\varepsilon}^{\infty} \frac{du}{F(u)} < \infty$ and $\int_{-\varepsilon}^{-\infty} \frac{du}{F(u)} < \infty$ for some $\varepsilon > 0$, or

(viii) $\int_{0}^{\varepsilon} \frac{du}{F(u)} < \infty$ and $\int_{-0}^{-\varepsilon} \frac{du}{F(u)} < \infty$ for some $\varepsilon > 0$.

2. Main Results

THEOREM 1. Let conditions (i) - (vii) hold. Assume there is a twice continuously differentiable function

$$\zeta : [T, \infty) \to (0, \infty)$$

such that

(ix) $\int^{\infty} \zeta(s)q(s)ds = \infty$.

Then each of the following conditions ensures the oscillation of every solution of equation (1):

(I) $\zeta^{\cdot}(t) \geq 0$ and $\gamma(t) = \left(\frac{\zeta^{\cdot}(t)}{\sigma^{n-2}(t)\sigma^{\cdot}(t)}\right)^{\cdot} \leq 0$ for $t \geq T$,

(II) $\zeta^{\cdot}(t) \geq 0$ and $\int_{T}^{\infty} |\gamma(s)|ds < \infty$ for $t \geq T$.

Proof. Let $x(t)$ be a nonoscillatory solution of equation (1). Assume $x(t) > 0$ for $t \geq t_0 \geq T$ and choose a $t_1 \geq t_0$ so that $g(t) \geq t_0$ for $t \geq t_1$. Be Lemma 1, there exists a $t_2 \geq t_1$ such that

$$x^{(n-1)}(t) > 0 \text{ and } \dot{x}(t) > 0 \text{ for every } t \geq t_2.$$

Choose $t_3 \geq t_2$ so that $\sigma(t) \geq 2t_2$ for $t \geq t_3$.
It is easy to check that we can apply Lemma 2 for $u = \dot{x}$, $\lambda = \frac{1}{2}$ and conclude that there exist a $M > 0$ and a $t_4 \geq t_3$ such that

$$x^{\cdot}[\tfrac{1}{2}\sigma(t)] \geq M\sigma^{n-2}(t)x^{(n-1)}[\sigma(t)] \tag{2}$$

$$\geq M\sigma^{n-2}(t)x^{(n-1)}(t) \quad \text{for} \quad t \geq t_4.$$

We conclude from conditions (iv) and (v) that

$$x^{(n)} + q(t)F(x[\sigma(t)]) \leq 0 \quad \text{for} \quad t \geq t_4.$$

Multiplying both sides of the above inequality by $\dfrac{\zeta(t)}{F(x[\frac{1}{2}\sigma(t)])}$ and integrating from t_4 to t we get

$$\frac{\zeta(t)x^{(n-1)}(t)}{F(x[\frac{1}{2}\sigma(t)])} \leq \frac{\zeta(t_4)x^{(n-1)}(t_4)}{F(x[\frac{1}{2}\sigma(t_4)])} - \int_{t_4}^{t} \zeta(s)q(s)ds \tag{3}$$

$$+ \int_{t_4}^{t} \frac{\zeta^{\cdot}(s)x^{(n-1)}(s)}{F(x[\frac{1}{2}\sigma(s)])}$$

$$- \int_{t_4}^{t} \frac{\zeta(s)x^{(n-1)}(s)x^{\cdot}[\frac{1}{2}\sigma(s)](\frac{1}{2}\sigma^{\cdot}(s))F'(x[\frac{1}{2}\sigma(s)])}{F^2(x[\frac{1}{2}\sigma(s)])} ds.$$

Now we consider the following two cases:

CASE 1. We assume that the assumption (I) hold. By condition (vi), inequality (3) becomes

$$\frac{\zeta(t)x^{(n-1)}(t)}{F(x[\frac{1}{2}\sigma(t)])} \leq c_1 - \int_{t_4}^{t} \zeta(s)q(s)ds + \int_{t_4}^{t} \frac{\zeta^{\cdot}(s)x^{(n-1)}(s)}{F(x[\frac{1}{2}\sigma(s)])} ds,$$

where c_1 is a constant which depends on t_4. An application of inequality (2) now leads to the following estimate:

$$\frac{\zeta(t)x^{(n-1)}(t)}{F(x[\frac{1}{2}\sigma(t)])} \le c_1 - \int_{t_4}^{t} \zeta(s)q(s)ds \tag{4}$$

$$+ \int_{t_4}^{t} \frac{\zeta^{\cdot}(s)x^{\cdot}[\frac{1}{2}\sigma(s)]\frac{1}{2}\sigma^{\cdot}(s)}{\frac{1}{2}M\sigma^{\cdot}(s)\sigma^{n-2}(s)F(x[\frac{1}{2}\sigma(s)])} ds.$$

We may note that

$$I_1 = \int_{t_4}^{t} \frac{\zeta^{\cdot}(s)x^{\cdot}[\frac{1}{2}\sigma(s)]\frac{1}{2}\sigma^{\cdot}(s)}{\frac{1}{2}M\sigma^{\cdot}(s)\sigma^{n-2}(s)F(x[\frac{1}{2}\sigma(s)])} ds,$$

$$= -\left(\frac{1}{\frac{1}{2}M}\right) \frac{\zeta^{\cdot}(t)}{\sigma^{\cdot}(t)\sigma^{n-2}(t)} G(x(t))$$

$$+ \left(\frac{1}{\frac{1}{2}M}\right) \frac{\zeta^{\cdot}(t_4)}{\sigma^{\cdot}(t_4)\sigma^{n-2}(t_4)} G(x(t_4))$$

$$+ \left(\frac{1}{\frac{1}{2}M}\right) \int_{t_4}^{t} \gamma(s)G(x(s))ds,$$

where

$$G(x(t)) = \int_{x[\frac{1}{2}\sigma(t)]}^{\infty} \frac{du}{F(u)} \quad \text{and} \quad \gamma(t) = \left(\frac{\zeta^{\cdot}(t)}{\sigma^{\cdot}(t)\sigma^{n-2}(t)}\right)^{\cdot}.$$

Since $\gamma(t) \le 0$, it follows that

$$0 \le I_1 \le K_1$$

for some constant K_1.

Consequently inequality (4) becomes

$$\frac{\zeta(t)x^{(n-1)}(t)}{F(x\frac{1}{2}(t))} \leq c_2 - \int_{t_4}^{t} \zeta(s)q(s)ds,$$

where $c_2 = c_1 + k_1$.

CASE 2. Let the assumption (II) hold. From condition (vii) it follows that

$$0 \leq G(x(t)) \leq M_1 \quad \text{for some constant } M_1,$$

and hence,

$$I_1 = -\frac{1}{\frac{1}{2}M}\frac{\zeta^{\cdot}(t)}{\sigma^{\cdot}(t)\sigma^{n-2}(t)} G(x(t)) + \frac{1}{(\frac{1}{2}M)}\frac{\zeta^{\cdot}(t_4)}{\sigma^{\cdot}(t_4)\sigma^{n-2}(t_4)} G(x(t_4))$$

$$+ \frac{1}{(\frac{1}{2}M)}\int_{t_4}^{t} \gamma(s)G(x(s))ds$$

$$\leq (\frac{1}{\frac{1}{2}M})\frac{\zeta^{\cdot}(t_4)}{\sigma^{\cdot}(t_4)\sigma^{n-2}(t)} G(x(t_4))$$

$$+ \frac{M_1}{\frac{1}{2}M}\int_{t_4}^{t} |\gamma(s)|ds = K_2.$$

Thus inequality (4) becomes

$$\frac{\zeta(t)x^{(n-1)}(t)}{F(x[\frac{1}{2}\sigma(t)])} \leq c_3 - \int_{t_4}^{t} \zeta(s)q(s)ds,$$

where $c_3 = c_1 + K_2$. Now it follows from the above two cases that

$$\frac{\zeta(t)x^{(n-1)}(t)}{F(x[\frac{1}{2}\sigma(t)])} \to -\infty \quad \text{as} \quad t \to \infty.$$

Consequently, we must have $x^{(n-1)}(t) < 0$ for $t \geq t_5 \geq t_4$, which is a contradiction. The proof of the case $x(t) < 0$ for $t \geq t_0$ is similar and is omitted.

THEOREM 2. Suppose that conditions (i) - (vii) hold, and

$$\text{(X)} \quad \int_T^{\infty} \sigma^{\cdot}(s)\sigma^{n-2}(s)\left(\int_s^{\infty} q(\tau)\,d\tau\right)ds = \infty.$$

Then equation (1) is oscillatory.

Proof. Let $x(t)$ be a nonoscillatory solution of equation (1), say $x(t) > 0$ for $t \geq t_0 \geq T$. It follows as in the proof of Theorem 2, that for any $b \geq t_4$ and $\zeta(t) = 1$,

$$\frac{x^{(n-1)}(t)}{F(x[\frac{1}{2}\sigma(t)])} \leq \frac{x^{(n-1)}(b)}{F(x[\frac{1}{2}\sigma(b)])} - \int_b^t q(s)\,ds.$$

Since $x^{(n-1)}(t) > 0$ for all $t \geq b$, we have

$$0 \leq \frac{x^{(n-1)}(b)}{F(x[\frac{1}{2}\sigma(b)])} - \int_b^{\infty} q(s)\,ds.$$

Hence for all $t \geq b$

$$\int_t^{\infty} q(s)\,ds \leq \frac{x^{(n-1)}(t)}{F(x[\frac{1}{2}\sigma(t)])} .$$

Using inequality (2) we get

$$\int_t^{\infty} q(s)\,ds \leq \frac{x^{\cdot}[\frac{1}{2}\sigma(t)]\frac{1}{2}\sigma^{\cdot}(t)}{\frac{1}{2}M\sigma^{\cdot}(t)\sigma^{n-2}(t)F(x[\frac{1}{2}\sigma(t)])} .$$

Thus,

$$\frac{1}{2}M\int_b^t \sigma^\cdot(s)\sigma^{n-2}(s)\left(\int_s^\infty q(\tau)d\tau\right)ds \le \int_{x(b)}^{x(t)} \frac{du}{F(u)}\ .$$

This contradicts condition (vii), since the integral on the left diverges. A similar proof holds when $x(t) < 0$ for $t \ge t_0$.

COROLLARY. If conditions (i) - (vi) and (x) hold, then all bounded solutions of equation (1) are oscillatory.

REMARK 1. Theorem 1 which is obtained for $\sigma^\cdot(t) > 0$, extends easily to the case $\sigma^\cdot(t) \ge 0$, if the weight function $\zeta(t)$ belongs to the set $\{\sigma^i(t) \mid i$ is any real number in $[0, n-1]$ and $t \in (0, \infty)\}$.

REMARK 2. For the case $f(x) = x^\gamma$, where γ is the ratio of two positive integers, $\gamma > 1$, $g(t) = t - \tau(t)$ with $0 < \tau(t) \le M_1$ and $-M_1 \le t - \tau(t)$ for some constant M_1. We let $\zeta(t) = t$, then Theorem 1 in [1] and our Theorem 1 are equivalent. We note that if $n > 2$, then our Theorem 1 is stronger than Theorem 1 in [1].

REMARK 3. Kiguradze's results in [6] can be obtained from Theorem 1 by taking $f(u) = ug(u)$, $u \in R_+$, $q(t) = F(t, c^2)$, $c > 0$ and $\zeta(t) = t^{n-2}$ (g and F are functions which appeared in [6]).

REMARK 4. For n even, our Theorem 1 is similar to and some sense more general than Theorem 1 in [7], Theorem in [11], Theorem in [3] since in our proof we do not need the existence of a nondecreasing function ϕ required in some of the above mentioned results.

THEOREM 3. Let conditions (i) - (vi), (viii) hold. If there is a twice continuously function $\zeta : [T, \infty) \to (0, \infty)$ such that $\zeta^\cdot(t) \ge 0$, $\gamma(t) = \left(\frac{\zeta^\cdot(t)}{\sigma^{n-2}(t)\sigma^\cdot(t)}\right)^\cdot$ is in $L(t, \infty)$ and

$$\int^\infty \zeta(s)q(s) = \infty,$$

then every bounded solution of equation (1) is oscillatory.

Proof. Let $x(t)$ be a bounded nonoscillatory solution of equation (4.1). Assume $x(t) > 0$ for $t \geq t_0 \geq T$. As in the proof of Theorem 1, we write I_1 as follows:

$$I_1 = \left(\frac{1}{\frac{1}{2}M}\right) \frac{\zeta^{\cdot}(t)}{\sigma^{\cdot}(t)\sigma^{n-2}(t)} G_1(x(t)) - \left(\frac{1}{\frac{1}{2}M}\right) \frac{\zeta^{\cdot}(t_4)}{\sigma^{\cdot}(t_4)\sigma^{n-2}(t_4)} G(x(t_4))$$

$$- \left(\frac{1}{\frac{1}{2}M}\right) \int_{t_4}^{t} \gamma(s) G_1(x(s))\,ds,$$

where we define

$$G_1(x(t)) = \int_0^{x[\frac{1}{2}\sigma(t)]} \frac{du}{F(u)} .$$

From condition (viii) we conclude that

$$0 \leq G_1(x(t)) \leq M_2,$$

and

$$\left|\frac{\zeta^{\cdot}(t)}{\sigma^{\cdot}(t)\sigma^{n-2}(t)}\right| \leq N$$

for some constants M_2 and N. Thus

$$I_1 \leq \frac{M_2 N}{\frac{1}{2}M} - \frac{\zeta^{\cdot}(t_4)}{\frac{1}{2}M\sigma^{\cdot}(t_4)\sigma^{n-2}(t_4)} G(x(t_4)) + \frac{M_2}{\frac{1}{2}M} \int_{t_4}^{\infty} |\gamma(s)|\,ds = K_3$$

and hence inequality (4) becomes

$$\frac{\zeta(t)\ x^{(n-1)}(t)}{F(x[\frac{1}{2}\sigma(t)])} \to -\infty \quad \text{as} \quad t \to \infty$$

contradicting the fact that $x^{(n-1)}(t) > 0$ for $t \geq t_4$. A similar proof holds when $x(t) < 0$ for $t \geq t_0$.

In the following Theorem, we take $f(x) = |x|^\gamma \operatorname{sgn} x$ and $g(t) \geq 0$ for $t \geq t_0$.

THEOREM 4. Let conditions (i), (ii) hold, $0 < \gamma < 1$. If

(xi) $\int^\infty g^{(n-1)\gamma}(s)q(s)ds = \infty$,

then equation (1) is oscillatory.

<u>Proof.</u> Let $x(t)$ be a nonoscillatory solution of equation (1), with $x(t) > 0$ for $t \geq t_0 \geq 0$. Choose a $t_1 \geq t_0$ sufficiently large so that $g(t) \geq 2t_0$ for $t \geq t_1$. By Lemmas 1 and 2, we have

$$x^{\cdot}(t) > 0 \quad \text{and} \quad x^{(n-1)}(t) > 0 \quad \text{for} \quad t \geq t_1,$$

and there exist a constant $M > 0$ and $t_2 \geq t_1$ such that

$$x[\tfrac{1}{2}t] \geq Mt^{n-1}x^{(n-1)}(t) \quad \text{for} \quad t \geq t_2.$$

Thus, we have

$$x[g(t)] \geq x[\tfrac{1}{2}g(t)] \geq Mg^{n-1}(t)x^{(n-1)}[g(t)] \geq Mg^{n-1}(t)x^{(n-1)}(t).$$

Hence,

$$Mg^{(n-1)}(t) \leq \frac{x[g(t)]}{x^{(n-1)}(t)},$$

or

$$M^\gamma g^{(n-1)\gamma}(t) \leq (x[g(t)])^\gamma [x^{(n-1)}(t)]^{-\gamma}.$$

Integrating from t_2 to t we obtain

$$M^\gamma \int_{t_2}^t q(s)g^{(n-1)\gamma}(s)ds \leq \frac{1}{1-\gamma}\,[(x^{(n-1)}(t_2))^{1-\gamma} - (x^{(n-1)}(t))^{1-\gamma}]$$

$$\leq \frac{1}{1-\gamma}\,[x^{(n-1)}(t_2)]^{1-\gamma} < \infty.$$

This inequality contradicts (xi). A similar proof holds when $x(t) < 0$ for $t \geq t_0$.

THEOREM 5. Let conditions (i) - (iii) hold,

(xii) f is positive, nondecreasing and submultiplicative function on $[0, \infty)$, $f(0) = 0$,

(xiii) $\displaystyle\int_{r_0}^{r} \frac{du}{f(u)} \to \infty$ as $r \to \infty$, $r \geq r_0 > 0$.

If equation (1) is oscillatory, then

$$\int^{\infty} q(s)f([g(s)]^{n-1})ds = \infty.$$

Proof. We assume that

$$\int^{\infty} q(s)f([g(s)]^{n-1})ds < \infty,$$

and demonstrate that equation (4.1) has a nonoscillatory solution $x(t)$ such that

$$\lim_{t\to\infty} \frac{x(t)}{t^{n-1}} = a \neq 0 \quad \text{for some constant } a.$$

We omit the proof, since it is similar to that of Theorem 2.1 in [12], except that we use Bellman-Bihari's Lemma to show that the function F_1 defined by

$$F_1(t) = c + \int_{t_0}^{t} q(s)f(x[g(s)])ds, \; c > 0,$$

is bounded.

THEOREM 6. Let conditions (ii) - (iii), (xii), (xiii) hold. Suppose $q(t) > 0$ and

$$\int^{\infty} s^{n-1}q(s)f([g(s)]^{n-1})ds < \infty \quad \text{for} \quad t \geq t_0.$$

Further suppose that e is continuous function such that

$$\frac{e(t)}{q(t)f([g(t)]^{n-1})}$$

approaches a limit as $t \to \infty$. Then a necessary and sufficient condition for all oscillatory solutions of equation

$$(**) \quad x^{(n)} + q(t)|x[g(t)]|^{\gamma}\operatorname{sgn}(x[g(t)] = 0, \ \gamma > 0$$

to approach zero asymptotically is that

$$(\text{xiv}) \quad \lim_{t\to\infty} \frac{e(t)}{q(t)f([g(t)]^{n-1})} = 0.$$

Proof (Necessity). Let $x(t)$ be an oscillatory solution of equation (**) approaching zero as $t \to \infty$. Suppose that

$$\lim_{t\to\infty} \frac{|e(t)|}{q(t)f([g(t)]^{n-1})} \geq \alpha > 0.$$

Dividing equation (4.2) by $q(t)f([g(t)]^{n-1})$ and taking the limit as $t \to \infty$ we find that $x^{(n)}(t)$ has one sign to the right of sufficiently large t. This forces $\dot{x}(t)$ to eventually assume a constant sign which contradicts the fact that $x(t)$ is oscillatory.

SUFFICIENCY. Suppose condition (xiv) holds. Then

$$t^{n-1}|e(t)| < t^{n-1}q(t)f([q(t)]^{n-1})$$

for sufficiently large t. Since $\int^{\infty} s^{n-1}q(s)f([g(s)]^{n-1})ds < \infty$ we have $\int^{\infty} s^{n-1}|e(s)|ds < \infty$, and the conclusion follows by an argument developed by Singh [Hiroshima Math. J. 10 (1980)].

REFERENCES

1. S. Dahiya and B. Singh, "On the oscillatory behaviour of even order delay equations", J. Math. Anal. Appl. 42 (1973), 183-190.

2. M. K. Grammatikopoulis, Y. G. Sficas and V. A. Stackos, "Oscillatory properties of strongly superlinear differential equations with deviating arguments", J. Math. Anal. Appl. 67 (1979), 171-187.

3. I. V. Kamanev, "On the oscillation of solutions of a nonlinear equation of high order", Differencial'nye Uravnenija, 7 (1971), 927-929.

4. A. G. Kartsatos, "Recent results on oscillation of solutions of forced and perturbed nonlinear differential equations of even order", Stability of Dynamical Systems, Theory and Applications, Lect. Notes Marcel Dekker, New York, 1977, 17-72.

5. I. T. Kiguradze, "The problem of oscillation of solutions of nonlinear differential equation", Differencial'nye Uravenija 1 (1965), 995-1006.

6. I. G. Kigurdaze, "On oscillatory solutions of some ordinary differential equation", Soviet. Math. Dokl. 144 (1962), 33-36.

7. T. Kussano and H. Onose, "Oscillation of functional differential equations with retarded arguments", J. Diff. Eqn. 15 (1974), 269-277.

8. I. Licko and M. Svec, "Le caractere oscillatoire des solutions de l'equation $y^{(n)} + f(x)y^{\alpha} = 0$, $n > 1$", Czech. Math. J. 13 (1963), 481-489.

9. W. E. Mahfoud, "Oscillation theorems for a second order delay differential equations", J. Math. Anal. Appl. 63 (1978), 339-346.

10. W. E. Mahfoud, "Oscillation and asymptotic behaviour of solutions of nth order nonlinear delay differential equations", J. Diff. Eqn. 24 (1977), 75-98.

11. P. Marushiak, "Oscillation property of solutions of an nth order differential equation with time lag", Differencial'nye Uravnenija, Vol. 14, No. 7, 1978, 842-846.

12. G. Ladas, "Oscillation and asymptotic behaviour of solutions of differential equations with retarded arguments", J. Diff. Eqn. 10 (1971), 281-290.

13. W. E. Mahfoud, "Remark on some oscillation theorems for nth order differential equations with a retarded argument", J. Math. Anal. Appl. 62 (1978), 68-80.

14. J. S. W. Wong, "Second order oscillation with retarded arguments".

NUMERICAL TREATMENT OF A NONLINEAR PROBLEM ARISING IN HEAT TRANSFER*

C. W. Groetsch

University of Cincinnati
Cincinnati, Ohio
U.S.A.

Abstract

A numerical method is proposed for solving a nonlinear Abel-Volterra integral equation which models the interface temperature between a gas and semi-infinite solid with a nonlinear heat transfer law. The method consists of applying the Abel inversion formula to produce an integro-differential equation in which the nonlinearity appears only in the free term. The unknown function is then replaced by a B-spline approximation and the coefficients are determined by collocation at the knots. This leads to a triangular system of fixed point equations for the unknown coefficients which is solved by fixed point iteration.

W. R. Mann and his colleagues Wolf and Roberts have studied the problem of determining the temperature history $U(0, t)$ of the interface between a solid $(x \geq 0)$ which is initially at temperature 0° and a gas $(x < 0)$ which is constantly at temperature 1° with a nonlinear heat transfer law at the interface given by

$$U_x(0, t) = -G(U(0, t)),$$

where G is a strictly decreasing continuous function on $[0, 1]$ with $G(1) = 0$. In [1] it is shown that the interface temperature

*This work was partially supported by the United Kingdom Science Research Council while the author was a Senior Visiting Fellow in the Department of Mathematics at the University of Manchester.

$f(t) := U(0, t)$ satisfies the integral equation

$$\sqrt{\pi}\, f(t) = \int_0^t \frac{G(f(\tau))}{\sqrt{t - \tau}}\, d\tau \,, \tag{1}$$

and that this equation has a unique nondecreasing solution if G satisfies a Lipschitz condition. In [2] the existence of a unique strictly increasing solution was established without the assumption that G is Lipschitz.

Our aim in this brief note is to propose a numerical method for the solution of (1) and present some preliminary numerical findings. The method begins by applying the Abel inversion formula [3] to (1), noting that $f(0) = 0$, to obtain the integro-differential equation

$$\sqrt{\pi}\, G(f(t)) = \int_0^t \frac{f'(\tau)}{\sqrt{t - \tau}}\, d\tau \tag{2}$$

in which the nonlinearity has been freed from the integral. For a convenient set of basis functions $\{\phi_i\}_{i=1}^N$ (with $\phi_i(0) = 0$), the approximation

$$f = \sum_{j=1}^{N} a_i \phi_i$$

is made and the coefficients $\{a_i\}$ are determined by collocation at certain points $\{t_j\}_{j=1}^N$:

$$\sqrt{\pi}\, G\left(\sum_{i=1}^{N} a_i \phi_i(t_j)\right) = \sum_{i=1}^{N} a_i \int_0^{t_j} \frac{\phi_i'(\tau)}{\sqrt{t_j - \tau}}\, d\tau, \quad j = 1, \ldots, N. \tag{3}$$

If an approximate solution is sought on the interval $[0, T]$ and the basis functions $\{\phi_i\}$ are chosen to be piecewise linear B-splines based on the knots $t_j = jh$, $h = T/N$, then (3) leads to the following system of equations for the coefficients $\{a_i\}_{i=1}^N$:

$$\frac{\sqrt{\pi h}}{2} G(a_j) + \sum_{i=1}^{j-1} s_{ji} a_i = a_j, \quad j = 1, \ldots, N \tag{4}$$

where

$$s_{ji} := -\sqrt{j - i + 1} + 2\sqrt{j - i} - \sqrt{j - i - 1}.$$

To put it another way, a_j is a fixed point of the function

$$H_j(x) := \frac{\sqrt{\pi h}}{2} G(x) + k_j, \quad \text{where} \quad k_j := \sum_{i=1}^{j-1} s_{ji} a_i.$$

From the properties of G it follows immediately that H_1 has a unique fixed point $a_1 \in [0, 1]$. If we suppose that H_i has a unique fixed point $a_i \in [0, 1]$ for $i \leq j - 1$, then since

$$0 \leq k_j < \sum_{i=1}^{j-1} s_{ji} = -\sqrt{j} + \sqrt{j-1} + 1 < 1,$$

we find that

$$H_j(0) = \frac{\sqrt{\pi h}}{2} G(0) + k_j > 0$$

and

$$H_j(1) = k_j < 1.$$

It then follows that H_j has a unique fixed point $a_j \in [0, 1]$.

Also, if G satisfies a Lipschitz condition with constant L, then H_j satisfies a Lipschitz condition with constant $\sqrt{\pi h}\, L/2$. We summarize these remarks in the following:

THEOREM. For any $h > 0$, the system of fixed point equations (4) has a unique solution $\{a_i\}_{i=1}^N \subset [0, 1]$. Moreover, if G satisfies a Lipschitz condition and h is sufficiently small, then each of the mappings defined by (4) is strictly contractive.

Guided by the theorem, we may apply fixed point iteration to each equation of (4) in turn to approximate the coefficients a_j. In the actual implementation it is found that the iteration converges very quickly for small h if a_{j-1} is used as the starting value in the iteration to determine a_j.

One could of course use higher order B-splines for the basis functions $\{\phi_i\}$. For example, using quadratic B-splines based on $\{t_j\}$, one obtains the system

$$\frac{3}{4} \sqrt{\pi h}\, G\left(\frac{a_{j-1} + a_j}{2}\right) + \sum_{i=1}^{j-1} (a_i - a_{i-1})\, \sigma_{ji} + a_{j-1} = a_j$$

where

$$\sigma_{ji} = -(j - i + 1)^{3/2} + 2(j - i)^{3/2} - (j - i - 1)^{3/2}.$$

Again computational experience indicates fast convergence for small h.

As a test of the method the linear equation with $G(x) = 1 - x$, corresponding to Newton's law of cooling, was solved. In this particular case, the exact solution is known to be

$$f(t) = 1 - e^t \operatorname{erfc}(\sqrt{t})$$

and hence, the error in the approximate solution can be determined. Table I gives the observed percentage error at various points in the interval $[0, 10]$ and for various N, using both linear and quadratic splines. In performing the computation, the fixed point iterations were terminated when successive iterates differed by less than 10^{-6}. Computed interface temperatures for the nonlinear problem with $G(x) = 1 - x^4$, corresponding to a cooling law of Stefan-Boltzmann type, and using quadratic splines are displayed in Table II.

A glance at these numerical results shows that the approximations at the first few knots are not reliable. This can be attributed to the fact that $f(t)$ generally has a singular derivative at $t = 0$, while the basis functions do not. In fact, if we assume that G has the form

$$G(x) = A(1 - H(x))$$

where H is continuous, increasing and bounded, then by (1)

$$f(t) = \frac{A}{\sqrt{\pi}} \int_0^t \frac{1 - H(f(\tau))}{\sqrt{t - \tau}} d\tau$$

$$f(t) = A\sqrt{\frac{t}{\pi}} \; \{2 - \int_0^1 \frac{H(f(tu))}{\sqrt{1 - u}} du\}$$

and hence by the Bounded Convergence Theorem $f(t) = 0(\sqrt{t})$ as $t \to 0^+$. Suitable approximations can, however, by found for small times simply by using a small T and sufficiently many knots so that the time in question does not occur at any of the first few knot points.

TABLE I

t \ N	QUAD 200	LIN 200	QUAD 400	LIN 400	QUAD 800	LIN 800
.1	5.4	11.0	4.5	5.4	2.4	2.6
.2	3.9	5.0	2.2	2.4	1.1	1.2
.5	1.5	1.6	.72	.78	.38	.38
2.	.25	.25	.12	.13	.06	.25
3.	.14	.14	.07	.15	.52	.50
4.	.09	.10	.04	.23	.29	.70
6.	.05	.12	.45	.43	.12	.85
9.	.28	.23	.12	.61	.08	.94
10.	.29	.27	.08	.64	.08	.96

TABLE II

t \ N	50	100	150	300
.06	.1627	.2495	.2496	.2641
.12	.3510	.3636	.3707	.3784
.30	.5547	.5673	.5712	.5750
.60	.7148	.7193	.7212	.7228
1.02	.8038	.8057	.8064	.8071
1.98	.8743	.8749	.8751	.8735
3.00	.9030	.9037	.9035	.9024

ACKNOWLEDGEMENTS

The author wishes to thank Dr. M. P. Benson for a number of helpful conversations and Mr. Donyen Chan for independently verifying some of the numerical calculations.

REFERENCES

1. W. R. Mann and F. Wolf, "Heat transfer between solids and gases under nonlinear boundary conditions", Quart. Appl. Math 9 (1951), 163-184.

2. J. H. Roberts and W. R. Mann, "On a certain nonlinear integral equation of Volterra type ", Pacific J. Math. 1 (1951), 431-445.

3. F. G. Tricomi, "Integral Equations", Intersciences, New York, 1957.

ITERATIONS DETERMINISTES ET STOCHASTIQUES SUR LES INTERVALLES

F. Guenard*

Ecole Normale Supérieure de l'Enseignement
Cachan
France

0. Introduction - Motivations

Considérons un ordinateur effectuant un algorithme itératif. Le fait que cet algorithme soit sensé aboutir peut souvent être modélisé mathématiquement par la notion de contraction asymptotique: si X est un espace topologique séparé, on dira qu'une application continue $f : X \to X$ est une *contraction asymptotique* si elle a un unique point fixe $c \in X$, et si, pour tout $x \in X$, la suite $[f^n[x]]_{n\in\mathbb{N}}$ converge vers c; on a noté $f^n[x] = f \circ f \circ \ldots \circ f[x]$ [n fois] la n-ième itérée de x par f, et posé: $f^o[x] = x$.

Hormis quelques cas exceptionnels comme celui de $X = \mathbb{Q}$ et de $f \in \mathbb{Q}[X]$, où f est un polynôme à coefficients rationnels, l'ordinateur ne fait jamais intervenir exactement f. A chaque étape de l'itération, il tronque ses calculs, de sorte que le processus mathématique défini par: $x_{n+1} = f[x_n]$ n'est pas respecté par la machine, et se trouve remplacé par un autre processus.

Il y a diverses modélisations possibles de ce dernier. Trois grands types de perturbations envisageables ont été énoncés par Collet et Eckmann ([2], p.9). Bien que l'optique de ces auteurs soit légèrement différente, on peut reprendre leur classification:

(1) perturbation du point initial: "Est-ce que les systèmes dont les états initiaux sont proches évoluent de facon similaire?"

*Current affiliation: Centre Pédagogigue Régional de l'Académie de Créteil, Creteil Cedex, France.

(2) perturbation de la fonction f : "f est seulement approximativement connue";

(3) perturbation stochastique: "La véritable équation n'est pas $x_{n+1} = f(x_n)$, mais il y a des termes de bruit que l'on peut modéliser en écrivant:

$$x_{n+1} = f(x_n) + r(x_n),$$

où $r(x)$ est un petit terme aléatoire, c'est-à-dire une variation de $f(x_n)$ suivant une loi de probabilité connue *a priori*".

Collet et Eckman citent dans leur ouvrage des travaux portant principalement sur le problème (1); le cadre est surtout celui des fonctions continues $f : [-1, 1] \to [-1, 1]$ ayant un unique maximum. En général, ces fonctions ne sont pas des contractions asymptotiques, ni même des applications *itérativement convergentes*, c'est-à-dire des fonctions f telles que, pour tout x, la suite $(f^n(x))_{n\in\mathbb{N}}$ converge. C'est ce qui motive leur étude; partant des points de vue ensemblistes, topologiques ou de la théorie de la mesure, ils s'intéressent à la nature de la suite $(f^n(x))$, à celle de ses valeurs d'adherence, à celle de l'ensemble des x dont la suite des itérées a telle ou telle propriété.

Revenons au problème de l'ordinateur qui tronque. Si l'on rand compte d'un algorithme convergent par une contraction asymptotique, le problème (1) disparaît. Il n'est à considérer que dans une modélisation par une application itérativement convergente.

Nous ne considérerons ici que des applications continues d'un intervalle I de $\mathbb{R}$ dans lui-même.

Dans ce cadre, la structure des contractions asymptotiques et des applications itérativement convergentes est connue (cf. [1] et [5]). Le problème (1) est implicitement résolu par les démonstrations caractérisant les applications itérativement convergentes que l'on trouve dans ces deux reféréncES.

Nous ne considérerons donc ici que deux modèles correspondant respectivement aux problèmes (2) et (3) de Collet et Eckmann:

- les balades aléatoires;
- le processus autorégressif d'ordre 1.

Les balades aléatoires

Intuitivement, une balade aléatoire est un processus itératif où à chaque pas, la fonction utilisée est tirée au sort.

Plus précisément, soient X un espace topologique séparé, Λ un ensemble, $(\mu_n)_{n\in\mathbb{N}}$ une suite de probabilités sur Λ, et $(f_\lambda)_{\lambda\in\Lambda}$ une famille de contractions asymptotiques de X. On appelle *balade aléatoire* associée à $((f_\lambda)_{\lambda\in\Lambda}; (\mu_n)_{n\in\mathbb{N}})$ l'application aléatoire $B : X \times \mathbb{N} \to X$ définie par:

$$B(x, n) = h_n \circ h_{n-1} \circ \dots \circ h_1(x) \quad \forall\, n \in \mathbb{N}$$

$$B(x, 0) = x$$

avec $h_n = f_\lambda$ où $\lambda \in \Lambda$ a pour loi μ_n.

On posera: $H_n = h_n \circ h_{n-1} \circ \dots \circ h_1$.

Les balades aléatoires posent différents types de problèmes, mais nous n'en considérerons ici qu'un dont nous donnerons la solution: recherche de conditions pour que, pour tout x, la suite $(H_n(x))_{n\in\mathbb{N}}$ converge, indépendamment de la suite $(\mu_n)_{n\in\mathbb{N}}$. Pour cela, il faut bien sûr supposer que les fonctions f_λ ont toutes le même point fixe. Mais il est faux de croire qu'en général, si les fonctions f_λ ont toutes le même point fixe, c, la suite $(H_n(x))_{n\in\mathbb{N}}$ convergera vers c.

Pour revenir à notre ordinateur qui tronque, cela signifie qu'en cas de "mauvaise" troncature, le programme ne marchere pas. Toujours en termes d'ordinateurs, cela peut avoir d'autres applications. Imaginons par exemple le couplage de plusieurs ordinateurs $0_1, \dots, 0_p$ disposant chacun d'un programme itératif permettant - disons pour fixer les idées - de résoudre des équations différentielles. Le traitement d'une équation est initialement confié à un ordinateur 0_{i_1} qui commence le calcul.

Pour des raisons de priorité dans le traitement des programmes, ce calcul est interrompu, abandonné par 0_{i_1}, et confié au premier ordinateur disponible, 0_{i_2}. Celui-ci reprend le calcul en l'état où l'avait laissé 0_{i_1}, c'est-à-dire à partir de la dernière approximation de la solution qu'avait calculée 0_{i_1}. Et ainsi de suite. En général, un tel couplage ne marchera pas, c'est-à-dire que l'on n'obtiendra pas une "bonne approximation" de la solution.

Mathématiquement, on dira qu'une famille $(f_\lambda)_{\lambda \in \Lambda}$ de contractions asymptotiques $X \to X$ ayant un même point fixe, c, est *universellement attractive* si, pour toute réalisation, et tout $x \in X$, la suite $(H_n(x))_{n \in \mathbf{N}}$ est convergente, et si la limite est le point c. La propriété d'être universellement attractive ne dépend que de la famille de contractions, pas de la suite $(\mu_n)_{n \in \mathbf{N}}$ de probabilités: c'est une propriété purement fonctionnelle, et non aléatoire.

Remarque et probléme ouvert

Collet et Eckmann, dans l'énoncé du problème (2) semblent plutôt penser à la stabilité des propriétés itératives d'une fonction f quand on la remplace tout au long de l'itération par une fonction g "proche" de f. Dans le cas d'un ordinateur qui tronque, le calcul (approximatif) de $f(x)$ pour un x donné se fait à l'aide d'un certain nombre d'opérations élémentaires où il y a troncature. Le résultat est un nombre $\bar{f}(x)$. En faisant varier x, on obtient une fonction $\bar{f}$. Peut-on espérer un théorème de sélection assignant à $\bar{f}$ des propriétés raisonnables? Si tel était le cas, le modèle déterministe auquel semblent penser Collet et Eckmann serait adapté au problème de l'ordinateur qui tronque.

Le processus autorégressif d'ordre 1

Il est défini, dans le cadre choisi de la droite réelle, par la formule:

$$x_{n+1} = f(X_n) + \tau_n,$$

où (τ_n) est une suite de variables aléatoires indépendantes équidistribuées (bruit blanc).

Les processus définis par la formule précédente ont été surtout étudiés

- d'une part dans le cas où la suite (τ_n) tend vers 0,

- d'autre part dans le cas où la loi des τ_n domine la measure de Lebesgue. Dans le premier cas, il ne peut être question d'équidistribution; cette hypothèse n'est pas justifiée pour notre problème: a-t-on déjà vu un ordinateur tronquant de moins en moins, indéfiniment, au fur et à mesure qu'il progresse dans une itération? Dans le deuxième cas, l'hypothèse utilisée signifie intuitivement que l'ordinateur peut faire des erreurs énormes. Cela peut paraitre excessif d'un point de vue physique, mais mathématiquement, la raison de cette hypothèse est claire: si la fonction f est raisonnable - disons continue -, pour tout $x \in \mathbb{R}$, et tout ouvert A, la probabilité $\mathbf{P}(X_1 \in A \mid X_o = x)$ est strictement positive. Dès lors, une hypothèse limitant la croissance de f à l'infini conduira à la récurrence de la chaine de Markov X_n. Dans ce dernier cas, l'aspect probabiliste prime le côté fonctionnel du problème.

Devant les bonnes dispositions de ce processus, on peut envisager d'aller plus loin et d'en étudier les aspects statistiques, en construisant des estimateurs. Revenons une dernière fois à notre ordinateur qui tronque en calculant par approximations successives la solution d'une équation différentielle. Chaque étape de l'approximation est le fruit d'un grand nombre, L, d'opérations élémentaires. Si à chaque opération élémentaire l'ordinateur travaille avec une précision de q chiffres, après L opérations, seuls les $p (p < q)$ premiers chiffres du résultat sont fiables. Les $q - p$ suivants ne sont individuellement pas significatifs. L'idée de la construction d'un estimateur pour le processus autorégressif est d'utiliser ces résultats à q chiffres dont on sait que seulement p sont bons pour calculer une estimation dont $p + r$ $(r < q - p)$ chiffres aient une forte probabilité d'être corrects.

1. L'attractivite Universelle Sur Les Segments

Soient $I = [a, b]$ un intervalle compact de $\mathbf{R}$, c un élément de I. Pour toute fonction continue $f : I \to I$, posons:

$$H(f) = \{(x, f(x)), x \in I\} \quad \text{(graphe de } f)$$

$$H_g(f) = \{(x, f(x)), x \in [a, c]\} \quad \text{(graphe gauche de } f)$$

$$\mathcal{H}_d(f) = \{(x, f(x)),\ x \in [c, b]\} \quad \text{(graphe droit de } f).$$

Désignons par Δ la diagonale de I^2, et posons:

$$\Delta_g = \Delta \cap [a, c]^2,\ \Delta_d = \Delta \cap [c, b]^2.$$

Si A est une partie de I^2, on désignera par A^{-1} l'ensemble $A^{-1} = \{(y, x) \mid (x, y) \in A\}$, par $\overline{A}$ l'adhérence de A, et par $\overset{\circ}{A}$ l'intérieur de A.

On dira que A *est séparé de* $\Delta - \{c\}$ si les deux conditions suivantes sont vérifiees:

- pour tout point (y, y) de $\Delta_g - \{(a, a); (c, c)\}$, il exists $x < y$ tel que l'intersection de l'intérieur du triangle $T_g(x, y)$ de sommets $((y, y); (x, x); (x, y))$ avec A soit vide:

$$\forall y \in]a, c[,\ \exists x \times y,\ A \cap \overset{\circ}{T_g}(x, y) = \emptyset$$

- pour tout point (z, z) de $\Delta_d - \{(b, b); (c, c)\}$, il existe $t > z$ tel que l'intersection de l'intérieur du triangle $T_d(z, t)$ de sommets $((z, z); (t, z); (t, t))$ avec A soit vide:

$$\forall z \in]b, c[,\ \exists t > z,\ A \cap \overset{\circ}{T_d}(z, t) = \emptyset.$$

Soient A et B deux parties de $\overset{\circ}{I}{}^2_- = \{(x, y) \in I^2 \mid x > y\}$. Soit $\overline{\overline{A}}$ (resp. $\overline{\overline{B}}$) l'adhérence de A (resp. B) dans $\overset{\circ}{I}{}^2_-$. On dira que A et B sont *droitement séparés* si $A \cap B = \emptyset$, et si, pour tout $(x, y) \in \overline{\overline{A}} \cap \overline{\overline{B}}$, il existe un carré $C(x, y, h)$ $(h > 0)$ de sommets $((x, y); (x + h, y); (x + h, y - h); (x, y - h))$ vérifiant l'une au moins des deux relations suivantes:

$$\overset{\circ}{C(x, y, h)} \cap \overline{\overline{A}} = \emptyset$$

$$\overset{\circ}{C(x, y, h)} \cap \overline{\overline{B}} = \emptyset$$

Avec ces notations, on peut caractériser les contractions asymptotiques de I, et les familles universellement attractives:

THEOREME 1 [5]. *Caractérisation des contractions asymptotiques.* Soit f une application continue $I \to I$. Les assertions suivantes sont équivalentes:

(1) f est une contraction asymptotique de I dont c est le point fixe.

(2) $\bigcap_{n \in \mathbf{N}} f^n(I) = \{c\}$.

(3) $H(f) \cap H^{-1}(f) = \{(c, c)\}$.

(4) $H^{-1}_g(f) \cap H_d(f) = \{(c, c)\}$.

THEOREME 2. *Caractérisation des familles universellement attractives.* Soit $(f_\lambda)_{\lambda \in \Lambda}$ une famille de contractions asymptotiques dont c est le point fixe commun. La famille $(f_\lambda)_{\lambda \in \Lambda}$ est universellement attractive si et seulement si les trois conditions suivantes sont verifiées:

$$(\bigcup_{\lambda \in \Lambda} H_g(f_\lambda))^{-1} \cap (\bigcup_{\lambda \in \Lambda} H_d(f_\lambda)) = \{(c, c)\}; \quad (1)$$

$$\bigcup_{\lambda \in \Lambda} H(f_\lambda) \text{ est séparé de } \Delta - \{(c, c)\}; \quad (2)$$

$$(\bigcup_{\lambda \in \Lambda} H_g(f_\lambda))^{-1} - \{(c, c)\} \text{ est droitement séparé de}$$

$$\bigcup_{\lambda \in \Lambda} H_d(f_\lambda) - \{(c, c)\}.$$

Démonstration Du Théoreme 2: *Première partie: Condition nécessaire.*

(1) La condition $(\bigcup_{\lambda \in \Lambda} H_g(f_\lambda))^{-1} \cap (\bigcup_{\lambda \in} H_d(f_\lambda)) = \{(c, c)\}$ est nécessaire.

Soit E l'ensemble défini par: $E = (\bigcup_{\lambda \in \Lambda} H^{-1}_g(f_\lambda)) \cap (\bigcup_{\lambda \in \Lambda} H_d(f_\lambda))$. Il résulte du théoreme 1 que (c, c) appartient à E. Supposons qu'un autre point, (x, y) appartienne à E. Il existe un couple (λ, θ) d'éléments de Λ tels que:

$y = f_\lambda(x), \; x = f_\theta(y).$

Considérons la réalisation $(H_n)_{n\in\mathbf{N}}$ de la balade, définie par:

$$\forall\, n \in \mathbb{N},\; h_{2n+1} = f_\lambda,\; h_{2n+2} = f_\theta$$

$$H_n = h_n \circ h_{n-1} \circ \ldots \circ h_1.$$

On a: $\forall\, n \in \mathbb{N},\; H_{2n+1}(x) = y$ et $H_{2n+2}(x) = x$. Pour que la suite $(H_n(x))_{n\in\mathbf{N}}$ converge, il faut $x = y$, ce qui implique $x = y = f_\lambda(x) = c$, car f_λ est une contraction asymptotique de point fixe c. Il faut donc $(x, y) = (c, c)$, et la condition est nécessaire.

(2) La condition: $\bigcup_{\lambda\in\Lambda} H(f_\lambda)$ est séparé de $\Delta - \{(c, c)\}$ est nécessaire:

Supposons que $\bigcup_{\lambda\in\Lambda} H(f_\lambda)$ ne soit pas séparé de $\Delta - \{(c, c)\}$, et soit (ℓ, ℓ) un point de $\Delta - \{(c, c)\}$ tel que l'intérieur de tout triangle $T_d(\ell, \ell+h)$ ($\ell > c$ et $h > 0$) rencontre $\bigcup_{\lambda\in\Lambda} H_d(f_\lambda)$. Si c'est $\bigcup_{\lambda\in\Lambda} H_g(f_\lambda)$ qui n'est pas séparé de $\Delta_g - \{(c, c)\}$, la démonstration est la même, *mutatis mutandis*. Posons, pour tout $n \in \mathbb{N}$, $h_n = 1/2^n$. Pour tout entier n, il existe $\lambda_n \in \Lambda$ tel qu'il existe $x_n \in \,]\ell, \ell+h_n[$ tel que:

$$(x_n, f_{\lambda_n}(x_n)) \in \overline{\overset{\circ}{T_d}(\ell, \ell+h_n)}.$$

On a en particulier:

$$f_{\lambda_n}(x_n) > \ell \quad \text{et} \quad f_{\lambda_n}(x_n) < x_n < \ell + h_n.$$

Mais comme $f_{\lambda_n}(\ell) < \ell$, il existe, d'après le théorème des valeurs intermédiaires, $y_n < x_n$ tel que:

$$f_{\lambda_n}(y_n) = \ell, \quad \text{et} \quad f_{\lambda_n}([y_n, x_n]) \subset T_d(\ell, \ell + h_n)$$

L'intervalle $[y_n, x_n]$ n'est pas réduit à un point. Définissons donc par récurrence, pour tout entier n, un tel intervalle $[y_n, x_n]$, en impossant de plus: $x_{n+1} < y_n$. Soit $p \in \mathbb{N}$. L'ensemble $f_{\lambda_{p-1}}^{-1}([y_p, x_p]) \cap [y_{p-1}, x_{p-1}]$ est un compact non vide de $[y_{p-1}, x_{p-1}]$; désignons - le par $K(p, p-1)$. Définissons par récurrence descendante sur n, les ensembles $K(p, n)$, $p - 1 > n > 0$, par:

$$K(p, n) = f_{\lambda_n}^{-1} (K(p, n+1)) \cap [y_n, x_n].$$

Pour tout n, l'ensemble $K(p, n)$ est un compact non vide. On a de plus, pour tout entier p, l'inclusion:

$$k(p+1, 1) \subset K(p, 1),$$

de sorte que $\bigcap_{p \in \mathbb{N}^*} K(p, 1)$ est un compact non vide. Soit z un élément de cette intersection. Considérons la réalisation suivante de la balade:

$$h_n = f_{\lambda_n}, \; H_n = h_n \circ \ldots \circ h_1.$$

Par définition de $K(p, 1)$, on a : $H_n(z) \in [y_n, x_n]$, ce qui montre que la suite $(H_n(z))_{n \in \mathbb{N}}$ converge vers ℓ, et que la condition considérée est nécessaire.

(3) La condition "$(\bigcup_{\lambda \in \Lambda} H_g(f_\lambda))^{-1} - \{(c, c)\}$ est droitement séparé de $\bigcup_{\lambda \in \Lambda} H_d(f_\lambda) - \{(c, c)\}$" est nécessaire. Supposons qu'il existe un point $M = (u, v) \in I^2$, $u > v$ tel que l'intérieur de tout carré $C(u, v, h)$ de sommets $((u, v); (u, v-h); (u+h, v-h); (u+h, v))$ $(h > 0)$ contienne simultanément des points de $\bigcup_{\lambda \in \Lambda} H_d(f_\lambda)$ et des points de $(\bigcup_{\lambda \in \Lambda} H_g(f_\lambda))^{-1}$.

Pour un point $M(u, v)$, $u < v$ vérifiant la condition correspondante à gauche, le raisonnement est le même, *mutatis mutandis*.

Remarquons dout d'abord que, pour tout $\lambda \in \Lambda$, on a:

$$f_\lambda(u) > v \tag{1}$$

$$f_\lambda(x) = u \quad x < v \tag{2}$$

Cela étant, posons: $H_n = 1/2^n$, $n \in \mathbb{N}$. Il existe $\lambda_o \in \Lambda$ tel que:

$$\mathcal{H}_d(f_{\lambda_o}) \cap C(u, v, h_o) \neq \emptyset \tag{3}$$

Il existe un intervalle compact non réduit à un point $[x_o, y_o]$ de $]u, u+h_o]$ vérifiant:

$$f_{\lambda_o}(x_o) = v \quad \text{(d'après (1), (3) et le théorème des valeurs intermédiaires)} \tag{4}$$

$$\forall x \in]x_o, y_o[\quad f_{\lambda_o}(x) \in]v-h_o, v[\tag{5}$$

Posons:

$$k_1 = \min\{h_1; x_o - u; v - \inf\{f_{\lambda_o}(z) \mid z \in [x_o, y_o]\}\} \tag{6}$$

Il existe $\lambda_1 \in \Lambda$ tel que:

$$\mathcal{H}_g^{-1}(f_{\lambda_1}) \cap \overline{\overset{o}{C}(u, v, k_1)} \neq \emptyset \tag{7}$$

Il existe un intervalle compact non réduit à un point $[\alpha_1, \beta_1]$ inclus dans $[v-k_1, v[$ vérifiant:

$$f_{\lambda_1}(\beta_1) = u \tag{8}$$

$$\forall \gamma \in]\alpha_1, \beta_1[\quad f_{\lambda_1}(\gamma) \in]u, u+k_1[\tag{9}$$

L'ensemble $K(0, 1) = [x_o, y_o] \cap f_{\lambda_o}^{-1}([\alpha_1, \beta_1])$ est un compact non vide.

De plus, pour tout $x \in K(0, 1)$, on a:

$$f_{\lambda_o}(x) \in [\alpha_1, \beta_1], \quad \text{et} \quad f_{\lambda_1} \circ f_{\lambda_o}(x) \in]u, u+h_1[\tag{10}$$

Construisons par récurrence pour chaque entier i:

- un indice $\lambda_i \in \Lambda$;
- un intervalle: $[x_i, y_i]$ si i est pair

  $[\alpha_i, \beta_i]$ si i est impair;
- un réel k_i

reliés par les relations suivantes:

$$[x_i, y_i] \subset u,]u+k_i] \quad \text{(i pair)}$$

$$[\alpha_i, \beta_i] \subset [v-k_i, v[\quad \text{(i impair)}$$

$$H_d(f_{\lambda_i}) \cap \overline{\overset{\circ}{C}(u, v, k_i)} \neq \emptyset \quad \text{(i pair)}$$

$$H_g^{-1}(f_{\lambda_i}) \cap \overline{\overset{\circ}{C}(u, v, k_i)} \neq \emptyset \quad \text{(i impair)}$$

$$k_i = \min\{h_i, x_{i-1} - u, v - \inf\{f_{\lambda_{i-1}}(z) \mid z \in [x_{i-1}, y_{i-1}]\}\}$$

si i est impair

$$k_i = \min\{h_i, v - \beta_{i-1}, \sup\{(f_{\lambda_{i-1}}(\gamma) \mid \gamma \in]\alpha_{i-1}, \beta_{i-1}]\} - u\}$$

si i est pair

$$f_{\lambda_i}(x_i) = v \quad \text{(i pair)}$$

$$f_{\lambda_i}(\beta_i) = u \quad \text{(i impair)}$$

$$\forall x \in]x_i, y_i], f_{\lambda_i}(x) \in]v - k_i, v[\quad \text{(i pair)}$$

$$\forall \gamma \in]\alpha_i, \beta_i], f_{\lambda_i}(\gamma) \in]u, u + k_i[\quad \text{(i impair)}$$

Tout ceci étant construit, posons:

$$K(2n, 2n + 1) = [x_{2n}, y_{2n}] \cap f_{\lambda_{2n}}^{-1}([\alpha_{2n+1}, \beta_{2n+1}])$$

$$K(2n-1, 2n) = [\alpha_{2n-1}, \beta_{2n-1}] \cap f_{\lambda_{2n-1}}^{-1}([x_{2n}, y_{2n}]).$$

Les ensembles $K(i, i+1)$ sont des compacts non vides. Par récurrence descendante sur p, on définit $k(p, n)$, $0 \le p \le n-1$ par:

$$K(p, n) = K(p, p+1) \cap K(p+1, n).$$

Les ensembles $K(p, n)$ sont des compacts non vides. Il en est particulièrement ainsi des $K(0, n)$, qui forment une suite décroissante, dont l'intersection n'est pas vide. Soit $x \in \bigcap_{n\in\mathbf{N}} K(0, n)$. La suite $(f_{\lambda_{2n}}(x), f_{\lambda_{2n+1}}(x))_{n\in\mathbf{N}}$ converge vers (v, u), ce qui montre que la famille $(f_\lambda)_{\lambda\in\Lambda}$ n'est pas universellement attractive, et achève la première partie de la démonstration du théorème.

Deuxième partie: Condition suffisante

Supposons que les conditions (1), (2) et (3) du théorème soient vérifiees. Soient $x \in I$, $(H_n)_{n\in\mathbf{N}}$ une réalisation de la balade:

$$H_n = h_n \circ h_{n-1} \circ \ldots \circ h_1$$
$$\forall n \in \mathbf{N}^*, \exists\lambda_n \in \Lambda, h_n = f_{\lambda_n}.$$

Considérons la suite $(H_n(x))_{n\in\mathbf{N}}$. Si cette suite est monotone au-delà d'un certain rang, elle est convergente. Supposons, quitte à changer de sens, qu'elle soit décroissante:

$$\exists n_o, \forall n > n_o, H_n(x) < H_{n-1}(x)$$

ce qui s'écrit aussi: $f_{\lambda_n}(H_{n-1}(x)) < H_{n-1}(x)$. Or f_{λ_n} est une contraction asymptotique de point fixe c; il résulte du théorème 1 que le graphe droit de f_{λ_n} est au-dessous de Δ, la diagonale, tandis que $\mathcal{H}_g(f_{\lambda_n})$ est au-dessus. La relation précédente suppose donc $c < H_{n-1}(x)$. La limite ℓ de la suite $(H_n(x))$ vérifie: $c \leq \ell$. Supposons $c < \ell$. On a: $\lim_{n\to\infty} (H_n(x), H_{n+1}(x)) = (\ell, \ell)$ et: $H_{n+1}(x) < H_n(x)$. Ceci suppose $(H_n(x), H_{n+1}(x)) \in T_d(\ell,.)$ pour n assez grand; ceci est exclu par la condition (2).

On a montré que si la suite $(H_n(x))_{n\in\mathbf{N}}$ est monotone au-delà d'un certain rang, elle est alors convergente vers c.

Si la suite $(H_n(x))_{n\in\mathbf{N}}$ n'est pas monotone au-delà d'un certain rang, introduisons les deux suites $(m_j)_{j\in\mathbf{N}}$ et $(p_k)_{k\in\mathbf{N}}$ des indices de changement de signe du terme $H_{n+1}(x) - H_n(x)$.

Supposons, quitte à décaler d'un cran les suites (m_j) et (p_k), que l'on ait: $H_1(x) < x$. On définit par récurrence les suites (m_j) et par les inégalités:

$$x > H_1(x) > H_2(x) > \dots > H_{m_o-1}(x) > H_{m_o}(x)$$

$$H_{m_o}(x) < H_{m_o+1}(x) < \dots < H_{p_o-1}(x) < H_{p_o}(x)$$

$$H_{p_o}(x) > H_{p_o+1}(x) > \dots > H_{m_1}(x)$$

...

$$H_{m_j}(x) < H_{m_j+1}(x) < \dots < H_{p_j}(x)$$

$$H_{p_j}(x) > H_{p_j+1}(x) > \dots > H_{m_{j+1}}(x)$$

...

On va montrer:

$$H_{p_j}(x) > H_{p_j+1}(x) > \dots > H_{m_{j+1}}(x) > H_{p_j-1}$$
$$H_{m_j}(x) < H_{m_j+1}(x) < \dots < H_{p_j}(x) < H_{m_j-1}$$

Il s'ensuivra que la suite $(H_{p_j}(x))_{j\in\mathbf{N}}$ est décroissante, et et que la suite $(H_{m_j}(x))_{j\in\mathbf{N}}$ est croissante. Supposons par l'absurde qu'il existe un entier j, et un entier k, vérifiant:

$$p_j + 1 \leq k \leq m_{j+1} \quad \text{et} \quad H_k(x) \leq H_{p_j-1}(x)$$

L'inégalité $H_{k-1}(x) > H_k(x)$ implique: $H_{k-1}(x) \in]c, H_{p_j}(x)]$. Le point $(H_{p_j}(x), H_{p_j-1}(x))$ appartiant à $\mathcal{H}_g^{-1}(f_{\lambda_{p_j}})$. Or, $\mathcal{H}_g^{-1}(f_{\lambda p_j})$ est un arc joignant (c, c) au segment [(a, a); (b, a)]. Il coupe le triangle T de sommets ((a, a); (b, a); (b, b)) en deux composantes connexes. Le segment [(b, a); (b, b)] appartient à l'une de ces composantes, tandis que le segment $[(a, a); (f_{\lambda_j}(a), a)]$ appartient à l'autre. De son côté l'arc $\mathcal{H}_d(f_{\lambda_k})$ coupe également le triangle en au moins deux composantes. Puisque par hypothese les arcs $\mathcal{H}_g^{-1}(f_{\lambda_{pj}})$ et $\mathcal{H}_d(f_{\lambda_k})$ ne se coupent qu'en (c, c), il faut qu'ils soient chacun dans une des composantes connexes du triangle T définies par l'autre. En conséquence, puisque $\mathcal{H}_g^{-1}(f_{\lambda_{p_j}})$ relie (c, c) au segment [(a, a); (b, a)], tout élément (z, y) de $\mathcal{H}_g^{-1}(f_{\lambda_{p_j}})$ doit vérifier: $y < f_{\lambda_k}(z)$.

Précisons ce point: les composantes connexes définies par $\mathcal{H}_d(f_{\lambda_k})$ sont incluses dans:

$$C_1 = \{(x, y) \in T \mid y \geq f_{\lambda_k}(x)\}$$

$$C_2 = \{(x, y) \in T \mid y \leq f_{\lambda_k}(x)\}$$

Si C_1 est toujours une composante connexe, il peut ne pas en être de même de C_2, s'il existe t tel que: $f_{\lambda_k}(t) = a$. Or, $\mathcal{H}_g^{-1}(f_{\lambda_{p_j}})$ est un arc joignant (c, c) à $(H_{p_j}(x), H_{p_j-1}(x))$ dans: $\{(z, t) \mid t > H_{p_j-1}(x)\}$. Il existe donc $v > H_{p_j-1}(x)$ tel que: $(H_{k-1}(x), v) \in \mathcal{H}_g^{-1}(f_{\lambda_{p_j}})$, $v > H_{p_j-1}(x)$.

Par hypothèse sur H_k, on a alors: $H_k(x) < H_{p_j-1}(x)$ de sorte que $(H_{k-1}(x), v)$ est un point de $C_1 \cap \mathcal{H}_g^{-1}(f_{\lambda_{p_j}})$, ce qui est impossible.

Mutatis mutandis, on montre la suite d'inégalités:

$$H_{m_j}(x) < H_{m_j+1}(x) < \dots < H_{p_j}(x) < H_{m_j-1}(x).$$

Bref, la suite $(H_{p_j}(x))_{j\in\mathbf{N}}$ est décroissante, minorée par c; elle est convergente, et sa limite ℓ_1 est supérieure à c. De même, la suite $(H_{m_k}(x))_{k\in\mathbf{N}}$ est croissante, majorée par c; elle est convergente, et sa limite ℓ_2 est inférieure à c.

Soit $\theta > 0$. Il existe n_o tel que, pour tout $n > n_0$, on ait: $H_n(x) \in [\ell_1, \ell_1 + \theta] \cup [\ell_2 - \theta, \ell_2]$. Il y a donc attraction vers deux points. A ce stade de la démonstration, seule la condition (1) a servi. (la condition (2) a servi dans le cas où la suite $(H_n(x))$ est monotone au-delà d'un certain rang). Montrons qu'avec les conditions (2) et (3), on a:

$$\ell_1 = \ell_2 = c.$$

Supposons $\ell_1 > c$. Il existe alors un triangle $T_d(\ell_1, \tau)$ dont l'intersection avec $\bigcup_{\lambda\in\Lambda} \mathcal{H}_d(f_\lambda)$ est vide.

Par suite, pour j assez grand, on doit avoir:

$$f_{\lambda_{p_j+1}}(H_{p_j}(x)) \in [\ell_2 - \tau, \ell_2]$$

Autrement dit, la balade ne reste pas près de ℓ_1 plus d'un coup. De même pour ℓ_2, si $\ell_2 < c$. Supposons $\ell_1 > c$, et considérons le point (ℓ_1, ℓ_2). Les relations:

$$H_{p_j}(x) \in [\ell_1, \ell_1 + \tau]$$

$$f_{\lambda_{p_j+1}}(H_{p_j}(x)) \in [\ell_2 - \tau, \ell_2]$$

$$H_{p_j-1}(x) \in [\ell_2 - \tau, \ell_2]$$

$$f_{\lambda_{p_j}}(H_{p_j-1}(x)) \in [\ell_1, \ell_1 + \tau]$$

impliquent:

$$\mathcal{H}_g^{-1}(f_{\lambda_{p_j}}) \cap C(\ell_1, \ell_2, \tau) \neq \emptyset$$

$$\mathcal{H}_d(f_{\lambda_{p_j+1}}) \cap C(\ell_1, \ell_2, \tau) \neq \emptyset$$

ce qui contredit la condition (3).

De même, si l'on suppose $\ell_2 < c$, et le théorème est démontré.

REMARQUE 1. Si l'ensemble Λ est fini, la condition (1) implique les conditions (2) et (3), du fait de la compacité des graphes.

REMARQUE 2. Si (c, c) est le seul point de Λ adhérent à $\bigcup_{\lambda\in\Lambda} \mathcal{H}(f_\lambda)$, et si les adhérences de $(\bigcup_{\lambda\in\Lambda} \mathcal{H}_g(f_\lambda))^{-1}$ et de $\bigcup_{\lambda\in\Lambda} \mathcal{H}_d(f_\lambda)$ ne se coupent qu'en (c, c), on peut préciser la vitesse de convergence de la balade en faisant intervenir dans les inégalités en $H_{p_j}(x)$ de la deuxième partie de la démonstration les fonctions $\inf_{\lambda\in\Lambda} f_\lambda$ et $\sup_{\lambda\in\Lambda} f_\lambda$. On montre alors en fait que la convergence est uniforme non seulement en x, mais aussi par rapport à la réalisation:

$$\forall\ \tau > 0,\ \exists n \in \mathbb{N},\ \forall x \in I,\ \forall (H_p),\ \forall\ q \in \mathbb{N},$$

$$(q > n) \Rightarrow (H_q(x) \in [c - \tau, c + \tau])$$

où "$\forall(H_p)$" signifie "pour toute réalisation de la balade".

2. Le Processus Autoregressif D'ordre 1

Convention: Soit f une contraction asymptotique de I. Si $f(x) > b - \tau$, on a: $f(x) + \tau > b$, de sorte que $f(f(x) + \tau)$ n'est pas défini. On conviendra d'étendre f à $\mathbf{R}$ par: $f(x) = f(a)$ si $x < a$, $f(x) = f(b)$ si $x > b$, lorsque cela sera nécessaire.

Nous avons vu dans l'introduction que les balades aléatoires et le processus autorégressif peuvent être interprétés comme deux modèles d'un même problème physique. Les liens entre ces deux modèles sont plus profonds, puisque sur le plan mathématique, on peut appliquer au processus autorégressif d'ordre 1 des résultats établis pour les balades:

COROLLAIRE 3. *Etat absorbant pour le processus autorégressif d'ordre* 1, *lorsque la fonction de régression est une contraction asymptotique, et que le bruit est borné.*

Soit f une contraction asymptotique de $I = [a, b]$, de point fixe $c \in I$. Soient δ un réel > 0, $(\theta_n)_{n\in\mathbf{N}}$ une suite de variables aléatoires réelles, indépendantes, équidistribuées, dont la loi F a son support dans $[-\delta, \delta]$ et charge tout intervalle de la forme $[-\gamma, \gamma]$, $0 < \gamma < \delta$. Soit $J = [\alpha, \beta]$ l'intervalle de I ayant la propriété d'être le plus petit intervalle $K = [z, t]$ contenant c et tel que pour tout $x \in ([z - \delta, z] \cup [t, t + \delta])$ on ait: $|f(x) - x| > \delta$. Soit $L = [\alpha - \delta, \beta + \delta]$. Alors, L est absorbant pour le processus défini, avec la convention précédente par: $X_{n+1} = f(X_n) + \theta_n$, au sens suivant:

- Si $X_n \in L$, alors, pour tout $m > n$, on a : $X_m \in L$
- Quel que soit $X_o \in I$, il existe presque sûrement $n \in \mathbf{N}$ tel que: $X_n \in L$.

DÉMONSTRATION. Joint à la convention précédents, le théorème 1 montre que L est défini.

Le graphe de f est compact, de même que la diagonale Δ de I^2. On a donc:

$$\nu = \inf \{|x - f(x)|, x \in [a, \alpha - \delta] \cup [\beta + \delta, b]\} > 0$$

Par ailleurs, d'après le théorème 1, on a: (d: distance euclidienne)

$$\omega = d(H^{-1}(f [a, \alpha - \delta]), H(f [\beta + \delta, b])) > 0$$

Posons: $\gamma = \frac{1}{2} \min(\nu, \omega)$. Soit $x_o \in I$. Si $x_o \in L$, on a presque sûrement:

$$\forall n \in \mathbf{N}, X_n \in L,$$

car presque sûrement, $|\theta_n| < \delta$, et parce que L est stable par l'operation: $x \to f(x) + \theta$, avec $|\theta| < \delta$.

Considérons à présent, pour tout $\theta \in [-\gamma, \gamma]$ la fonction f_θ définie par:

$$f_\theta(x) = f(x) + \theta \quad \text{si} \quad x \in I - L$$

$$f_\theta(x) = f(x) \quad \text{si} \quad x \in J$$

$$f_\theta(\alpha - t\delta) = f(\alpha - t\delta) + t.\theta \quad \text{si} \quad t \in [0, 1]$$

$$f_\theta(\beta + t\delta) = f(\beta + t\delta) + t.\theta \quad \text{si} \quad t \in [0, 1]$$

Le théorème 2 s'applique a la famille $(f_\theta)_{\theta \in [-\gamma, \gamma]}$, de sorte que si l'on a: $|\theta_1| < \gamma$, $|\theta_2| < \gamma$, ..., $|\theta_p| < \gamma$, le processus coincide sur $0, 1, \ldots, p$ avec une balade aléatoire sur les f_θ. On sait qu'une telle balade converge vers c.

Donc, si $x \in I - L$, il existe $p \in \mathbf{N}$ tel que: $H_p(x) \in L$. De plus, de par la remarque suivant le théorème 2, cet entier p ne dépend pas de x, ni de la réalisation de la balade.

Or, la probabilité pour que l'on ait:

$$|\theta_1| < \gamma, |\theta_2| < \gamma, \ldots, |\theta_p| < \gamma$$

est strictement positive.

Par conséquent, il existe $n \in \mathbf{N}$, et $\tau > 0$, tels que, pour tout

$x \in I$, on ait: $\mathbf{P}\{X_n \in L \text{ si } X_0 = x\} > \tau > 0$.

Ceci, joint à la stabilité de L, montre que cet intervalle est absorbant pour la chaine de Markov $(X_n)_{n\in\mathbf{N}}$.

A partir d'ici, les résultats énonces sont dus à Paul Doukhan. Le processus autorégressif $X_{n+1} = f(X_n) + \theta_n$ est étudie dans le cas où le bruit blanc a une loi équivalente à la mesure de Lebesgue: le support est alors $\mathbf{R}$ tout entier. On se limitera ici au cas d'une loi normale $N(0, \sigma^2)$.

THEOREME 4 [3]. *Propriétés probabilistes du processus autorégressif d'ordre 1.* Soient $f : \mathbf{R} \to \mathbb{R}$ une fonction mesurable bornée, et $(\theta_n)_{n\in\mathbf{N}}$ une suite de variables aléatoires indépendantes, équidistribuées, dont la loi est la loi normale $N(0, \sigma^2)$. Le processus (X_n) défini par: $X_{n+1} = f(X_n) + \theta_n$ est une chaine de Markov apériodique, irréductible, et géométriquement ergodique.

Sous les hypothèses du théorème 4, on définit:

$$u : u(x) = \frac{1}{2} e^{-|x|} \quad \text{(noyau double exponentielle)}$$

$$\beta_n = n^{-1/3}$$

$$\hat{\pi}_n(x) = \frac{1}{n\beta_n} \sum_{k=0}^{n-1} u\left(\frac{X_k - x}{\beta_n}\right) \quad \text{(densité } \pi \text{ de la loi limite de } (X_n))$$

$$\hat{B}_n(x) = \frac{1}{n\beta_n} \sum_{k=0}^{n-1} X_{k+1}\, u\left(\frac{X_k - x}{\beta_n}\right)$$

Si a est un reel > 0, on définit un réel b vérifiant:

$$b \leq \frac{1}{2} \frac{1}{\sqrt{2\pi}\,\sigma} \exp\left(-(a + \|f\|_\infty)^2 / 2\sigma^2\right)$$

A partir d'un échantillon $(X_1, \ldots, X_n)$, on construit une estimation $\hat{f}_n$ de f sur $[-a, a]$ donnée par:

$$\hat{f}_n(x) = \frac{\hat{B}_n(x)}{\max(b, \hat{\pi}_n(x))}$$

On estime la variance σ^2 du bruit blanc par:

$$\hat{\sigma}2 = \frac{1}{p} \sum_{k=n+p_o+1}^{n+p_o+p} (X_{k+1} - \hat{f}(X_k))^2$$

Ce procédé d'estimation est justifié par le théorème suivant:

THEOREME 5 [4]. *Estimation dans le processus autorégressif d'ordre 1.* Sous les hypothèses et notation précédentes, et si de plus f vérifie: $\|f\|_\infty + \|f'\|_2 < \infty$, il existe des constantes C_π et C_f ne dépendant que de σ^2, $\|f\|_\infty$ et $\|f'\|_2$ telles que pour toute loi initiale ν:

$$\int \mathbf{E}_\nu(\hat{\pi}_n(x) - \pi(x))^2 \, dx < C_\pi \, n^{-2/3}$$

$$\int_{[-a,\, a]} \mathbf{E}_\nu(\hat{f}_n(x) - f(x))^2 \, dx < C_f \frac{n^{-2/3}}{b^2}$$

Ces constantes C_π et C_f sont en outre majorées par des expressions indépendantes de σ^2 de la forme:

$$C_\pi < \alpha + \frac{\beta}{\sigma^4} \qquad C_f < \alpha.\sigma^2 + \frac{\beta}{\sigma^4}$$

On montre par ailleurs que lorsque σ^2 tend vers 0, les valeurs d'adhérence de π_σ sont a support dans $\bigcap_{n\in\mathbf{N}} f^n(\mathbf{R})$.

En particulier, le théorème 5 montre que si la fonction de régression, f, est une contraction asymptotique, la mesure π_σ converge vers la mesure de Dirac au point fixe de f, lorsque σ tend vers 0.

Le corollaire 3 et le théorème 4 montrent que lorsque la fonction de régression est une contraction asymptotique, les propriétés probabilistes du processus autorégressif sont fort différentes, selon que le bruit est borné ou non. La mise en oeuvre sur ordinateur du théorème 5 semble conduire au "paradoxe" suivant:

"Considérons une contraction asymptotique f, de point fixe c. Grace à un programme de génération de nombres pseudoaléatoires, faisons engendrer par un ordinateur une suite $(\theta_n)_{n\leq N}$ simulant les réalisations indépendantes d'une loi normale $N(0, \sigma^2)$; enfin, faisons calculer l'estimateur de f défini plus haut. Les théormès 4 et 5 affirment que si N est assez grand, le résultat sera proche de f. D'un autre côté, un programme de nombres pseudo-aléatoires ne peut jamais engendrer qu'un nombre fini de tels nombres, de sorte que les θ_n sont bornés, et que c'est le corollaire 3 qui devrait s'appliquer, empêchant toute estimation en dehors de l'état absorbant".

Il n'y a bien sûr pas véritablement de paradoxe, et les résultats d'une expérience sont exposés dans la figure 1. C'est bien le corollaire 3 qui s'applique. Mais en pratique, on n'est intéressé par l'estimation de la fonction f que sur un intervalle de longueur fixée. Si la variance du bruit est assez grande, le programme de nombres pseudo-aléatoires produit des nombres répartis sur un intervalle $[-\delta, \delta]$ qui conduit dans le corollaire 3 à un intervalle d'absorption plus grand que l'intervalle sur lequel on cherche à estimer la fonction f; l'estimation fonctionne alors. Si au contraire la variance est trop petite, le processus est "piégé" au voisinage du point fixe, et l'estimation ne marche pas.

Dans le cas de la figure 1, la fonction f est itérativement convergente, avec deux attracteurs. Lorsque la variance est trop petite, chaque réalisation du processus est piégée par l'un ou l'autre des attracteurs. L'estimation est bonne sur la moitié de l'intervalle considéré, exactement dans le bassin de l'attracteur qui a attiré le processus. La fonction est estimée par zéro dans l'autre bassin. En moyenne, chaque point fixe attire le processus une fois sur deux, de sorte que l'estimation moyenne fournit $\frac{1}{2}f$, au lieu de f.

Pour revenir à notre point de départ, on a donc construit un outil fonctionnant bien pour des bruits assez grands, mal s'ils sont trop petits. Sans doute est-ce là le vrai paradoxe: trouver un moyen de pallier les erreurs des ordinateurs qui ne marche que si ces erreurs sont assez grandes, et qui est inopérant pour les petites erreurs...

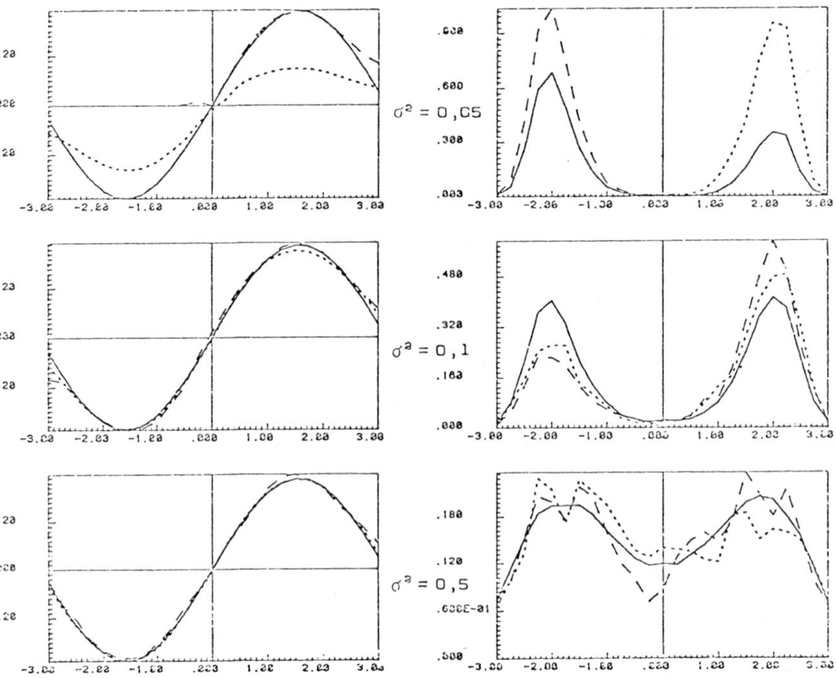

Figure 1:
Résultats d'une simulation du processus autorégressif d'ordre 1.

fonction de régression: $f : x \to \frac{\alpha}{\sin \alpha} \sin x$, où $\alpha \in \left]\frac{\pi}{2}, \pi\right[$ est l'unique solution de: $\alpha + \mathrm{tg}\, \alpha = 0$.

A gauche: en traits continus: graphe de f
en tirets: estimation à partir d'une réalisation
en pointillés: estimation moyenne à partir de 20 réalisations.

A droite: courbes de densité de probabilites (noter le changement d'échelle): en pointillés et en tirets, courbes obtenues à partir d'une réalisation du p processus (pas la même qu'à gauche)
en trait continu: courbe obtenue à partir des 20 réalisations utilisées à gauche.

Pour chaque réalisation: processus simulé sur 10^4 itérations.

BIBLIOGRAPHIE

1. V. V. Bashurov and V. N. Ogibin, "Conditions for the convergence of iterative processes on the real axis", U.S.S.R. Comp. Math. and Math. Phys. 6, 5 (1966), 178-184.

2. P. Collet and J. P. Eckmann, "Iterated maps on the interval as dynamical systems", PPh no. 1 Birkhäuser éd. 1980.

3. P. Doukhan and M. Ghindès, "Etude probabiliste du processur autorégressif d'ordre 1", C. R. Acad. Sc. Paris Série A, t. 290, pp. 921-923.

4. P. Doukhan, "Simulations in the general first order autoregressive process (unidimensional case)", A paraitre dans: Actes des Rencontres de Statisticiens Franco-Belges - Louvain-La-Neuve 1981 - Springer éd.

5. F. Guénard, "Caractérisations des contractions asymptotiques et des applications iterativement convergentes des intervalles", C. R. Acad. Sc. Paris, t. 292, série I, 1981, pp. 55-58.

THE METHOD OF UPPER AND LOWER SOLUTIONS FOR DIFFERENTIAL EQUATIONS IN A BANACH SPACE

V. Lakshmikantham

University of Texas at Arlington
Arlington, Texas
U.S.A.

One of the most useful techniques in proving existence of multiple solutions of initial and boundary value problems of differential equations is the monotone iterative technique. For differential equations in finite dimensions, the method of upper and lower solution yields existence of at least one solution and then monotone technique can be utilized to obtain multiple solutions by constructive method. If one wants to extend these considerations to differential equations in a Banach space, the method of upper and lower solutions alone is not good enough even to obtain existence of one solution. However, the monotone technique coupled with the method of upper and lower solutions offers a good mechanism to discuss existence problems by constructive methods. In the paper we shall report some recent results.

First, consider the IVP

$$u' = f(t, u), \; u(0) = u_0, \tag{1.1}$$

where $f \in C[I \times E, E]$, $I = [0, T]$ and E a real Banach space. Let K be a normal cone in E and let the inequalities be induced by K. Suppose that $v_0, w_0 \in C^1[I, E]$ and

$$v_0' \leq f(t, v_0), \; w_0' \geq f(t, w_0) \quad \text{on} \quad I. \tag{1.2}$$

The function f is said to be quasimonotone relative to K, if $v \leq u$ and $\phi(v - u) = 0$, $\phi \in K^*$, then $\phi(f(t, v) - f(t, u)) \leq 0$ where K^* is the dual cone. We can prove the following result using the monotone iterative technique [1].

THEOREM A. Assume that

(A_1) for any bounded set B in E, $\alpha(f(I \times B)) \leq L\,\alpha(B)$, $L > 0$ and α denotes the Kuratowski's measure of noncompactness;

(A_2) $f(t, u_1) - f(t, u_2) \geq -M(u_1 - u_2)$ whenever $v_0(t) \leq u_2 \leq u_1 \leq w_0(t)$ for some $M > 0$. If v_0, w_0 satisfy (1.2) such that $v_0(t) \leq w_0(t)$ on I, then there exist monotone sequences $\{v_n\}$, $\{w_n\}$ which converge uniformly and monotonically to the minimal and maximal solutions $\rho(t)$, $r(t)$ respectively of (1.1) such that $\rho, r \in [v_0, w_0] = [u \in C[I, E]: v_0 \leq u \leq w_0$ on $I]$.

We note that (A_2) implies quasimonotonicity of f. In case f is not quasimonotone, we can not assume (A_2). In order to prove a result similar to Theorem A, condition on v_0, w_0, namely (1.2), has to be strenghened in terms of linear functionals from K^* as follows:

(A_3) $v_0, w_0 \in C^1[I, E]$ with $v_0 \leq w_0$ on I such that there exists $M > 0$ satisfying $\phi[v_0' - f(t, \sigma) + M(v_0 - \sigma)] \leq 0$, $\phi[w_0' - f(t, \sigma) + M(w_0 - \sigma)] \geq 0$ for all σ such that $v_0(t) \leq \sigma \leq w_0(t)$, $t \in I$ and $\phi \in K^*$. Then we have the following result [2].

THEOREM B. Let the assumptions (A_1) and (A_3) hold. Suppose that f is Lipschitzian on $\Omega = [(t, u) \in I \times E : t \in I$ and $v_0(t) \leq u \leq w_0(t)]$. Then there exists a solution u of (1.1) such that $u \in [v_0, w_0]$ provided $v_0(0) \leq u_0 \leq w_0(0)$.

If f is assumed to be uniformly continuous, then (A_1) is superfluous. If σ in (A_3) is such that for every $\phi \in K^*$, $\phi(v_0(t) - \sigma) = 0$ and $\phi(w_0(t) - \sigma) = 0$, then ($A_3$) reduces to

$$\phi(v_0' - f(t, \sigma)) \leq 0, \ \phi(w_0' - f(t, \sigma)) \leq 0. \qquad (A_3^*)$$

Theorem B need not be true even in $E = R^n$ when K is an arbitrary cone if (A_3^*) holds. However, if $K = R^n_+$, Theorem B is true and it is the classical Muller's result, see [3, 4].

One can combine Theorems A and B to prove a result that unifies all the existing results by assuming that f admits a decomposition of the form

$$f(t, u) = f_0(t, u) + f_1(t, u) + f_2(t, u)$$

where f_1 is nondecreasing and f_2 is nonincreasing. For details, see [5].

Let us next consider the BVP

$$u'' = f(t, u, u'), \quad B^i u = b_i, \quad i = 0, 1, \tag{3}$$

where $B^i u = \alpha_i u(i) + (-1)^{i+1}\beta_i u'(i)$, $\alpha_i, \beta_i \geq 0$, $\alpha_i^2 + \beta_i^2 > 0$ and $f \in C[I \times E \times E, E]$. Here $I = [0, 1]$. Let us list the following assumptions:

(B_1) $v_0, w_0 \in C^2[I, E]$ with $v_0 \leq w_0$ on I and

$$v_0'' \geq f(t, v_0, v_0'), \quad B^i v_0 \leq b_i,$$

$$w_0'' \leq f(t, w_0, w_0'), \quad B^i w_0 \geq b_i, \quad i = 0, 1;$$

(B_2) $f(t, x, y) - f(t, \bar{x}, \bar{y}) \leq M_1(C)(x - \bar{x}) + M_2(C)(y - \bar{y})$ whenever $v_0(t) \leq \bar{x} \leq x \leq w_0(t)$, $t \in I$ and $\|y\|, \|\bar{y}\| \leq C$, C being a positive constant;

(B_3) $\|f(t, x, y)\| \leq J(\|y\|)$ whenever $t \in I$, $v_0(t) \leq x \leq w_0(t)$ where J is continuous on $[0, \infty]$;

(B_4) J, M_1 and M_2 are increasing and positive functions such that $b(s) = J(s) + RM_1(s) + 2M_2(s)s = 0(s^2)$ as $s \to \infty$, where $R = \max[\max\|v_0\|, \max\|w_0\|, \max\|v_0 - w_0\|]$;

(B_5) $\alpha(f(I \times A \times B)) \leq L \max(\alpha(A), \alpha(B))$ for any bounded sets A and B in E.

We can then prove the following result which corresponds to Theorem A for BVP's, see [6].

THEOREM C. Let the assumptions (B_1) to (B_5) hold. Then if $Q(L + M_1 + M_2) < 1$ where $Q = \max [\max \|G(t, s)\|, \max \|G_+(t, s)\|]$, $G(t, s)$ being the green's function, then there exist monotone sequences v_n, w_n which converge uniformly and monotonically to the minimal and maximal solutions , r respectively of BVP (1.3) such that $v_0 \leq \rho \leq r \leq w_0$ on I.

The conclusions of Theorem C are valid under the assumptions (B_1) to (B_4) if $E = R^n$ and $K = R^n_+$. Then the assumptions (B_5) and $Q(L + M_1 + M_2) < 1$ are superfluous.

The assumption (B_2) implies that f is linear in x'. Because of this, it was possible to obtain a linear procedure which yields that the elements of the sequences $\{v_n\}$, $\{w_n\}$ are solutions of linear BVP's. For the price we paid, we obtained a rich dividend. If, on the other hand, we allow nonlinearities of f in x' so that we assume

(B_2^*) $f(t, x, y) - f(t, \bar{x}, y) \leq M_1(C)(x - \bar{x})$ whenever $v_0 \leq \bar{x} \leq x \leq w_0$ and $\|y\| \leq C$, then we will have the problem of proving existence of solution of auxiliary BVP

$$u'' = f(t, \eta, u') + M_1(C)(u - \eta), \quad B^i u = b_i.$$

The usual methods of upper and lower solutions is not valid here since we are in E. Other methods which might demand more assumptions have to be considered. Furthermore, the advantage of linear procedure no longer exists.

When f does not satisfy (B_2) or (B_2^*), that is, when f does not possess any quasimonotone property, results corresponding to Theorem B are still unknown.

REFERENCES

1. Du, W. W. and Lakshmikantham, V. "Monotone iterative technique for differential equations in a banach space", (to appear in J. Math. Anal. & Appl.).

2. Lakshmikantham, V. and Leela, S. "The method of upper and lower solutions in abstract cones", (to appear in Ann. Polon. Math.).

3. Volkmann, P. (1978), "Ausdehnung eines satzes von Max Muller auf unendliche systeme von gewohnlichen differentialgleichungen", Funkcialaj Ekvacioj 21, 81-96.

4. Müller, M., (1926), "Uber das fundamental theorem in der theorie der gewohnlichen differentialgleichungen", Math. Z. 26, 619-645.

5. Lakshmikantham, V., Leela, S. and Vatsala, A. S., "Method of quasi-upper and lower solutions in abstract cones", (to appear in Nonlinear Analysis).

6. Bernfeld, S. R. and Lakshmikantham, V., "Linear monotone method for nonlinear boundary value problems in Banach spaces", (to appear in Rocky Mountain J.

APPLICATIONS OF FIXED POINT THEOREMS TO PROBLEMS IN OPTIMIZATION AND BEST APPROXIMATION

D. V. Pai and P. Veeramani
Indian Institute of Technology
Bombay
India

1. Introduction

Applications of fixed point theorems to problems in optimization and best approximation are well known (e.g., cf. [4] and [13]). In approximation theory, the problems that can be tackled using fixed point theorems include: existence of elements of best approximation, existence of mutually nearest points of pairs of convex sets, solarity of Chebyshev sets and the generalized Kolmogorov criterion. Brosowski [1] has employed Fan's fixed point theorem as well as a fixed point theorem of Nadler [8] for studying existence of elements of best approximation. Cheney and Goldstein [2] have investigated existence of mutually nearest points of a pair of convex sets in a Hilbert space. In game theory, Glicksberg [5], Fan [3] and others have employed fixed point theorems for multimappings for studying existence of Nash equilibrium points, for n-person games played over n compact Hausdorff pure strategy spaces.

Given a continuous convex function f defined on a separated locally convex space X, we are concerned, in this exposition, with f-best approximation of an element x from a given nonempty closed subset K of X as well as with f-mutually nearest points of a pair of convex subsets K_1, K_2 of X (the so-called multioptimum of f, cf. [9]). In Section 2, a fixed point theorem of Himmelberg [7] is employed to study existence of f-best approximation (Theorem 2.3). In Section 3, a fixed point result (Theorem 3.4) is obtained for f-nearest points. This yields as corollaries

(3.7 and 3.9) results for best approximations from (i) approximatively compact subsets of a metric space, (ii) a closed convex subset of a reflexive Banach space.

2. On Applications Of Himmelberg Fixed Point Theorem

Let X be a Hausdorff locally convex space, f be a continuous real valued convex function defined on X and let K be a nonempty closed subset of X. For each x in X, let

$$f_K(x) = \inf\{f(x - k) : k \in K\}$$

(possibly equal to $-\infty$).

An element $k \in K$ is said to be <u>f-best approximation</u> to x in K if $f(x - k) = f_K(x)$.

Let

$$P_K(x) = \{k \in K : k \text{ is f-best approximation to } x \text{ in } K\}$$

(eventually void).

The mapping $x \to P_K(x)$ is called <u>f-projection supported on K</u>. K is said to be <u>f-proximinal</u> (resp. <u>f-Chebyshev</u>) if $P_K(x) \neq \phi$ (resp. $P_K(x)$ is singleton) for each x in X (cf. [6]).

DEFINITION 2.1. (i) (cf. [6]) f is said to be <u>inf-compact</u> (resp. <u>inf-bounded</u>) if the sub-level sets $S_\alpha = \{x \in X : f(x) \leq \alpha\}$ are compact (resp. bounded) for each real number α.

(ii) (cf. [6]) The set K is said to be inf-compact if for each $x \in X$, each minimizing net k_α in K $(f(x - k_\alpha) \to f_K(x))$ has a convergent subnet converging in K.

(iii) (cf. [9]) Let K_1, K_2 be a pair of convex sets in X. A pair $(\bar{k}_1, \bar{k}_2) \in K_1 \times K_2$ is called <u>multioptimum</u> for f if

$$f(\bar{k}_1 - \bar{k}_2) = \inf_{k_2 \in K_2} f(\bar{k}_1 - k_2) = \inf_{k_1 \in K_1} f(k_1 - \bar{k}_2)$$

(i.e., the points $\bar{k}_1, \bar{k}_2$ are f-mutually nearest points of the sets K_1,

K_2).

We mostly employ here the following fixed point theorem of Himmelberg (cf. [7], Theorem 2) which is an extension of the well known theorem of Fan.

THEOREM 2.2. (cf. [7]), Theorem 2). Let K be a nonempty convex subset of a Hausdorff locally convex space X. Let $T : K \to cc(K)$ (Here $cc(K)$ denotes the family of nonempty closed convex subsets of K) be upper semi-continuous (u.s.c.) and let $T(K)$ be contained in a compact subset C of K. Then T has a fixed point.

THEOREM 2.3. Let K be a nonempty closed convex subset of a Hausdorff locally convex space X and let f be a continuous convex functional on X. Let x in K be fixed and let $\psi_x : K \to cc(K)$ be defined by

$$\psi_x(k) = K \cap \{x - S((f_K(x) + f(x - k))/2)\}.$$

(As before, $S(r) = \{x : f(x) \leq r\}$, $r \in \mathbb{R}$, denote sub-level subsets of f.) Assume that $\psi_x(K)$ is contained in a compact subset of K.

Then there exists f-best approximation to x in K.

Proof. Let A be a nonempty closed subset of K. In order to prove that ψ_x is u.s.c., it suffices to show that the set $\psi_x^{-1}(A) = \{k \in K : \psi_x(k) \cap A \neq \phi\}$ is closed.

Let $\{u_\alpha\}$ be a net in $\psi_x^{-1}(A)$ such that u_α tends to u_o. Since $u_\alpha \in \psi_x^{-1}(A)$, $\psi_x(u_\alpha) \cap A \neq \phi$. Let $k_\alpha \in \psi_x(u_\alpha) \cap A$. Then

$$f(x - k_\alpha) \leq \frac{f_K(x) + f(x - u_\alpha)}{2}. \tag{1}$$

Since $\{k_\alpha\} \subset \psi_x(K)$ and $\psi_x(K)$ is contained in a compact subset of K, $\{k_\alpha\}$ has a convergent subnet $\{k_\beta\}$ converging to k_o. K being closed, $k_o \in K$. Passing to limits in (1), continuity of f yields

$$f(x - k_o) \leq \frac{f_K(x) + f(x - u_o)}{2}.$$

Since $k_\beta \in A$ and A is closed, $k_o \in A$. Thus $k_o \in \psi_x(u_o) \cap A$, whence $u_o \in \psi_x^{-1}(A)$ and ψ_x is u.s.c.

Now by using Theorem 2.2, we get an element $k \in K$ such that $k \in \psi_x(k)$. Hence

$$f(x - k) \leq \frac{f_K(x) + f(x - k)}{2}$$

and this gives $f(x - k) = f_K(x)$.

3. Some Results On Best Approximations

Let X be a Hausdorff topological space, f be a real valued function on $X \times X$ and let K be a nonempty closed subset of X.

We denote by

$$f(x, K) = \inf\{f(x, k) : k \in K\}$$

(possibly equal to $-\infty$).

An element $k \in K$ is said to be f-nearest to x in K if

$$f(x, k) = f(x, K).$$

We denote

$$P(x, K) = \{k \in K : f(x, k) = f(x, K)\}$$

(eventually void).

DEFINITION 3.1. (i) The set K is said to be inf-compact at a point $x \in X$, if each minimizing net k_α in K $(f(x, k_\alpha) \to f(x, K))$ has a convergent subnet converging in K.

(ii) K is said to be inf-compact if it is inf-compact at each point $x \in X$.

DEFINITION 3.2. A real valued convex function f defined on a Hausdorff linear topological space X is said to be rotund if the sub-level sets S_α $(\alpha \in \mathbf{R})$ of f are rotund, i.e., $f(x) = f(y) = \alpha$ implies $f(\beta x + (1 - \beta)y) < \alpha$ for $0 < \beta < 1$.

PROPOSITION 3.3. Let K be a nonempty subset of Hausdorff topological space X and let f be a real valued function on $X \times X$. Suppose that K is inf-compact at a point $x \in X$ and let $x \to f(y_0, x)$ be l.s.c. (lower semi-continuous). Then $P(x, K)$ is a nonempty compact subset of K.

Proof. By the definition of $f(x, K)$, we can extract a net $\{k_\alpha\}$ in K such that $\lim_\alpha f(x, k_\alpha) = f(x, K)$. Now K being inf-compact at x, $\{k_\alpha\}$ has a convergent subnet $\{k_\beta\}$ converging in K.

Let $k_\beta \to k_0$ in K. Then $f(x, k_0) \leq \underline{\lim}_\alpha f(x, k_\alpha) = f(x, K)$. Hence $f(x, k_0) = f(x, K)$; and this gives $k_0 \in P(x, K)$. Compactness of $P(x, K)$ follows trivially from the definition of inf-compactness of K at x.

THEOREM 3.4. Let K be a nonempty subset of a Hausdorff topological space X and let f be a real valued function on $X \times X$. Assume K is inf-compact at a point $y_0 \in X$ and that the mapping $x \to f(y_0, x)$ is l.s.c. Let T be a mapping from K into itself satisfying:

(i) $f(y_0, Tx) \leq f(y_0, x)$, $x \in K$;

(ii) For each compact subset F of K which contains more than one element and is mapped into itself by T, there is an element x_0 in F such that $f(y_0, F) < f(y_0, x_0)$.

Then T has a fixed point y in K which is f-nearest to y_0 in K.

Proof. By Proposition 3.3, $P(y_0, K)$ is a nonempty compact subset of K and by hypothesis (i) it is mapped into itself by T. Let $\mathcal{F}$ denote the family of all nonempty closed subsets of $P(y_0, K)$ each of which is mapped into itself by T. Ordering $\mathcal{F}$ by set inclusion, by compactness of $P(y_0, K)$ and Zorn's lemma, we obtain a minimal element F of $P(y_0, K)$.

Suppose that F contains more than one element. For $x \in P(y_0, F)$,

$$f(y_0, Tx) \leq f(y_0, x) = f(y_0, F)$$

implies $Tx \in P(y_0, F)$. Hence $P(y_0, F) \in \mathcal{F}$. But by hypothesis (ii), $P(y_0, F)$ is a proper subset of F, which contradicts the minimality of F. Hence F contains a unique element y such that $Ty = y$; and since F is a subset of $P(y_0, K)$, we get that

$f(y_o, y) = f(y_o, K)$.

REMARK 3.5. In case X is a metric space and $f = d$, the metric on X, the notion of inf-compactness of K coincides with the well known notion of approximative compactness of K (cf. [10], p. 382). In this case, f-nearest elements to x in K are usually called elements of best approximation to x in K (cf. [10]).

REMARK 3.6. Conditions similar to (i) of Theorem 3.4 have been employed recently be Subrahmanyam [12] and Singh [11] for studying fixed points that are best approximations of a given point.

COROLLARY 3.7. Let K be a nonempty approximatively compact subset of a metric space (X, d). Let $y_o \in X \setminus K$ and let T be a mapping of K into itself satisfying:

(i) $d(y_o, Tx) \leq d(y_o, x)$, $x \in K$;

(ii) For each compact subset F of K which contains more than one element and is mapped into itself by T, there is an element x_o in F such that $d(y_o, F) < d(x_o, y_o)$.

Then T has a fixed point y which is a best approximation to y_o in K.

COROLLARY 3.8. Let f be a rotund l.s.c. convex function on a Hausdorff linear topological space X. Assume that K is a nonempty convex subset of X such that K is inf-compact at $y_o \in X$. If T is a mapping from K into itself such that

$$f(y_o - Tx) \leq f(y_o - x), \quad x \in K.$$

Then T has a fixed point y in K which is a f-best approximation to y_o in K.

Proof. It is easily verified that f being rotund and K being convex the hypothesis (ii) of Theorem 3.4 is satisfied. Now the corollary follows immediately from Theorem 3.4.

COROLLARY 3.9. Let K be a nonempty closed convex subset of a reflexive

Banach space X. Let $y_o \in X \setminus K$ and let T be a mapping from K into itself such that:

(i) $\|y_o - Tx\| \leq \|y_o - x\|$, $x \in K$;

(ii) For each bounded closed convex subset F of K which contains more than one element and is mapped into itself by T, there is an element x_o in F such that $d(y_o, F) < \|y_o - x_o\|$.

Then T has a fixed point y which is a best approximation to y_o in K.

Proof. Let $x \in X$. Fix up an element $k_o \in K$ and consider the set $K \cap \{k : \|x - k\| \leq \|x - k_o\|\}$. This set is bounded, closed and convex, and X being reflexive, it is weakly compact. Thus every minimizing net for x has a weakly convergent subnet converging in K, and this proves that K is inf-compact in the weak topology. Also $\|\cdot\|$ is weakly l.s.c. Now the corollary follows easily from Theorem 3.4.

COROLLARY 3.10. Let K be a nonempty closed convex subset of a reflexive and strictly convex Banach space X. Let $y_o \in X \setminus K$ and let T be a mapping from K into itself satisfying

$$\|y_o - Tx\| \leq \|y_o - x\|, \quad x \in K.$$

Then T has a fixed point y in K which is also the unique best approximation to y_o in K.

REFERENCES

1. B. Brosowski, "Fixpunktsatze in der approximations theorie", Mathematica, (Cluj), 11 (1969), 195-220.

2. W. Cheney and A. A. Goldstein, "Proximity maps for convex sets", Proc. Amer. Math. Soc., 10 (1959), 448-450.

3. K. Fan, "Fixed point and minimax theorems in locally convex linear spaces", Proc. Nat. Acad. Sci., U.S.A., 38 (1952), 121-126.

4. K. Fan, "Convex sets and their applications", Arogonne National Laboratory Applied Mathematics Division, Summer Lectures, Argonne, U.S.A. (1959).

5. I. L. Glicksberg, "A further generalization of the Kakutani fixed point theorem, with application to Nash equilibrium points", Proc. Amer. Math. Soc., 3 (1952), 170-174.

6. P. Govindarajulu and D. V. Pai, "On properties of sets related to f-projections", J. Math. Anal. Appl. 73(2) (1980), 457-465.

7. C. J. Himmelberg, "Fixed points of compact multi-functions", J. Math. Anal. Appl., 38 (1972), 205-207.

8. S. B. (Jr.) Nadler, "Multivalued contraction mappings", Pacific J. Math., 30 (1969), 475-488.

9. D. V. Pai, "Multioptimum of a convex functional", J. Approximation Theory, 19(1) (1977), 83-99.

10. I. Singer, "Best approximation in normed linear spaces by elements of linear subspaces", Springer-Verlag, Berlin, Heidelberg, New York (1970), (English translation).

11. S. P. Singh, "Some results on best approximation in locally convex spaces", J. Approximation Theory, 28 (1980), 329-332.

12. P. V. Subrahmanyam, "An application of a fixed point theorem to best approximation", J. Approximation Theory, 20 (1977).

13. L. P. Vlasov, "Approximative properties of sets in normed linear spaces", Russian Math. Surveys, 28(6), (1973), 1-66.

APPROXIMATION THEORY AND RELATED RESULTS

EXISTENCE AND CONTINUITY OF THE CHEBYSHEV OPERATOR, II

C. B. Dunham

University of Western Ontario
London, Ontario
Canada

Let X be a compact space and $C(X)$ be the space of continuous functions on X with norm

$$||h|| = \sup \{|h(x)| : x \in X\}.$$

Let G be a non-empty subset of $C(X)$. The Chebyshev problem is, given $f \in C(X)$, to find an element g^* of G for which $||f - g||$ attains its infimum $\rho(f)$ over $g \in G$. Such an element g^* is called a best approximation to f on X.

An examination of several existence proofs for non-linear Chebyshev approximation suggested that the following compactness property played a key role in existence.

DEFINITION. A non-empty subset G of $C(X)$ is <u>dense compact</u> if every bounded sequence of elements of G has a subsequence converging pointwise on a dense subset Y of X to an element g of G.

In [3] is proven

THEOREM 1. Let G be dense compact on X. Then a best approximation exists to all $f \in C(X)$.

Finite dimensional linear families are dense compact. Families unisolvent of degree n on an interval $[\alpha, \beta]$ are dense compact on any compact subset of $[\alpha, \beta]$ containing n points [7, 72]. Ordinary ra-

Ordinary rational families $R^n_m[\alpha, \beta]$ are dense compact on $[\alpha, \beta]$. Exponential families E_n are dense compact on $[\alpha, \beta]$ [8, 449]. Some dense compact familes G containing elements of the form $a + b\phi(cx)$ are given in [5,6].

Necessity Of Dense Compactness In Approximation On An Interval

DEFINITION. G has property Z of degree n at $g \in G$ if $h \in G$ and $g - h$ having n zeros implies that $g = h$.

THEOREM 2. Let X be an interval. Let $G \subset H \subset C(X)$. Let all bounded sequences from H have a subsequence converging pointwise on a dense subset Y of X to an element h of H (and convergence be uniform on a non-degenerate closed subinterval I of X). Let H have property Z of some degree at all approximants. A necessary and sufficient condition that each $f \in C(X)$ have a best approximation in G is that G is dense compact.

Proof. Sufficiency is known. Suppose dense compactness fails for a sequence $\{g_k\}$ from G, then we can assume that $\{g_k\} \to h \in H \sim G$ pointwise on a dense set and uniformly on $I = [\gamma, \delta]$. Let H have property Z of degree n at h and let $\gamma < x_0 < \ldots < x_{2n} < \delta$. Let us choose M such that $\|g_k - h\| < M$ for all k. Let us define f such that

$$\begin{aligned} f(x_i) - h(x_i) &= (-1)^i\, 2M \\ f(\gamma) - h(\gamma) &= f(\delta) - h(\delta) = 0 \end{aligned} \qquad i = 0, \ldots, 2n.$$

There is a continuous extension of f such that

$$\begin{aligned} |f(x) - h(x)| &\le 2M & x \in I \\ f(x) - h(x) &= 0 & x \notin I. \end{aligned}$$

By standard arguments of the alternating theory, h is uniquely best in H to f. We have

$$\begin{aligned} |f(x) - g_k(x)| &< M & x \notin I \\ |f(x) - g_k(x)| &\to |f(x) - h(x)| & x \in I, \end{aligned}$$

hence

$$\|f - g_k\| \to \|f - h\| .$$

As h is not in G, there is no best approximation to f in G.

COROLLARY. A necessary and sufficient condition that every $f \in C[\alpha, \beta]$ have a best approximation from $G \subset R_m^n[\alpha, \beta]$ on $[\alpha, \beta]$ is that G be dense compact.

By using the results of Schmidt [8], we obtain:

COROLLARY A necessary and sufficient condition that every $f \in C[\alpha, \beta]$ have a best approximation from $G \subset E_n$ on $[\alpha, \beta]$ is that G be dense compact.

By using results on property Z and uniform convergence for other dense compact families cited, the same necessary and sufficient condition can be obtained.

Positive Continuous Weight Functions

Let s be a positive element of C(X). The problem of Chebyshev approximation with weight function s is given $f \in C(X)$, to find g^* to minimize $\|s*(f - g)\|$. An equivalent approach is to consider Chebyshev approximation of $s * f$ by

$$s * G = \{s * g : g \in G\},$$

where $*$ denotes multiplication.

THEOREM 3. Let s be a positive element of C(X) and G be dense compact on X. Then $s * G$ is dense compact on X.

Proof. Let $\sigma = \inf \{s(x) : x \in X\}$, then $\sigma > 0$. Let $\|s * g_k\| < M$, then $\|g_k\| < M/\sigma$ and we apply the definition of dense compactness.

In [11], is given an example of non-existence with s non-negative and G a family of ordinary rational functions.

Transformations of G

Let ϕ be a continuous mapping of the real line into the extended real line such that $|\phi(t)| \to \infty$ as $|t| \to \infty$. Define

$$\phi(G) = \{\phi(g) : g \in G, \|\phi(g)\| < \infty\}.$$

Transformations of this type are considered in [2; 6].

THEOREM 4. Let G be dense compact and $\phi(G)$ be non-empty, then $\phi(G)$ is dense compact.

Proof. Let $\|\phi(g_k)\| < M$ then by hypothesis on ϕ, $\{g_k\}$ is a bounded sequence. There is a subsequence (assume it is $\{g_k\}$) converging pointwise on Y dense in X to $g \in G$. Hence $\{\phi(g_k)\}$ converges pointwise on Y to $\phi(g)$. As $|\phi(g(x))| \leq M$ for $x \in Y$, $\phi(g) \in \phi(G)$.

The full transformations of [2; 6] involve multiplication by a positive continuous weight function, but this is taken care of by the previous section.

With other transformations (such as $\phi(y) = \exp(y)$) we may not have existence.

Approximation With Non-continuous Weight Functions

Let s be a positive lower semi-continuous function on X. Lower semi-continuity ensures that $\sigma = \inf \{s(x) : x \in X\}$ is attained by s at some x, hence the infimum is positive. The problem of Chebyshev approximation with weight function s is given $f \in C(X)$, to find g^* for which $\|s * (f - g)\|$ attains its infimum $\rho(s, f)$.

THEOREM 5. Let s be a positive lower semi-continuous function on X. Then a best approximation with respect to weight s exists to all $f \in C(X)$.

Proof. Let $\|s * (f - g_k)\|$ be a decreasing sequence with limit $\rho(s, f)$. If $\|s * (g - g_1)\| > 2 \|s * (f - g_1)\|$ then $\|s * (f - g)\| >$

$\|s * (f - g_1)\|$. Hence, we have $\|s * (g_k - g_1)\| \leq 2\|s * (f - g_1)\|$ and $\{\|s * g_k\|\}$ is a bounded sequence. But $\|g_k\| \leq \|s * g_k\|/\sigma$. Hence there is a subsequence of $\{g_k\}$, assume without loss of generality it is $\{g_k\}$ converging pointwise on a dense subset Y to $g_0 \in G$. For $x \in Y$,

$$s(x)|f(x) - g_0(x)| \leq \limsup_{h\to\infty} s(x)|f(x) - g_k(x)| \leq \rho(s, f). \qquad (1)$$

By lower semi-continuity, (1) holds for all $x \in X$.

Approximation With Restricted Ranges

Let u be upper semi-continuous into the extended real line and v be lower semi-continuous into the extended real line, $u \leq v$. A special case of this is where u, v are continuous into the extended real line. The problem of Chebyshev approximation with restricted ranges is to find g^* to minimize $\|f - g\|$ under the constraint $u \leq g(x) \leq v(x)$ [9, 497-498]. Special cases are one-sided approximation from below with $u = -\infty$, $v = f$, and one-sided approximation from above with $u = f$, $v = \infty$.

THEOREM 6. Let G be dense compact then $\{g : g \in G, u \leq g \leq v\}$ is dense compact if non-empty.

Proof. Let $\{g_k\}$ be a bounded sequence of elements of G with $u \leq g_k \leq v$. There exists $g \in G$ such that a subsequence, assume it is g_k, converges pointwise to g on Y dense in X. We have

$$u(x) \leq g(x) \leq v(x) \qquad x \in Y.$$

As $g - u$ is lower semi-continuous and ≥ 0 on a dense set Y, $g - u \geq 0$. Similarly $v - g \geq 0$.

Simultaneous Approximation

Let f^- be a lower semi-continuous function on X and f^+ be an upper semi-continuous function on X, $f^- \leq f^+$. The Chebyshev problem of simultaneous approximation of f^- and f^+ [1] is to find g^* in G for which $e(g)$ $\max\{\|f^- - g\|, \|f^+ - g\|\}$ attains its infimum $\rho(f^-, f^+)$. Such an element g^* is called a best simultaneous approximation to f. It is shown in [1, bottom 476] that the problem of Chebyshev approximation of a bounded function on X can be replaced by a problem of simultaneous approximation.

THEOREM 7. Let G be dense compact. A best simultaneous approximation to f^-, f^+ exists.

Proof. Let $e(g_k)$ be a decreasing sequence with limit $\rho(f^-, f^+)$. By semi-continuity of f^-, f^+, $\|f^-\|$ and $\|f^+\|$ are finite.

$$\|g_k\| \leq \|f^- - g_k\| + \|f^-\| \leq e(g_k) + \|f^-\| \leq e(g_1) + \|f^-\| .$$

Hence $\{g_k\}$ has a subsequence, assume it is $\{g_k\}$, converging pointwise to $g \in G$ on Y dense in X. We have

$$\liminf_{k\to\infty} f^-(x) - g_k(x) \geq -\rho(f^-, f^+) \qquad x \in Y$$

$$\limsup_{k\to\infty} f^+(x) - g_k(x) \leq \rho(f^-, f^+) \qquad x \in Y,$$

hence

$$-\rho(f^-, f^+) \leq f^-(x) - g(x) \leq f^+(x) - g(x) \leq \rho(f^-, f^+) \qquad x \in Y.$$

By density of Y in X and semi-continuity, the above inequality holds for all $x \in X$.

Biased Approximation

Let fixed positive r be given and define

$$d_r(y) = y \qquad y \leq 0$$

$$= ry \qquad y > 0$$

$$\|h\|_r = \|d_r(h)\| .$$

The problem of r-biased Chebyshev approximation is, given $f \in C(X)$, to find an element g^* of G for which $\|f - g\|_r$ attains its infimum $\rho_r(f)$ over $g \in G$. Biased approximation was introduced in [10, 224] in terms of generalized weight functions.

THEOREM 8. Let G be dense compact. There exists a best approximation to all $f \in C(X)$.

Proof. Let $\{\|f - g_k\|_r\}$ be a decreasing sequence with limit $\rho(f)$. As

$$\|h\|_r \leq \|h\| \max\{r, 1\}$$

it follows that $\{\|f - g_k\|\}$ is bounded and by classical arguments, such as those of the corresponding result in [3], $\{\|g_k\|\}$ is bounded. By dense compactness we take a subsequence (assume it is $\{g_k\}$) converging pointwise to g_0 on dense subset Y. Then

$$|d_r(f(x) - g_0(x))| \leq \rho_r(f) \qquad x \in Y.$$

By density and continuity the inequality holds for all $x \in X$.

Approximation With Interpolation

Unfortunately dense compactness is not sufficient for existence of best approximations under Chebyshev approximation with Lagrange-type interpolation (Taylor [9, 495ff]). Loeb has an example cited by Taylor [9, bottom 496] of non-existence by ordinary rational functions on [0, 1]. The author has extended this non-existence to Chebyshev approximation with Hermite-type interpolation by ordinary rational functions on $[0, \beta]$.

Behaviour Of The Best Approximation Operator

In case there is a unique best approximation g to f, we can denote it by $T(f)$, defining the Chebyshev operator. In [3] appeared:

DEFINITION. An element g of a family G is called <u>normal</u> if any bounded sequence of elements of G which converges pointwise to g on a dense subset of X must converge uniformly on X.

It is easily seen that if g is a normal element of G then

(i) $s * g$ is a normal element of $s * G$, s a positive element of $C(X)$.

(ii) if $\phi(g) \in \phi(G)$ and g is a normal element of G, $\phi(g)$ is a normal element of $\phi(G)$.

(iii) if $u \leq g \leq v$ and g is a normal element of G, g is a normal element of $\{h : h \in G, u \leq h \leq v\}$.

In [3] is proven:

THEOREM 9. Let G be dense compact. The operator T is continuous at all $f \in C(X)$ possessing a unique and normal best approximation and at all $f \in G$.

In the case of simultaneous approximation, we can denote the unique best simultaneous approximation to f^-, f^+ by $T(f^-, f^+)$. It is readily seen that a continuity result similar to the above holds. In biased approximation a comparable result holds.

We now give sufficient conditions for non-normality to imply discontinuity of T.

THEOREM 10. Let X be an interval. Let bounded sequences from G have a subsequence converging uniformly on a non-degenerate closed interval I in X to an element of G. Let G have property Z of degree n at all elements. If g is not normal, there exists $f \in C(X)$ such that $T(f) = g$ and T is not continuous at f.

<u>Proof</u>. Let $\|g_k\| < M$ and $\{g_k\} \to g$ uniformly on $I = [\gamma, \delta]$, but not uniformly on X. Let $\gamma < x_0 < \ldots < x_{2n} < \delta$ and define

$$f(x_i) = g(x_i) + (-1)^i 2M$$

$$f(\gamma) = g(\gamma) \qquad f(\delta) = g(\delta).$$

There is a continuous extension of f to X such that

$$f(x) = g(x) \qquad x \notin I$$

$$\|f - g\| = 2M$$

$f - g$ alternates $2n$ times on X.

By standard arguments from the alternating theory, g is uniquely best to f. Define

$$f_k(x) = g_k(x) + f(x) - g(x) \qquad x \in I$$
$$= g_k(\gamma) + f(x) - g(\gamma) \qquad x < \gamma$$
$$= g_k(\delta) + f(x) - g(\delta) \qquad x > \delta,$$

then $\|f - f_k\| = \max\{|g(x) - g_k(x)| : x \in I\} \to 0$. For $x \notin I$,

$$|f_k(x) - g_k(x)| \leq |f_k(x) - f(x)| + |f(x) - g(x)| + |g(x) - g_k(x)| .$$

For all k sufficiently large, the right-hand side is less than $2M$. Further $f_k - g_k$ alternates $2n$ times on I with amplitude $2M$, so for all k sufficiently large, g_k is uniquely best to f_k. We have

$$\{f_k\} \to f \text{ and } \{T(f_k)\} \not\to T(f).$$

Discontinuity of the best approximation operator in simultaneous and biased approximation can be proven by similar arguments involving alternation.

REFERENCES

1. C. B. Dunham, "Simultaneous Chebyshev approximation of functions on an interval", Proc. Amer. Math. Soc. 18 (1967), 472-477.

2. C. B. Dunham, "Transformed rational Chebyshev approximation", Numer. Math. 10 (1967), 147-152.

3. C. B. Dunham, "Existence and continuity of the Chebyshev operator", SIAM Rev. 10 (1968), 444-446.

4. C. B. Dunham, "Chebyshev approximation by A+B*log(1+CX).II", J. Inst. Math. Applics. 10 (1972), 369-372.

5. C. B. Dunham, "Existence of best Chebyshev approximations by A+B*ϕ(CX)", Util. Math. 4 (1973), 87-93.

6. C. B. Dunham, "Transformed linear Chebyshev approximation", Aeq. Math. 12 (1975), 6-11.

7. J. Rice, "The approximation of functions", Volume 1, Addison-Wesley, Reading, Mass., 1964.

8. E. Schmidt, "Zur Kompaktheit bei Exponentialsummen", J. Approx. Theory 3 (1970), 445-454.

9. G. D. Taylor, "Uniform approximation with side conditions", in G. G. Lorentz (ed.) Approximation Theory, Academic Press, New York, 1973, 495-503.

10. C. B. Dunham, "Chebyshev approximation with respect to a weight function", J. Approx. Theory 2 (1969), 223-232.

11. C. B. Dunham, "Rational approximation with a vanishing weight function and with a fixed value at zero", Math. Comp. 30 (1976), 45-47.

APPROXIMATION OF INTEGRABLE FUNCTIONS BY BOREL MEANS

A. S. B. Holland
University of Calgary
Calgary, Alberta
Canada

R. N. Mohapatra
American University of Beirut
Beirut
Lebanon

B. N. Sahney
University of Calgary
Calgary, Alberta
Canada

1. Introduction

Approximation of a continuous function by means of the Fourier series representation of the function, has been studied by several authors. In particular, Alexits [1], studied error bounds for (C, α) means, Kathal, Holland and Sahney [5] generalized this result and Holland, Sahney and Tzimbalario [4] studied error bounds for Norlund means using the modulus of continuity, $\omega(\delta; f)$. This paper makes a detailed study of order or degree of approximation, using the Borel means of the Fourier series for f and, in particular, using the integral modulus of continuity $\omega_p(t; f)$, $p \geq 1$.

2. Let $f \in L[0, 2\pi]$ and be periodic with period 2π. The Fourier series associated with f at the point x is given by

$$f(x) \sim \frac{a_0}{2} + \sum_{k=1}^{\infty} (a_k \cos kx + b_k \sin kx). \tag{2.1}$$

Let $s_n(x)$ denote the nth partial sum of the series (2.1).

The series (2.1) is said to be Borel summable to s if $B^r(f; x)$, the Borel transform of the series (2.1) tends to s as $r \to \infty$. $B^r(f; x)$ is given by

$$B^r(f;\, x) = e^{-r} \sum_{n=0}^{\infty} \frac{r^n}{n!} s_n(x). \tag{2.2}$$

All norms considered in the paper will be L_p norms unless otherwise stated and will be with respect to the variable x. When $p = \infty$, L_p will be regarded as $C_{2\pi}$, the space of continuous functions with period 2π. $\mathrm{Lip}(\alpha, p)$ $0 < \alpha \leq 1$, $1 \leq p < \infty$, will be as usual (see [3], p. 612). When $p = \infty$, $\mathrm{Lip}(\alpha, p)$ will be regarded as $\mathrm{Lip}\ \alpha$.

Throughout the paper we shall write

$$\phi_x(t) = \frac{1}{2}\{f(x + t) + f(x - t) - 2f(x)\}, \tag{2.3}$$

$$K(r, t) = \frac{\sin(\frac{t}{2} + r \sin t)}{\sin \frac{t}{2} \exp\{r(1 - \cos t)\}}, \tag{2.4}$$

$$w_p(t;\, f) = \|f(x + t) - f(x)\|_p \quad (1 \leq p < \infty)$$

$$w(t;\, f) = \sup_{|h| \leq t} |f(x + h) - f(x)|$$

$$\rho = \pi(r + \tfrac{1}{2})^{-1}, \quad 0 < r < \infty.$$

When $p \to \infty$, we regard $w_p(t;\, f) = w(t;\, f)$.

3. We require the following lemmas.

LEMMA 3.1 Let $\rho = \dfrac{\pi}{(r + \frac{1}{2})}$, then

$$K(r, t) = \begin{cases} O(\frac{rt}{t}) = O(r) & (\text{as } r \to \infty) \quad (0 < t < \rho), \\ O(t^{-1} e^{-Art^2}) & (\rho \leq t < \pi), \end{cases}$$

where A is a positive constant.

Proof. Since in $0 < t < \rho$, $\sin nt = O(nt)$, and $\sin \frac{t}{2} \geq \frac{t}{\pi}$ for $0 < t < \pi$; the first estimate follows.

The second estimate follows by taking $\sin nt = O(1)$ uniformly in t and noting that the exponential function is an increasing function and

$\sin \frac{t}{2} \geq \frac{t}{\pi}$.

LEMMA 3.2. (See: Hardy, Littlewood, Polya, "Inequalities", p. 148, 6.13.9). If $h(x, t)$ is a function of two variables, x and t defined over $0 \leq t \leq \pi$, $0 \leq x \leq 2\pi$, then

$$\| \int h(x, t)dt \|_p \leq \int \| h(x, t) \|_p dt , \quad (p > 1).$$

(This is the generalized Minkowski inequality.)

THEOREM 1. Let $f \in L_p$ $(p > 1)$. Then

$$\| f - B^r(f) \|_p = 0(\omega_p(r^{-1}; f)) + 0(r^{\beta} e^{-r^{1-2\beta}}) \tag{3.3}$$

$$+ 0(\int_{\rho}^{\rho^{\beta}} \frac{\omega^*(t; f)}{t} dt).$$

where $0 < \beta < \frac{1}{2}$, and $\omega^*(t; f)$ is a positive nondecreasing function of t with the property that

$$\omega_p(t; f) = 0(\omega^*(t; f)).$$

Proof. Clearly

$$B^r(f; x) - f(x) = \frac{e^{-r}}{\pi} \int_0^{\pi} \phi_x(t) \sum_{n=0}^{\infty} \frac{r^n}{n!} \frac{\sin(n + \frac{1}{2})t}{\sin \frac{t}{2}} dt .$$

Following Hardy [2] ("Divergent Series", p. 364), we have

$$B^r(f; x) - f(x) = \frac{1}{2\pi} [\int_0^{\rho} + \int_{\rho}^{\rho^{\beta}} + \int_{\rho^{\beta}}^{\pi}] \phi_x (t) K(r, t) dt \tag{3.5}$$

$$= I_1 + I_2 + I_3, \text{ say,}$$

where

$$\rho = \pi(r + \tfrac{1}{2})^{-1} \quad \text{and} \quad 0 < \beta < \tfrac{1}{2}.$$

By the Minkowski inequality,

$$\|B^r(f;\ x) - f(x)\| \le \|I_1\| + \|I_2\| + \|I_3\| . \tag{3.6}$$

We now use Lemma (3.2) and the first estimate of $K(r,\ t)$ from Lemma (3.1),

$$\|I_1\| = O\left(\int_0^\rho \|\phi_x(t)\|\ |K(r,\ t)|dt\right)$$

$$= O\left(r\int_0^\rho w_p(t;\ f)dt\right)$$

$$= O(w_p(\rho;\ f)) = O(w_p(r^{-1};\ f)). \tag{3.7}$$

By Lemma (3.2) and the second estimate of $K(r,\ t)$ from Lemma (3.1), we get

$$\|I_3\| = O\left(\int_{\rho^\beta}^\pi \frac{\|\phi_x(t)\|}{t}\ e^{-Art^2}\ dt\right)$$

$$= O\left(\rho^{-\beta}\ e^{-Ar\rho^{2\beta}} \int_{\rho^\beta}^\pi w_p(t;\ f)dt\right)$$

$$= O\left(r^\beta\ e^{-Ar^{1-2\beta}}\right), \tag{3.8}$$

since $w_p(t,\ f)$ is a nondecreasing function of t.

Now we consider I_2. For t in $[\rho,\ \rho^\beta]$, we use the second estimate of Lemma (3.1), after using Lemma (3.2) in I_2. We obtain

$$\|I_2\| \le \int_\rho^{\rho^\beta} \frac{\|\phi_x(t)\|}{t}\ e^{-Art^2} dt$$

$$= O\left(e^{-A/r} \int_\rho^{\rho^\beta} \frac{w_p(t;\ f)}{t}\ dt\right)$$

$$= O\left(\int_\rho^{\rho^\beta} \frac{w^*(t;\ f)}{t}\ dt\right),\quad 0 < \beta < \frac{1}{2}. \tag{3.9}$$

On collecting the estimates from (3.7), (3.8), (3.9), and substituting in (3.6), the proof of the theorem is completed.

THEOREM 2. Let $f \in \mathrm{Lip}(\alpha,\ p)$, $0 < \alpha \le 1$, $p > 1$, then

$$\|B^r(f;\ x) - f(x)\| = O(r^{-\alpha\beta}),\quad 0 < \beta < \frac{1}{2}. \tag{3.10}$$

An immediate corollary of Theorem 2 is the following:

COROLLARY 1. If $f \in \text{Lip}\,\alpha$, $(0 < \alpha \leq 1)$, then

$$\sup_{0 \leq x \leq 2\pi} |B^r(f; x) - f(x)| = 0(r^{-\alpha\beta})$$

where $0 < \beta < \frac{1}{2}$.

The corollary follows from Theorem 2 by taking $p = \infty$ in it and noting the fact that in this case $\text{Lip}(\alpha, p)$ reduces to $\text{Lip}\,\alpha$.

<u>Proof of Theorem 2.</u> We make use of the fact that when $f \in \text{Lip}(\alpha, p)$, $0 < \alpha < 1$, $p > 1$, then

$$w_p(t; f) = 0(t^\alpha)$$

in (3.7). Hence the estimate in (3.7) reduces to $0(r^{-\alpha})$. Since $0 < \beta < \frac{1}{2}$,

$$r^\beta e^{-Ar^{1-2\beta}} = 0(r^{-\alpha}), \quad (0 < \alpha \leq 1).$$

Hence I_3 in the proof of Theorem 1, is of order $(r^{-\alpha})$.

For estimating I_2, note that $f \in \text{Lip}(\alpha, p)$ implies that $w^*(t; f) = 0(t^\alpha)$, so I_2 in the proof of Theorem 1 is

$$0\left(\int_\rho^{\rho^\beta} t^{\alpha-1}dt\right) = 0(\rho^{\beta\alpha}) + 0(\rho^\alpha)$$

$$= 0(r^{-\beta\alpha}), \quad (0 < \beta < \tfrac{1}{2}).$$

On substituting these estimates in (3.6), the theorem follows.

THEOREM 3. If $f \in \text{Lip}(\alpha, p)$, $0 < \alpha \leq 1$, $p > 1$, $\alpha p > 1$, then

$$|B^r(f; x) - f(x)| = 0(r^{(-\alpha + \frac{1}{p})\beta})$$

uniformly in x in $0 \leq x \leq 2\pi$ almost everywhere.

This theorem can be proved using the following lemma.

LEMMA 3.11. ([3], Theorem 5(ii), p. 627). Suppose that $f \in \text{Lip}(\alpha, p)$ where $p \geq 1$, $0 < \alpha \leq 1$ and $\alpha p > 1$. Then f is equivalent to a function, g in $\text{Lip}(\alpha - \frac{1}{p})$, almost everywhere.

Proof of Theorem 3. Under the assumptions of the lemma, $\phi_x(t) = 0(t^{\alpha - \frac{1}{p}})$. We use this, and the result will follow as in Theorem 1.

4. Now we study approximation properties of the Borel means, by imposing an integrability condition on the function $\phi_x(t)$. Our first theorem in that direction is

THEOREM 4. Let $f \in L_p$ $(p \geq 1)$ and let $w_p(t, f) = 0(w^*(t; f)$ where $w^*(t; f)$ is a positive nondecreasing function of t in $0 \leq x \leq \pi$ such that for $0 < \delta \leq 1$,

$$\frac{w^*(t)}{t^\delta}$$

is nonincreasing in t. Further, let

$$\tilde{I} = \int_\rho^{\rho^\beta} \frac{\|\phi_x(t) - \phi_x(t + \rho)\|}{t} e^{\frac{-rt^2}{2}} dt = 0(w^*(r^{-1}; f)) \tag{4.1}$$

where $0 < \beta < \frac{1}{2}$. Then

$$\|B^r(f; x) - f(x)\|p = 0(w^*(r^{-1}; f)) + 0(r^\beta e^{-Ar^{1-2\beta}}) \tag{4.2}$$

$$+ 0(r^{-\beta} w^*(r^{-\beta}; f)).$$

Proof of Theorem 4. The first two estimates on the right of (4.2) are estimates for I_1 and I_3 of Theorem 1 (see the proof). It remains to prove that I_2 has estimates which are like $0(w^*(r^{-1}; f))$ or $0(r^{-\beta} w^*(r^{-\beta}; f))$. Thus, we check I_2 of Theorem 1.

We can write I_2 in the following manner.

$$I_2 = \frac{-1}{2\pi} \int_0^{\rho^\beta - \rho} \frac{\phi_x(t+\rho)}{\sin\frac{(t+\rho)}{2}} \frac{\sin t(r + \frac{1}{2})}{\exp[r(1-\cos(t+\rho)]} dt$$

$$= I_{21} + I_{22} + I_{23} + I_{24} + I_{25},$$

where

$$I_{21} = \frac{1}{4\pi} \int_{\rho}^{\rho^{\beta}} \frac{\phi_x(t) - \phi_x(t+\rho)}{\sin\frac{t}{2}} \frac{\sin t(r+\frac{1}{2})}{\exp[r(1-\cos t)]} dt ,$$

$$I_{22} = \frac{1}{4\pi} \int_{\rho}^{\rho^{\beta}} \frac{\phi_x(t+\rho)}{\sin\frac{t}{2}} \{\frac{1}{\exp[r(1-\cos t)]} - \frac{1}{\exp[r(t-\cos(t+\rho)]}\}$$

$$\times \sin t(r+\tfrac{1}{2}) dt,$$

$$I_{23} = \frac{1}{4\pi} \int_{\rho}^{\rho^{\beta}} \phi_x(t+\rho) \left| \frac{1}{\sin\frac{t}{2}} - \frac{1}{\sin\frac{(t+\rho)}{2}} \right| \frac{\sin t(r+\frac{1}{2})}{\exp[r(1-\cos(t+\rho)]} dt,$$

$$I_{24} = -\frac{1}{4\pi} \int_{0}^{\rho} \frac{\phi_x(t+\rho)}{\sin\frac{(t+\rho)}{2}} \frac{\sin t(r+\frac{1}{2})}{\exp[r(1-\cos(t+\rho)]} dt,$$

and

$$I_{25} = \frac{1}{4\pi} \int_{\rho^{\beta}-\rho}^{\rho^{\beta}} \frac{\phi_x(t+\rho)}{\sin\frac{(t+\rho)}{2}} \frac{\sin t(r+\frac{1}{2})}{\exp[r(1-\cos(t+\rho))]} dt .$$

By the Minkowski inequality

$$\|I_2\| \le \|I_{21}\| + \|I_{22}\| + \|I_{23}\| + \|I_{24}\| + \|I_{25}\| . \tag{4.3}$$

Now by Lemma (3.2), and the fact that $\sin\frac{t}{2} \geq \frac{t}{\pi}$, $0 < t \leq \pi$,

$$\|I_{21}\| \le \frac{1}{2} \int_{\rho}^{\rho^{\beta}} \frac{\|\phi_x(t) - \phi_x(t+\rho)\|}{t} e^{-rt^2/2} dt$$

$$= 0(w^*(r^{-1}; f)),$$

by hypothesis.

By Lemma (3.2), and the fact that the difference of the exponentials in I_{22} are or order t, we have

$$\|I_{22}\| = 0(\int_0^{\rho^{\beta}} \frac{\|\phi_x(t+\rho)\|}{t} \cdot t \; dt)$$

$$= O\left(\int_0^{\rho^\beta} w_p(t + \rho; f)dt\right)$$

$$= O\left(\int_0^{\rho^\beta} w^*(t; f)dt\right)$$

$$= O(\rho^\beta w^*(\rho^\beta; f))$$

$$= O(r^{-\beta} w^*(r^{-\beta}; f)). \tag{4.4}$$

By Lemma (3.2), and estimating the terms in I_{23},

$$\|I_{23}\| = O\left(\rho \int_\rho^{\rho^\beta} \frac{\phi_x(t + \rho)}{t(t + \rho)} dt\right)$$

$$= O\left(\rho \int_\rho^{\rho^\beta} \frac{w^*(t; f)}{t(\rho + t)} dt\right)$$

$$= O\left(\frac{\rho w^*(\rho, f)}{\rho^\delta} \int_\rho^\infty \frac{dt}{t^{2-\delta}}\right)$$

$$= O\left(w^*(\rho; f) \cdot \frac{\rho^{1-\delta}}{\rho^{1-\delta}}\right) = O(w^*(r^{-1}; f)), \quad (0 < \delta < 1). \tag{4.5}$$

By Lemma (3.2),

$$\|I_{24}\| \leq \int_0^\rho \frac{\|\phi_x(t + \rho)\|}{O \sin\frac{(t + \rho)}{2}} \frac{|\sin t(r + \frac{1}{2})|}{\exp[r(1 - \cos(t + \rho))]} dt$$

$$= O\left(\frac{1}{2\rho} \int_\rho^{2\rho} \|\phi_x(t)\| dt\right)$$

$$= O(w^*(r^{-1}; f)).$$

Finally, we have by a similar argument

$$\|I_{25}\| \leq \frac{1}{2} \int_0^\rho \frac{\|\phi_x(t + \rho^\beta)\|}{t + \rho^\beta} dt$$

$$= \frac{1}{2} \int_{\rho^\beta}^{\rho+\rho^\beta} \frac{\|\phi_x(t)\|}{t} dt$$

$$= O\left(\int_{\rho^\beta}^{\rho+\rho^\beta} \frac{\omega_p(t; f)}{t} dt \right)$$

$$= O\left(\frac{\omega^*(\rho; f)}{\rho^\delta} \int_{\rho^\beta}^{\rho+\rho^\beta} \frac{dt}{t^{1-\delta}} \right)$$

$$= O\left(\frac{\omega^*(\rho; f)}{\rho^\delta} \{(\rho + \rho^\beta)^\delta - \rho^{\delta\beta}\} \right)$$

$$= O(\omega^*(\rho; f)).$$

Since

$$\frac{(\rho + \rho^\beta)^\delta - \rho^{\delta\beta}}{\rho^\beta} = O\left(\left(\frac{\rho}{\rho^\beta}\right)^{1-\eta}\right) = O(1).$$

On collecting the estimates the result follows.

5. Parallel to Theorem 2 and Theorem 3 we can deduce the following two theorems:

THEOREM 5. Let $f \in \mathrm{Lip}(\alpha, p)$, $0 < \alpha < 1$, $p > 1$. Then

$$\|B^r(f; x) - f(x)\| = O(r^{-\alpha})$$

whenever the integral

$$I = \int_{\rho}^{\rho^\beta} \frac{\|\phi_x(t) - \phi_x(t + \rho)\|}{t} e^{-r\frac{t^2}{2}} dt = O(r^{-\alpha}),$$

and

$$\frac{\alpha}{(1 + \alpha)} \leq \beta < \frac{1}{2}.$$

THEOREM 6. Let $f \in \mathrm{Lip}(\alpha, p)$, $0 < \alpha < 1$, $p > 1$, and $\alpha p > 1$. Then whenever $I = O(r^{-\alpha + \frac{1}{p}})$, we obtain

$$B^r(f; x) - f(x) = O(r^{-\alpha + \frac{1}{p}})$$

uniformly for $0 \leq x \leq 2\pi$ almost everywhere.

REFERENCES

1. G. Alexits, "Uber die Annaherung einer stetigen Function durch die Cesàroschen Mittel ihrer Fourrierreihe", Math. Annalen, 100, (1928), 264-277.
2. G. H. Hardy, "Divergent Series", Clarendon Press, Oxford (1973).
3. G. H. Hardy and J. E. Littlewood, "A convergence criterion for Fourier series", Math. Zeit., 28 (1928), 612-634.
4. A. S. B. Holland, B. N. Sahney and J. Tzimbalario, "On degree of approximation of a class of functions by means of Fourier series", Acta Sci. Math. 38 (1976), 69-72.
5. P. D. Kathal, A. S. B. Holland and B. N. Sahney, "A class of continuous functions and their degree of approximation", Acta Math. Acad. Sci. Hung. 30(3-4), (1977), 227-331.

SOME RESULTS ON L_p APPROXIMATION OF FUNCTIONS BY THEIR FOURIER SERIES*

R. N. Mohapatra**

American University of Beirut
Beirut
Lebanon

S. P. Singh

Memorial University of Newfoundland
St. John's, Newfoundland
Canada

Abstract

Given a measurable, periodic, integrable function of period 2π and a lower triangular infinite matrix of scalars. With the help of the Fourier expansion of the given function, a linear operator is defined by considering the matrix transformation of the partial sum of the Fourier series. For functions of class L_p and $\mathrm{Lip}(\alpha, p)$, we have studied the rapidity of convergence of the operator to the given function. Study has also been made to determine the degree of convergence of the Borel and Euler means of Fourier series of the function to itself.

1. Definitions And Notations

Let $f \in L[-\pi, \pi]$ be a periodic function with period 2π. Let the Fourier series of f be given by

AMS classification 42A 10.

Key words: Lipschitz class, Summability methods, L_p and Sup norms, modulus of continuity and integral modulus of continuity.

*This work was supported by NSERC and NATO grants.

**The author was visiting Memorial University from the American University of Beirut.

$$\frac{1}{2}a_0 + \sum_{n=1}^{\infty}(a_n \cos nx + b_n \sin nx) \equiv \sum_{n=0}^{\infty}A_n(x). \tag{1.1}$$

Let $S_n(x)$ denote the nth partial sum of the series (1.1).

Let $A = (a_{nk})(k = 0, 1, \ldots; n = 0, 1, \ldots)$ be an infinite matrix of real numbers. We denote by $A_n(f; x)$ the A transform of the series (1.1). $A_n(f, x) \equiv A_n(f)$ is given by

$$A_n(f, x) = \sum_{k=0}^{\infty} a_{nk} S_k(x), \tag{1.2}$$

whenever the series on the right of (1.2) converges for each $n = 0, 1, \ldots$.

Throughout the paper, L_p $(1 \leq p \leq \infty)$ spaces will be as usual. When $p = \infty$, $L_p[-\pi, \pi]$ will be regarded as the space $C_{2\pi}$ of all continuous functions on $[-\pi, \pi]$ which are periodic with period 2π. All norms considered in this paper will be L_p norms unless otherwise mentioned.

All norms will be taken with respect to the variable x and will be denoted by $\|\cdot\|_p$ $(1 \leq p \leq \infty)$. When $p = \infty$, $\|\cdot\|_p$ will be regarded as the sup norm. The classes of functions Lip α, Lip(α, p), Lip$^*\alpha$, Lip$^*(\alpha, p)$ $(1 \leq p < \infty)$ will have their usual significance (for definitions, see [6], p.p. 612. Also, see [18], p.p. 42 and 45). When $p = \infty$, the classes Lip(α, p) and Lip$^*(\alpha, p)$ will be regarded as the classes Lip α and Lip$^*\alpha$, $0 < \alpha \leq 1$, respectively.

For $1 \leq p < \infty$ and $\delta > 0$, we write $w(g, \delta)$ and $w_p(g; \delta)$ for modulus of continuity and integral modulus of continuity of a function g respecitvely. (See [18], p.p. 42 and p.p. 45.)

Throughout the paper, we shall let K denote an absolute positive constant which may be different at different occurences. We shall also use the following notations:

$$\phi_x(t) = f(x + t) + f(x - t) - 2f(x), \tag{1.3}$$

$$\overline{a}_{nk} = \sum_{r=0}^{k} a_{nr}, \tag{1.4}$$

$$\overline{a}_n(k) = \overline{a}_{n,k}, \tag{1.5}$$

$$a'_{nk} = \sum_{r=k}^{n} a_{nr}, \tag{1.6}$$

$$\tau = [\tfrac{\pi}{t}] \quad (0 < t \leq \pi), \tag{1.7}$$

$$E(n, t) = (1 + q)^{-n}\sum_{k=0}^{n} \binom{n}{k}q^{n-k} \sin (k + \tfrac{1}{2})t \, (q \geq 0), \tag{1.8}$$

(1.9) $M(p, t) = e^{-p} \sum_{n=0}^{\infty} \frac{p^n}{n!} \sin (n + \frac{1}{2})t.$ *

2. The problem of obtaining the degree of approximation of a function in Lip α $(0 < \alpha \leq 1)$ by the Fejér means of its Fourier series was first considered by Alexits [1]. Subsequently, Prem Chandra [2], and Holland, Sahney and Tzimbalario [8] generalized the result of Alexits by replacing the Fejér means by $(\overline{N}, p_n)$ and (N, P_n) means. (See [5] for definitions of $(\overline{N}, p_n)$ and (N, p_n)). Independently, Kathal, Holland and Sahney [12], and Khan and Wafi [13] have attempted to generalize the result of [8] by replacing the Nörlund matrix by a lower triangular matrix (a_{nk}) satisfying some conditions. Subsequently, Mohapatra and Sahney [16] unified the results of [2] and [8] by estimating $\|f - A_n(f)\|_{\infty}$. Degrees of approximation of functions belonging to Lip α by the Euler and Borel means of their Fourier series have been studied in [3] and [4]. Quade [17] has studied the approximation of functions belonging to class Lip (α, p) by their Fourier series. Quade's results concerning L_p approximations have been deduced from results in Orlicz spaces ([17], Theorem 1, Theorem 3). In [10] and [11] order of approximation of functions of L_p by the Euler and Borel means of their Fourier series have been studied. L_p approximation of functions by the Taylor means of their Fourier series has been considered in [9]. Mohapatra and Chandra [15] have determined the degrees of approximation of continuous functions by their Borel, Euler and Taylor means in the Hölder metric. (See [15] for definitions.)

The purpose of this paper is to obtain the degree of convergence of $A_n(f)$ to f in the L_p norm. Such results are extensions of the results obtained in [16], [2] and [8]. By specializing the matrix $A = (a_{nk})$ in §5, we obtain many interesting results of independent interest. We also obtain estimates for $\sup_{-\pi \leq x \leq \pi} |f(x) - A_n(f; x)|$ for f in L_p $(1 < p < \infty)$. In §6, we shall study the L_p approximation of functions by the Euler and Borel means of their Fourier series and deduce results concerning the approximation of functions of class Lip (α, p) $(0 < \alpha \leq 1, 1 \leq p \leq \infty)$.

We begin by stating some general results involving the operator $A_n(f; x)$.

THEOREM 1. Let $A = (a_{nk})$ be a lower triangular matrix satisfying the following:

*Whenever there is no confusion, we shall write $w_p(\delta)$ or $w(\delta)$ for $w_p(g, \delta)$ and $w_p(g, \delta)$ respectively.

(2.1) $a_{nk} \geq 0$ $(k = 0, 1, \ldots, n;\ n = 0, 1, \ldots)$ and $\sum_{k=0}^{n} a_{nk} = 1.$

(2.2) $a_{nk} \geq a_{n,k+1}$ $(k = 0, 1, \ldots, n - 1;\ n = 0, 1, \ldots).$

Then for $f \in L_p[-\pi, \pi]$ $(1 \leq p \leq \infty)$,

$$(2.3)\quad \|f - A_n(f)\|_p = O\Big(\sum_{k=1}^{n} \frac{w_p(\frac{1}{k})\bar{a}_n(k + 1)}{k}\Big).$$

THEOREM 2. Let $A = (a_{nk})$ be a lower triangular matrix satisfying (2.1) and

(2.4) $a_{nk} \leq a_{n,k+1}$ $(k = 0, 1, \ldots, n - 1;\ n = 0, 1, \ldots).$

Then, for $f \in L_p[-\pi, \pi]$ $(1 \leq p < \infty)$,

$$(2.5)\quad \|f - A_n(f)\|_p = O\Big(\sum_{k=1}^{n} \frac{w_p(\frac{1}{k})a'_{n,n-k}}{k}\Big).$$

THEOREM 3. Let $A = (a_{nk})$ satisfy (2.1) and (2.4). Then, for $f \in \mathrm{Lip}$ (α, p) $(0 < \alpha \leq 1,\ 1 \leq p \leq \infty)$,

$$\|f - A_n(f)\|_p = \begin{cases} O(a_{nn}^{\alpha}) & (0 < \alpha < 1); \\ O(a_{nn}(1 + \log \frac{1}{a_{nn}})) & \end{cases}$$

From the definitions of $\mathrm{Lip}\ \alpha$ and $\mathrm{Lip}\ (\alpha, p)$ $(0 < \alpha \leq 1,\ p > 1)$, we know that $f \in \mathrm{Lip}\ \alpha$ implies $f \in \mathrm{Lip}\ (\alpha, p)$. To complement studies on $\sup_{-\pi \leq x \leq \pi} |f(x) - A_n(f; x)|$ when $f \in \mathrm{Lip}\ \alpha$ we prove the following results:

THEOREM 4. Let $A = (a_{nk})$ be a lower triangular matrix satisfying (2.1) and (2.2). Then, for $f \in \mathrm{Lip}\ (\alpha, p)$, $0 < \alpha \leq 1$, $p > 1$, $\alpha p > 1$,

$$\sup_{-\pi \leq x \leq \pi} |f(x) - A_n(f; x)| = O\Big(\sum_{k=1}^{n} \frac{\bar{a}_n(k+1)}{k^{\alpha+1-1/p}}\Big).$$

THEOREM 5. Let $A = (a_{nk})$ be a lower triangular matrix satisfying (2.1) and (2.4). Then, for $f \in \mathrm{Lip}\ (\alpha, p)$, $0 < \alpha \leq 1$, $p > 1$, $\alpha p > 1$,

$$\sup_{-\pi\leq x\leq\pi} |f(x) - A_n(f; x)| = O\left(\sum_{k=1}^{n} \frac{a'_{n,n-k}}{k^{\alpha+1-1/p}}\right).$$

THEOREM 6. Let $A = (a_{nk})$ satisfy (2.1) and (2.4). Then, for $f \in \mathrm{Lip}(\alpha, p)$, $0 < \alpha \leq 1$, $p > 1$, $\alpha p > 1$,

$$\sup_{-\pi\leq x\leq\pi} |f(x) - A_n(f; x)| = O(a_{nn}^{\alpha-1/p}).$$

3. We shall need the following lemmas for the proof of our results.

LEMMA 1. (See [16], Lemma 2.) Let the lower triangular matrix $A = (a_{nk})$ satisfy $a_{nk} > 0$ $(k = 0, 1, \ldots, n;\ n = 0, 1, \ldots)$ and (2.4). Then

$$\sum_{k=0}^{n} a_{nk} \sin(k + \tfrac{1}{2})t = O(a_{nn}/t)$$

uniformly in $0 < t \leq \pi$.

This can be proved by using Abel's lemma.

LEMMA 2. (See [16], Lemma 1.) Let the lower triangular matrix $A = (a_{nk})$ satisfy $a_{nk} \geq 0$ $(k = 0, 1, \ldots, n;\ n = 0, 1, \ldots)$ and (2.2). Then

$$\sum_{k=0}^{n} a_{nk} \sin(k + \tfrac{1}{2})t = O(\bar{a}_n(\tau))$$

where $\tau = [\frac{\pi}{t}]$.

This lemma can be proved by using arguments similar to that of McFadden [14, p.p. 182].

LEMMA 3. (See [16], Lemma 3.) Let the lower triangular matrix $A = (a_{nk})$ satisfy the conditions of Lemma 2. Then

$$\sum_{k=0}^{n} a_{nk} \sin(k + \tfrac{1}{2})t = O(a'_{n,\, n-\tau+1})$$

where $\tau = [\frac{\pi}{t}]$.

LEMMA 4. (See [7], p.p. 148, 6.13.9.) If $h(x, t)$ is a function of two variables x and t defined for $0 \leq t \leq \pi$, $-\pi \leq x \leq \pi$, then for

$1 \leq p < \infty$,

$$\left\| \int h(x, t)dt \right\|_p \leq \int \|h(x, t)\|_p \, dt.$$

LEMMA 5. (See [6], Theorem 5(ii), p.p. 627.) Suppose that $f \in \text{Lip}(\alpha, p)$ where $p \geq 1$, $0 < \alpha \leq 1$, $\alpha p > 1$. Then f is equivalent to a function $g \in \text{Lip}(\alpha - \frac{1}{p})$ almost everywhere.

LEMMA 6. (See [15], Lemma 1.) Let $q \geq 0$ and $E(n, t)$ be given by (1.8). Then the following holds:

(3.1) $E(n, t) = [P(q, t)]^{n/2} \sin \{\frac{1}{2}t + nQ(q, t)\}$, where

(3.2) $P(q, t) = 1 + q^2 + 2q \cos t$, and

(3.3) $Q(q, t) = \tan^{-1} \{\sin t/(q + \cos t)\}$.

Further, $E(n, t)$ has the following two estimates:

(3.4) $E(n, t) = 0(\exp \{-2nqt^2/(1 + q)^2\pi^2\}]$, $(0 < t < \pi)$ and

$$E(n, t) = 0[t^{-1} n^{-1/2}) \quad (0 < t < \pi).$$

4. In this section we prove the theorems.

Proof of Theorem 1. From (1.2), (2.1), (1.3), we get

$$(4.1) \quad A_n(f; x) - f(x) = \frac{1}{\pi} \int_0^\pi \frac{\phi_x(t)}{2 \sin \frac{1}{2}t} \left(\sum_{k=0}^n \sin (k + \tfrac{1}{2})t\right)dt.$$

Since $\sin x \geq \frac{2}{\pi}x$, $0 \leq x \leq \frac{\pi}{2}$, we get, after writing

$$\int_0^\pi = \int_0^{\pi/n} + \int_{\pi/n}^\pi$$

in (4.1) and using Minkowski inequality.

(4.2) $\|f - A_n(f)\|_p \leq \|I_1\|_p + \|I_2\|_p$, where

$$I_1 = \frac{1}{2} \int_0^{\pi/n} \frac{\phi_x(t)}{t} \left(\sum_{k=0}^n a_{nk} \sin (k + \tfrac{1}{2})t\right)dt$$

and

$$I_2 = \frac{1}{2}\int_{\pi/n}^{\pi} \frac{\phi_x(t)}{t} \sum_{k=0}^{n} a_{nk} \sin(k + \tfrac{1}{2})t \, dt$$

By using Lemma 4 and the fact that $\|\phi_x(t)\|_p \leq 2\, w_p(t)$, we see that

$$(4.3)\quad \|I_1\|_p \leq \int_0^{\pi/n} \frac{w_p(t)}{t} \left| \sum_{k=0}^{n} a_{nk} \sin(k + \tfrac{1}{2})t \right| dt.$$

Since $\sin nt \leq nt$, $w_p(t)$ is a non-decreasing of t, we get

$$(4.4)\quad \|I_1\|_p = O(w_p(\tfrac{\pi}{n})), \text{ by } (2.1).$$

By Lemma 2 and the fact that $\bar{a}_n(x)$ increases with x, we get

$$\|I_2\|_p \leq K \int_{\pi/n}^{\pi} \frac{w_p(t)}{t} \bar{a}_n(\tfrac{\pi}{t}) dt$$

$$= K \sum_{k=1}^{n-1} \int_{\pi/(k+1)}^{\pi/k} \frac{w_p(t)}{t} \bar{a}_n(\tfrac{\pi}{t}) dt$$

$$(4.5)\quad \leq K \sum_{k=1}^{n} \frac{w_p(\frac{\pi}{k}) \bar{a}_n(k+1)}{k}.$$

Finally, observe that by (2.1), (2.2) and the non-decreasing nature of $w_p(t)$,

$$\sum_{k=1}^{n} \frac{w_p(\frac{1}{k}) \bar{a}_n(k+1)}{k} \geq w_p(\tfrac{\pi}{n}) \sum_{k=1}^{n} \frac{1}{k} \sum_{r=0}^{k+1} a_{nr}$$

$$\geqq \ell w_p(\tfrac{\pi}{n}),$$

where ℓ is a non-zero constant.

From (4.4), (4.5), (4.6) and (4.2), the required estimate follows.

<u>Proof of Theorem 2.</u> Observe that by Lemma 3,

$$\int_{\pi/n}^{\pi} \frac{w_p(t)}{t} \left| \sum_{k=0}^{n} a_{nk} \sin(k + \tfrac{1}{2})t \right| dt$$

$$\leq k \int_{\pi/n}^{\pi} \frac{w_p(t)}{t} a'_{n,n-\tau+1} \, dt$$

$$= k \sum_{k=1}^{n-1} \int_{\pi/(k+1)}^{\pi/k} \frac{w_p(t)}{t} a'_{n,n-\tau+1} \, dt$$

$$(4.7) \quad \leq k \sum_{k=1}^{n} \frac{w_p(\frac{\pi}{k}) a'_{n,n-k}}{k}$$

since $w_p(t)$ and $a'_{n,n-\tau}$ are nondecreasing functions of t.

Following the method of proof of Theorem 1 and replacing the estimate (4.5) by (4.7), we shall get

$$(4.8) \quad \|f - A_n(f)\|_p = 0(w_p(\frac{1}{n})) + 0(\sum_{k=1}^{n} \frac{w_p(\frac{\pi}{k}) a'_{n,n-k}}{k}).$$

Using (1.6), the monotonic nondecreasing nature of $w_p(t)$, (2.1) and (2.4), we can show that

$$(4.9) \quad w_p(\frac{1}{n}) = 0(\sum_{k=1}^{n} \frac{w_p(\frac{\pi}{k}) a'_{n,n-k}}{k})$$

Hence, the required result follows from (4.8) and (4.9).

Proof of Theorem 3. Proceeding in a manner similar to that of Theorem 1, we can get

$$(4.10) \quad \|f - A_n(f)\|_p \leq E_1 + E_2,$$

where

$$(4.11) \quad E_1 = \int_0^{a_{nn}} \frac{w_p(t)}{t} \left| \sum_{k=0}^{n} a_{nk} \sin (k + \tfrac{1}{2})t \right| dt,$$

and

$$(4.12) \quad E_2 = \int_{a_{nn}}^{\pi} \frac{w_p(t)}{t} \left| \sum_{k=0}^{n} a_{nk} \sin (k + \tfrac{1}{2})t \right| dt.$$

Since $f \in \text{Lip}(\alpha, p)$, $w_p(t) = 0(t^\alpha)$ $(0 < \alpha \leq 1)$, using (2.1) we get

(4.13) $E_1 = 0(a^\alpha_{nn})$ $(0 < \alpha \leq 1)$.

By Lemma 1 and the fact that $w_p(t) = 0(t^\alpha)$, we get

(4.14) $E_2 = 0(a_{nn} \int_{a_{nn}}^{\pi} t^{\alpha-2}\, dt)$.

If $0 < \alpha < 1$, then

$$E_2 = 0(a_{nn}) + 0(a^\alpha_{nn}),$$

(4.15) $\quad = 0(a^\alpha_{nn})$

since by (2.1) $0 \leq a_{nn} \leq 1$. If $\alpha = 1$, then from (4.14),

(4.16) $E_2 = 0(a_{nn}(\log \pi + \log \frac{1}{a_{nn}}))$

$$= 0(a_{nn}(1 + \log \frac{1}{a_{nn}}))$$

On collecting the estimates from (4.13), (4.15), and (4.16) and substituting in (4.10), the required result is obtained.

Theorem 4 can be proved by using Lemma 5 and modifying the proof of Theorem 1. Similarly, we can prove Theorem 5 and Theorem 6 by modifying the proofs of Theorem 2 and Theorem 3 respectively with the help of Lemma 5.

5. In this section, we apply our theorems stated in §2 to deduce a number of interesting corollaries.

Let $\{p_n\}$ be a non-negative sequence of real numbers with $P_n = p_0 + p_1 + \ldots + p_n \neq 0$. If $a_{nk} = p_{n-k}/P_n$ $(k \leq n)$ and $a_{nk} = 0$ $(k > n)$, then the matrix A is said to be the Nörlund matrix (N, p_n). In this case, let us write $A_n(f; x)$ as $N_n(f; x)$.

If $p_n = E_n^{\delta-1}$ $(\delta \geq 1)$ where $E_n^\delta = \binom{n+\delta}{\delta}$, then the matrix (N, p_n) reduces to the matrix (C, δ) usually called the Cesaro matrix of order $\delta(\delta \geq 0)$. If $p_n = \frac{1}{n+1}$ $(n = 0, 1, \ldots)$, then the matrix (N, p_n) reduces to the matrix of the Harmonic mean (see [5]). For $N_n(f; x)$ we shall write $\sigma_n^\delta(f; x)$ or $H_n(f; x)$ according as the matrix (N, p_n) is the Cesaro or the Harmonic method.

If $a_{nk} = p_k/P_n$ $(0 \leq k \leq n)$ and $a_{nk} = 0$ $(k > n)$, then $A = (a_{nk})$ is said to be $(\overline{N}, p_n)$ matrix and in this case we denote $A_n(f; x)$ by $\overline{N}_n(f; x)$. If $p_n = \frac{1}{n+1}$ $(n = 0, 1, \ldots)$, then we shall write $\overline{H}_n(f; x)$ for $\overline{N}_N(f; x)$.

From Theorem 2, we shall obtain

COROLLARY 1. Let $\{p_n\}$ be a non-negative, non-increasing sequence. For $f \in L_p[-\pi, \pi]$ $(1 \leq p \leq \infty)$,

$$\|f - N_n(f)\|_p = 0\left(\frac{1}{P_n} \sum_{k=1}^{n} \frac{w_p(\frac{1}{k})\, p_k}{k}\right).$$

The case $p = \infty$ of Corollary 1 was proved in [5]. On putting $p_n = E_n^{\delta-1}$ $(\delta \geq 0)$ in Corollary 1, we get

COROLLARY 2. Let $1 \leq p \leq \infty$, then

$$\|f - \sigma_n^\delta(f)\|_p = \begin{cases} 0(n^{-\delta}) & (0 < \delta < \alpha \leq 1); \\ 0\left(\frac{\log n}{n^\delta}\right) & (0 \leq \delta \leq \alpha \leq 1). \end{cases}$$

The case $p = \infty$ was proved by Alexits [1]. On putting $p_n = \frac{1}{n+1}$ $(n = 0, 1, \ldots)$ in Corollary 1, we get

COROLLARY 3. Let $f \in \mathrm{Lip}(\alpha, p)$, $0 < \alpha \leq 1$, $1 \leq p \leq \infty$. Then

$$\|f - H_n(f)\|_p = 0\left(\frac{1}{\log n}\right).$$

From Theorem 3, we can deduce

COROLLARY 4. Let $f \in \mathrm{Lip}(\alpha, p)$, $0 < \alpha \leq 1$, $1 \leq p \leq \infty$. Let $\{p_n\}$ be a non-negative and non-decreasing sequence. Then

$$\| f - \bar{N}_n(f) \|_p = 0\left(\frac{p_n}{P_n}\right)^{\alpha} \quad (0 < \alpha < 1);$$

$$0\left\{\frac{p_n}{P_n} \log \frac{P_n}{p_n}\right\} \quad (\alpha = 1).$$

The case $p = \infty$ of Corollary 4 was proved in [2].

When $p_n = \frac{1}{n+1}$ $(n = 0, 1, \ldots)$, Corollary 4 cannot be used since in this case $\{p_n\}$ is a decreasing sequence. However, from Theorem 1 we can obtain the following corollaries:

COROLLARY 5. Let $\{p_n\}$ be a non-negative, non-increasing sequence. If $f \in L_p$ $(1 \leq p \leq \infty)$, then

$$\| f - \bar{N}_n(f) \|_p = 0\left(\frac{1}{P_n} \sum_{k=1}^{n} \frac{p_{k+1}\, w_p(\frac{1}{k})}{k}\right).$$

If, in Corollary 5, we put $p_n = \frac{1}{n+1}$, then we get

COROLLARY 6. Let $f \in \mathrm{Lip}(\alpha, p)$, $0 < \alpha \leq 1$, $1 \leq p \leq \infty$. Then

$$\| f - \bar{H}_n(f) \|_p = 0\left(\frac{1}{\log n}\right).$$

In the same manner we can obtain corollaries from Theorems 4, 5 and 6. To illustrate, we state the following result which can be deduced from Theorem 5.

COROLLARY 7. Let $0 < \alpha \leq 1$, $1 < p < \infty$, $\alpha p > 1$ and $f \in \mathrm{Lip}(\alpha, p)$. If $0 < \delta \leq 1$, then

$$\sup_{-\pi \leq x \leq \pi} |f(x) - \sigma_n^{\delta}(f; x)| = \begin{cases} 0\left(\frac{1}{n^{\delta}}\right) & (\alpha > \frac{1}{p} + \delta); \\ 0\ \frac{\log n}{n^{\delta}} & (\alpha = \frac{1}{p} + \delta). \end{cases}$$

6. In this section we shall study some approximation properties of Euler and Borel means.

If $A = (a_{nk})$ is such that for $q \geq 0$,

$$(6.1)\quad a_{nk} = \begin{cases} (1+q)^{-n} \binom{n}{k} q^{n-k} & (0 \leq k \leq n); \\ 0 & (k > n); \end{cases}$$

then the matrix A is called the Euler matrix. In this case, let us denote the operator $A_n(f;\, x)$ by $E^q_n\,(f;\, x)$.

Let us also define the Borel transform of the series (1.1) by

$$(6.2)\quad B^p(f;\, x) = e^{-p} \sum_{n=0}^{\infty} \frac{p^n}{n!}\, s_n(x),$$

where p is a real number. Series (1.1) is said to be Euler summable or Borel summable to sum s according as

$$\lim_{n\to\infty} E_n^{q}(f;\, x) = s \quad \text{or} \quad \lim_{p\to\infty} B^p\,(f;\, x) = s.$$

Concerning the L^p approximation by Euler and the Borel means the following results are known:

THEOREM A ([10], Theorem 1). Let $F \in \text{Lip}\,(\alpha,\, p)$, $0 < \alpha \leq 1$, $1 \leq p \leq \infty$. Then

$$(6.3)\quad \|f - E^q_n(f)\|_p = 0(n^{-\beta\alpha}) \quad (0 < \beta < \tfrac{1}{2}).$$

THEOREM B ([11], Theorem 2). Let $f \in \text{Lip}\,(\alpha,\, p)$, $0 < \beta < 1$, $1 \leq p \leq \infty$. Then

$$(6.4)\quad \|f - B^p(f)\|_p = 0(p^{-\alpha\beta}),\ 0 < \beta < \tfrac{1}{2}.$$

Our object here is to prove the following theorems with a view to improving the estimates (6.3) and (6.4).

THEOREM 7. Let $f \in \text{Lip}\,(\alpha,\, p)$ $(0 < \alpha \leq 1)$, $1 \leq p \leq \infty$. Then

(6.5) $\|f - E^q_n(f)\|_p = 0(n^{-\alpha/2})$.

THEOREM 8. Let $f \in \mathrm{Lip}(\alpha, p)$ $(0 < \alpha \leq 1)$, $1 \leq p \leq \infty$. Then

(6.6) $\|f - B^p(f)\|_p = 0(n^{-\alpha/2})$.

<u>Proof of Theorem 7</u>. Using the definition of $E^q_n(f; x)$ and the fact that $\sin \frac{1}{2} t \geq \frac{t}{\pi}$ $(0 \leq t \leq \pi)$, we have

(6.7) $$|E^q_n(f; x) - f(x)| \leqq \int_0^{\pi} \frac{|\phi_x(t)|}{t} |E(n, t)| dt$$

where $E(n, t)$ is given by (1.8). Writing the integral on the right of (6.7) as

$$\int_0^{\pi/\sqrt{n}} + \int_{\pi/\sqrt{n}}^{\pi}$$

we have by Minkowski inequality and Lemma 4

(6.8) $\|f - E^q_n(f)\|_p \leqq \|I_1\|_p + \|I_2\|_p$,

where

$$I_1 = \int_0^{\pi/\sqrt{n}} \frac{\|\phi_x(t)\|_p}{t} \mid E(n, t) \mid dt$$

and

$$I_2 = \int_{\pi/\sqrt{n}}^{\pi} \frac{\|\phi_x(t)\|_p}{t} \mid E(n, t) \mid dt.$$

Since $|E(n, t)| \leq 1$ and $\|\phi_x(t)\|_p \leq 2w_p(t) \leq k\, t^{\alpha}$, $0 < \alpha \leq 1$, we get

(6.9) $$I_1 = 0\,(\int_0^{\pi/\sqrt{n}} k^{\alpha-1}\, dt) = 0(n^{-\alpha/2}).$$

By using Lemma 6 and following a procedure similar to the estimate of I_2 in [15], we can get

(6.10) $I_2 = 0(n^{-\alpha/2})$ (See [15], (4.4.)

From (6.8), (6.9) and (6.10), the required result follows.

Proof of Theorem 8. From (6.4) and the inequality $\sin\frac{1}{2}t > \frac{t}{\pi}$ $(0 \leq t \leq \pi)$, we get

$$(6.11)\quad |B^p(f;\, x)| - f(x) \leqq \int_0^{\pi} \frac{|\phi_x(t)|}{t} |M(p,\, t)|\, dt,$$

where

$$(6.12)\quad M(p,\, t) = e^{-p} \sum_{n=0}^{\infty} \frac{p^n}{n!} \sin\,(n + \tfrac{1}{2})t.$$

Writing the integral on the right of (6.1) as the sum of two integrals viz. from 0 to $1/\sqrt{p}$ and $1/\sqrt{p}$ to π, and using Minkowski inequality and Lemma 4, we can get

$$\|f - B^p(f)\|_p \quad \int_0^{1/\sqrt{p}} \|\phi_x(t)\|_p\, t^{-1} \mid M(p,\, t) \mid dt$$

$$+ \int_{1/\sqrt{p}}^{\pi} \|\phi_x(t)\|_p\, t^{-1} \mid M(p,\, t) \mid dt$$

$\varepsilon_1 + \varepsilon_2$, say.

Since $|M(p,\, t)| \leq 1$ and $\|\phi_x(t)\|_p = 0(t^{\alpha})$ whenever $f \in \mathrm{Lip}\,(\alpha,\, p)$, we get

$$(6.13)\quad \varepsilon_1 = 0\left(\int_0^{1/\sqrt{p}} t^{\alpha-1} dt\right) = 0(p^{-\alpha/2}).$$

Before estimating E_2, note that we can write

$$(6.14)\quad M(p,\, t) = \sin\,(\tfrac{t}{2} + p \sin t) e^{-2p} \sin^2 \tfrac{1}{2}\, t.$$

In view of (6.14),

$$\varepsilon_2 = 0\left(\int_{1/\sqrt{p}}^{\pi} t^{\alpha-1} e^{-2p \sin^2 \frac{1}{2} t}\, dt\right)$$

$$(6.15) \quad = 0(1) \int_{1/\sqrt{p}}^{\pi} \frac{t^{\alpha-1}}{p \sin t} - \frac{d}{dt}(\exp(-2p \sin^2 \tfrac{1}{2} t))\, dt.$$

Now write the integral on the right of (6.15) as the sum of the integrals

$$(6.16) \quad \int_{1/\sqrt{p}}^{\pi} + \int_{\pi/2}^{\pi} \frac{t^{\alpha-1}}{p \sin t} \frac{d}{dt}\{-\exp(-2p \sin^2 \tfrac{1}{2} t))\}\, dt.$$

Let the two integrals of (6.16) be called ε_{21} and ε_{22}.

$$\varepsilon_{21} = 0(p^{-1}) \int_{1/\sqrt{p}}^{\pi/2} t^{\alpha-2} \frac{d}{dt}\{-\exp\,(-2p \sin^2 \tfrac{1}{2} t)\}\, dt$$

$$(6.17) \quad = 0(p^{-\alpha/2}), \quad \text{since}$$

$$\sin t \geq 2t/\pi \quad \text{for} \quad 0 \leq t \leq \frac{\pi}{2}.$$

We also see that

$$(6.18) \quad \varepsilon_{22} = \int_{\pi/2}^{\pi} \frac{t^{\alpha-1}}{p \sin t} \frac{d}{dt}\{-\exp(-2p \sin^2 \tfrac{1}{2} t)\}\, dt = 0(p^{-1}).$$

On collecting the estimates from (6.13), (6.17) and (6.18), we get the required estimate for $\|f - B^p(f)\|$.

REMARK. One can have analogues of the result of the type considered in Theorems 4, 5, 6, for Euler and Borel means. We state the results in

THEOREM 9. If $0 < \alpha \leq 1$, $1 < p < \infty$, $\alpha p > 1$ and $f \in \text{Lip}\,(\alpha, p)$, then

$$\sup_{-\pi \leq x \leq \rho} |f(x) - E^q_n(f;\, x)| = 0(n^{-\frac{1}{2}(\alpha - \frac{1}{p})}).$$

and

$$\sup_{-\pi\leq x\leq\pi} |f(x) - B^p(f; x)| = O(p^{-\frac{1}{2}(\alpha - \frac{1}{p})}).$$

REMARK. A cursory examination of the proofs of our results reveals that when $f \in \mathrm{Lip}^*(\alpha, p)$ the 'O' in all estimates will be replaced by 'o'.

REFERENCES

1. G. Alexits, "Über die Annaherung einer Stetigen Funktions durch die Cesaròschen Mittel ihrer Fourierreihe", Math. Annalen, 100 (1928), 264-277.
2. Prem Chandra, "On the degree of approximation of functions belonging to Lipschitz class", Nanta Math., 8 (1975), 88-91.
3. Prem Chandra, "On the degree of approximation of functions belonging to the Lipschitz class", Labdev Jour. Science and Technology, 13A, (1975), 181-183.
4. Prem Chandra, "On the degree of approximation of continuous functions", Commun. Fac. Sci. Univ. Ankara, Ser A, 29 (1980) (to appear).
5. G. H. Hardy, "Divergent series", (Oxford, 1949).
6. G. H. Hardy and J. E. Littlewood, "A convergence criterion for Fourier series", Math. Zeitschr., 28 (1928), 612-634.
7. G. H. Hardy, J. E. Littlewood and G. Polya, "Inequalities", (Cambridge, 1967).
8. A. S. B. Holland, B. N. Sahney and J. Tzimbalario, "On the degree of approximation of a class of functions by means of Fourier series", Acta. Sci. Math. (Szeged), 38 (1976), 69-72.
9. A. S. B. Holland, R. N. Mohapatra and B. N. Sahney, "Functions of Class Lip (α, p) and their Taylor Mean" (to appear).
10. A. S. B. Holland, R. N. Mohapatra and B. N. Sahney, "L_p approximation of functions by Euler means", (to appear).
11. A. S. B. Holland, R. N. Mohapatra and B. N. Sahney, "Approximation properties of Borel means of the Fourier series of a function of class L_p", (to appear in the Proceedings of the Conference on Nonlinear Analysis and Applications held in St. John's, 1981).
12. P. D. Kathal, A. S. B. Holland and B. N. Sahney, "A class of continuous functions and their degree of approximation", Acta Math. Sci. Hungaricae, 30 (1977), 227-231.
13. Huzoor H. Khan and A. Wafi, "On the degree of approximation by matrix means", Studia Sci. Math. Hungarica, 12 (1977), 185-192.
14. L. McFadden, "Absolute Norlund Summability", Duke Math. Jour., 9 (1942), 168-207.

15. R. N. Mohapatra and Prem Chandra, "Holder continuous functions and their Euler, Borel and Taylor means", (to appear).

16. R. N. Mohapatra and B. N. Sahney, "Approximation of continuous functions by their Fourier Series", (to appear in Jour. L'Analyse Numerique et la Theorie de l'Approximation (Romania).

17. E. S. Quade, "Trigonometric approximation in the mean", Duke Math. Jour., 3 (1937), 529-543.

18. A. Zygmund, "Trigonometric series", Vol. 1 and Vol. II, (Second Edition, Cambridge, 1968).

EXTENSION OF A THEOREM OF J. L. WALSH ON DIFFERENCES OF INTERPOLATING POLYNOMIALS

A. Sharma
University of Alberta
Edmonton, Alberta
Canada

E. B. Saff
University of South Florida
Tampa, Florida
U. S. A.

R. S. Varga
Kent State University
Kent, Ohio
U. S. A.

Let A_ρ denote the set of functions $f(z)$ analytic in the disk $|z| < \rho$, where $1 < \rho < \infty$. Let π_m denote the set of all polynomials of degree $\leq m$ and let $p_n(z; f) \in \pi_n$ be the Lagrange polynomial interpolant of $f(z) \in A_\rho$ in the $(n + 1)$ roots of unity. Writing $f(z) = \sum_{j=0}^{\infty} a_j z^j$ for $|z| < \rho$, let $P_n(z; f) = \sum_{j=0}^{n} a_j z^j$ be the associated n^{th} partial sum of f, so that $P_n(z; f) - f(z) = 0\ (z^{n+1})$ as $z \to 0$. Walsh proved in 1931 that

$$\lim_{n\to\infty} [p_n(z; f) - P_n(z; f)] = 0 \quad \text{for} \quad |z| < \rho^2,$$

the convergence being uniform and geometric on any closed set of the circle $|z| < \rho^2$. Here we extend this result to functions of the class $M_\rho(\nu)$ which are meromorphic in $|z| < \rho$ and have precisely ν poles in

$|z| < \rho$. The Taylor polynomial is replaced by the (n, ν) Padé approximant and the $p_n(z; f)$ above by the generalized (n, ν) Padé approximant on the $(n + \nu + 1)$ roots of unity. Other extensions to more general point sets and Jordan domains are also given.

The details will appear elsewhere.

ON BEST SIMULTANEOUS CHEBYSHEV APPROXIMATION WITH ADDITIVE WEIGHT † FUNCTIONS AND RELATED RESULTS

B. N. Sahney

University of Calgary
Calgary, Alberta
Canada

S. P. Singh

Memorial University of Newfoundland
St. John's, Newfoundland
Canada

The classical problem of simultaneous approximation of a function and its derivatives has been studied by several authors like D. Johnson [29], Meir and Sharma [38], G. G. Lorentz [35], K. Lim [32] and others. Here, the problem is presented for the best simultaneous approximation of two or more functions and their extensions. We start with the supremum norm and then include extensions given in recent years.

Two Real Functions

Let $a < b$ be real numbers and on $a \leq x \leq b$, we define the real valued functions $f_1(x)$ and $f_2(x)$. The ideas of C. B. Dunham [17] have been followed up by Diaz and McLaughlin [13], who gave the following definition.

DEFINITION 1. If there exists a $s^* \in S$, where S is a nonempty family of real valued functions on $[a,b]$, such that

$$\inf \max\{ \sup_{a\leq x\leq b} |f_1(x) - s(x)|, \sup_{a\leq x\leq b} |f_2(x) - s(x)|\}$$

$$= \max\{ \sup_{a\leq x\leq b} |f_1(x) - s^*(x)|, \sup_{a\leq x\leq b} |f_2(x) - s^*(x)|\} ,$$

then we say that s^* is a best simultaneous approximation to $f_1(x)$ and $f_2(x)$.

†This work was supported by NSERC and a N A T O grant.

They proved the following theorem based on this definition.

THEOREM 1. If S is a nonempty family of real valued functions on [a, b], then

$$\left\| \left|\frac{f_1 + f_2}{2} - s\right| + \frac{1}{2}\left|f_1 - f_2\right| \right\|$$

$$= \max\{\|f_1 - s\|, \|f_2 - s\|\},$$

for any $s \in S$ and

$$\inf_{s\in S} \left\| \left|\frac{f_1 + f_2}{2} - s\right| + \frac{1}{2}\left|f_1 - f_2\right| \right\| \tag{A}$$

$$= \inf_{s\in S} \max\{\|f_1 - s\|, \|f_2 - s\|\}.$$

It is obvious that the best approximation to $\frac{1}{2}|f_1 + f_2|$ is not necessarily the best simultaneous approximation to f_1 and f_2 except in the case when $f_1 = f_2$. The authors also raised the question about f_1 and f_2 being complex and that the line of proof must be different even if such a result is true.

The proof of Theorem 1 follows by using the identity

$$|m + n| + |m - n| = 2\max(|m|, |n|).$$

The following definition is due to Ling [33], who considered the sum norm.

DEFINITION 2. If there exists an $s^* \in S$ such that

$$\| |f_1 - s^*| + |f_2 - s^*| \|_\infty = \inf_{s\in S} \| |f_1 - s| + |f_2 - s| \|_\infty .$$

Then s^* is said to be a best simultaneous approximation to f_1 and f_2 in the supremum norm. Using this definition, the following results are proved.

THEOREM 2. If S is a nonempty family of real valued functions on $[a, b]$, then

$$\| |f_1 - s| + |f_2 - s| \|_\infty \le 2 \max\{\|\frac{f_1 + f_2}{2} - s\|_\infty, \|\frac{f_1 - f_2}{2}\|_\infty\}$$

and

$$\inf_{s \in S} \| |f_1 - s| + |f_2 - s| \|_\infty \le 2 \inf_{s \in S} \max\{\|\frac{f_1 + f_2}{2} - s\|_\infty, \|\frac{f_1 - f_2}{2}\|_\infty\}.$$

The proof follows on the same lines as Theorem 1.

The following inequalities were also given. If

$$\inf_{s \in S} \|\frac{f_1 + f_2}{2} - s\|_\infty \ge \frac{1}{2} \|f_1 - f_2\|_\infty$$

then

$$\inf_{s \in S} \| |f_1 - s| + |f_2 - s| \|_\infty = 2 \inf_{s \in S} \|\frac{f_1 + f_2}{2} - s\|_\infty \tag{1}$$

which implies that under (1), s^* is the best simultaneous approximation to f_1 and f_2 in the above sense if and only if s^* is a best approximation to $\frac{f_1 + f_2}{2}$ from S.

Further, if

$$\inf \|\frac{f_1 + f_2}{2} - s\|_\infty < \frac{1}{2} \|f_2 - f_1\|_\infty$$

then

$$\inf_{s \in S} \| |f_1 - s| + |f_2 - s| \|_\infty = \|f_2 - f_1\|_\infty \tag{2}$$

(See for detailed proof [33]).

The proof of (1) follows by the identity

$$|m + n| + |m - n| = 2(|m|)$$

provided

$$|m| \geq |n|.$$

Also, the proof of (2) follows by the identity

$$|m + n| + |m - n| = 2(|n|)$$

provided

$$|n| > |m|.$$

Phillips and Sahney [43] have given the following definition for L_1 and L_2 norms.

DEFINITION 3. If there exists an $s^* \in S$ such that

$$\inf_{s \in S} \int_0^1 \max\{|f_1(x) - s(x)|, |f_2(x) - s(x)|\}dx$$

$$= \int_0^1 \max\{|f_1(x) - s^*(x)|, |f_2(x) - s^*(x)|\}dx,$$

then we say that s^* is a best simultaneous approximation to $f_1(x)$ and $f_2(x)$ in the L_1 norm.

The following results are given by them [43].

THEOREM 3. For any f_1 and f_2 defined on $[0, 1]$

$$\inf_{s \in S} \int_0^1 \left|\frac{f_1 + f_2}{2} - s\right| dx + \int_0^1 \left|\frac{f_1 - f_2}{2}\right| dx$$

$$= \inf_{s \in S} \int_0^1 \max\{|f_1(x) - s(x)|, |f_2(x) - s(x)|\} dx.$$

Thus, s^* is the best simultaneous approximation to f_1 and f_2 in the L_1 norm if and only if it is the best approximation to their arithmetic mean.

DEFINITION 4. If there exists an $s^* \in S$ such that

$$\inf_{s \in S}\{||f_1 - s||_1 + ||f_2 - s||_1\} = ||f_1 - s^*||_1 + ||f_2 - s^*||_1,$$

then s^* is said to be a best simultaneous approximation to f_1 and f_2 in L_1 norm.

THEOREM 4. For any f_1 and f_2 defined on [0, 1] and $s^* \in S$, then

$$\inf_{s \in S}\{||f_1 - s||_1 + ||f_2 - s||_1\}$$

$$= 2 \inf_{s \in S} \int_0^1 \max\{|\frac{f_1(x) + f_2(x)}{2} - s(x)|, |\frac{f_1(x) - f_2(x)}{2}|\} dx.$$

DEFINITION 5. If there exists a $s^* \in S$ such that

$$\inf_{s \in S}\{||f_1 - s||_2^2 + ||f_2 - s||_2^2\} = ||f_1 - s^*||_2^2 + ||f_2 - s^*||_2^2,$$

then we say that s^* is a best simultaneous approximation to f_1 and f_2 in the L_2 norm.

The following theorem is given using Definition 5.

THEOREM 5. For any f_1 and f_2

$$\inf_{s \in S} ||\frac{f_1 + f_2}{2} - s||_2^2 + ||\frac{f_1 + f_2}{2}||_2^2$$

$$= 2 \inf_{s \in S} \{ \| \frac{f_1 - s}{2} \|_2^2 + \| \frac{f_2 - s}{2} \|_2^2 \} .$$

This is equivalent to the statement that s^* is a best simultaneous approximation to f_1 and f_2 in the L_2 norm if and only if it is a best approximation to $\frac{1}{2}(f_1 + f_2)$.

Proofs of Theorem 3 and Theorem 4 follow on the identity

$$|m + n| + |m - n| = 2 \max(|m|, |n|),$$

and the proof of Theorem 5 depends on the Apollonius identity

$$|\alpha + \beta|^2 + |\alpha - \beta|^2 = 2\{|\alpha|^2 + |\beta|^2\},$$

which holds for any real or complex α and β.

Holland and Sahney [25] gave the following.

DEFINITION 6. Let f_1 and f_2 and $s \in S$, all are L_p integrable such that

$$\inf_{s \in S} [\| f_1 - s \|_p^p + \| f_2 - s \|_p^p] = \| f_1 - s^* \|_p^p + \| f_2 - s^* \|_p^p .$$

Then s^* is the best simultaneous approximation to the functions f_1 and f_2 in L^p norm.

THEOREM 6. If f_1, f_2 and S are as above and p is an even natural number, then

$$\inf [\| f_1 - s \|_p^p + \| f_2 - s \|_p^p] = 2 \inf_{s \in S} \{ \sum_{k=0}^{p/2} \binom{p}{2k} \int_a^b [\frac{f_1 + f_2}{2} - s]^{p-2k} .$$

$$[\frac{f_1 - f_2}{2}]^{2k} \} dx ,$$

i.e., the best simultaneous approximation to the two functions (in the L_p norm) is equivalent to the best approximation to the mean value of the functions in a certain sense such that weight functions become involved as multipliers.

If s^* is the best simultaneous approximation to f_1 and f_2 in the L_p norm and if s^{**} is the best approximation to $\frac{f_1 + f_2}{2}$, then the error bounds are given by

$$\|f_1 - s^*\|_p^p + \|f_2 - s^*\|_p^p \leq [\|\frac{f_1 + f_2}{2} - s^{**}\|_p + \|\frac{f_1 - f_2}{2}\|_p]^p$$

$$+ [\|\frac{f_1 + f_2}{2} - s^{**}\|_p - \|\frac{f_1 - f_2}{2}\|_p]^p .$$

By using the identity

$$(a + b)^p + (a - b)^p = 2 \sum_{k=0}^{p/2} \binom{p}{2k} a^{p-2k} b^{2k}$$

for p even and taking $a + b = f_1 - s$, $a - b = f_2 - s$.

Goel, Holland, Nasim and Sahney [24] proved the following results.

THEOREM 7. Let K be a subspace of a strictly convex normed linear space X, then there is at most one best simultaneous approximation from the elements of K to any two elements f_1 and f_2 in X.

THEOREM 8. Let K and X be as above, then any two elements $f_1, f_2 \in X$ have a unique best simultaneous approximation from the elements of K.

A generalized version to L_p norm has been given by them as follows.

THEOREM 9. Let X and K be as above and $f_1, f_2 \in X$ such that at least one of f_1 and f_2 does not belong to K. Then there is at most one best L^p simultaneous approximation to f_1 and f_2.

THEOREM 10. Let X be a Banach space and K a closed convex subset of X. If X is uniformly convex, then any two elements $f_1, f_2 \in X - K$ have a unique best L^p simultaneous approximation from the elements of

K for $1 \leq p \leq \infty$.

THEOREM 11. Let X be a normed linear space and G a subspace of X. Consider any two elements $f_1, f_2 \in X$ such that $f_1 \neq f_2 \neq f(f \in \overline{G})$ and $f_0 \in G$. Then f_0 is a best ℓ^p simultaneous approximation to f_1 and f_2 if there exist elements $g_1, g_2 \in X^*$ with the following properties.

$$\|g_1\|^q + \|g_2\|^q = 1, \frac{1}{p} + \frac{1}{q} = 1$$

$$g_1(f) + g_2(f) = 0 \quad \text{for all} \quad f \in G,$$

and

$$g_1(f - f_0) + g_2(f - f_0) = [\|f_1 - f_0\|^p + \|f_2 - f_0\|^p]^{\frac{1}{p}} .$$

An elegant general approach of the above result has been given by Brøndsted [8].

In order to answer the question about n functions in L^2 norm, Holland and Sahney have given the following result [25].

THEOREM 12. If there exists an element $s^* \in S$ such that a best simultaneous approximation to the n functions $f_1, f_2, \ldots, f_n$ is defined as

$$\inf_{s \in S}[\sum_{i=1}^{n} \|f_i - s\|_2^2] = \sum_{i=1}^{n} \|f_i - s^*\|_2^2 .$$

Then

$$\inf_{s \in S} \sum_{i=1}^{n} \|f_i - s\|_2^2 = \inf_{s \in S} n\|\frac{1}{n} \sum_{i=1}^{n} f_i - s\|_2^2 + \frac{1}{n} \sum_{i<J} \|f_i - f_J\|_2^2 .$$

The proof follows easily by using the identity

$$[\sum_{i=1}^{n} a_i]^2 + \sum_{i<J} [a_i - a_J]^2 = n \sum_{J=1}^{n} a_J^2 .$$

McCabe and Phillips [37] gave the following:

DEFINITION 7. Given the elements $f_1, f_2, \ldots, f_n$ of $C[a, b]$ and S a subset of $C[a, b]$, we say that $s^* \in S$ is a best simultaneous Chebyshev approximation to $f_1, f_2, \ldots, f_n$ if

$$\| |f_1 - s^*| + |f_2 - s^*| + \ldots + |f_n - s^*| \|_\infty \leq \| |f_1 - s| + |f_2 - s| + \ldots + |f_n - s| \|_\infty$$

for all $s \in S$.

The main result is the following.

THEOREM 13. If $s^* \in S$ satisfies

$$\|\frac{1}{n} \sum_{j=1}^{n} f_j - s^*\|_\infty \leq \|\frac{1}{n} \sum_{j=1}^{n} f_j - s\|_\infty$$

for all $s \in S$ and if

$$\|\frac{1}{n} \sum_{j=1}^{n} f_j - f_k\|_\infty \leq \|\frac{1}{n} \sum_{j=1} f_j - s^*\|_\infty$$

for $1 \leq k \leq n$, then s^* is a best simultaneous approximation to $f_1, f_2, \ldots, f_n$ in the sense of Definition 7.

Holland, McCabe, Phillips and Sahney [26] have shown that Theorem 3 cannot be extended to more than two functions, as such, in L_1 norm. They, however, proved the following.

THEOREM 14. If s^* is a best approximation to $\frac{1}{n} \sum f_i$ from S and $n > 2$, then in general s^* is not a best simultaneous approximation to $f_1, f_2, \ldots, f_n$.

THEOREM 15. Let $f_1, f_2, \ldots, f_n$ be elements of $C[a, b]$ and S be a subset of $C[a, b]$. Then $s^* \in S$ is a best simultaneous approximation to $f_1, f_2, \ldots, f_n$ if and only if it is a best approximation to g_1 and g_2.

THEOREM 16. Let $f_1, f_2, \ldots, f_n$ be elements of $C[a, b]$ and S be a subset of $C[a, b]$. Then $s^* \in S$ is a best simultaneous approximation to $f_1, f_2, \ldots, f_n$ if and only if it is a best L_1 approximation to the arithmetic mean of $\max_k\{f_k(x)\}$ and $\min_k\{f_k(x)\}$.

The work of Cheney, McCabe and Phillips [11] entitled "A mixed-norm bivariate approximation problem with application to Lewanowicz operators" is worth mentioning.

The Case Of Complex Functions

Diaz and McLaughlin [3] raised the question whether there is an analogous result to Theorem 1 for the complex case. In answering this question in negative Chui, Rahman, Sahney and Smith [12] gave the following example.

EXAMPLE. Let $f_1(x) = -1$ and $f_2(x) = 1$ $a \le x \le b$. If S is the set of functions $\{s_n\}$ where for $a \le x \le b$,

$$s_1(x) = 1 + ic \quad c \text{ is real, } 0 < |c| < \sqrt{3}$$

$$s_n(x) = n \quad n = 2, 3, 4, \ldots,$$

then the left hand side of (A) Theorem 1 is equal to $(1 + c^2)^{\frac{1}{2}} + 1$ where as the right hand side is equal to $(4 + c^2)^{\frac{1}{2}}$.

They proved the following.

THEOREM 17. Let f_1, f_2 and S be complex valued functions on $[a, b]$. Then

$$\max(\|f_1 - s\|, \|f_2 - s\|) = \|\{|\frac{f_1 + f_2}{2} - s| + (|\frac{f_1 - f_2}{2}|)^2$$

$$+ |\frac{|f_1 - s|^2 - |f_2 - s|^2}{2}|$$

$$- \left|\frac{(f_1 - s)^2 - (f_2 - s)^2}{2}\right|\}^{\frac{1}{2}}\| .$$

Diaz and McLaughlin have considered simultaneous Chebyshev approximations of a set of bounded complex valued functions in [14].

Some further results in this area have been given by Bredendiek [7], Carroll [9], [10], Dunham [18], Phillips, McCabe and Cheney [42], Goel, Holland, Nasim and Sahney [23], Gillotte and McLaughlin [22], and others. This list is, however, far from complete.

BEST SIMULTANEOUS APPROXIMATION IN FUNCTIONS SPACES II

This is in continuation of section I. This section is again an expository survey, where a set of finitely many elements is replaced by a closed, convex set.

Holland, Sahney and Tzimbalario [27, 28] studied the problem in a more general setting by replacing two element set by a closed, convex set. Mach [36], Bosznay [6], Milman [39] and Owens [41] gave more general results. Some further results were given by Dunham [19]. The problem of nonlinear best simultaneous approximation was considered by Blatt [5].

Recently a few mathematicians have studied relative Chebyshev centers, a concept closely related to the best simultaneous approximation, see e.g. Amir and Ziegler [2, 3], Franchetti and Cheney [20], Lambert and Milman [30], Smith and Ward [49, 50], Ward [51], Zamijatin [52, 53] and others.

Several interesting related results could not be included in this survey and therefore it is again far from complete as section I.

The following results are given in [27].

LEMMA 1. Let $K \subset X$, and F be a bounded subset of X. Then

$$\phi(k) = \sup_{f \in F} \|f - k\|$$

is a continuous functional on X.

LEMMA 2. If K is a finite dimensional subspace of a normed linear space X, then there exists a best simultaneous approximation $k^* \in K$ to any given compact subset $F \subset X$.

Proof. Since F is compact, there is a finite constant M such that

$$||f|| \leq M \quad \text{for every} \quad f \in F.$$

We define the subset S of K by

$$S = S(0, 2M),$$

then

$$\inf_{k \in S} \sup_{f \in F} ||f - k|| = \inf_{k \in K} \sup_{f \in F} ||f - k|| \leq M.$$

Since S is compact, the continuous functional $\phi(k)$ attains its minimum over S for some k^* which will be the best simultaneous approximation to F.

LEMMA 3. Let K be a convex subset of X and $F \subset X$. If k_1 and k_2 in K are best simultaneous approximations to F by elements of K, then

$$k^* = \lambda k_1 + (1 - \lambda)k_2 \quad 0 \leq \lambda \leq 1$$

is also a best simultaneous approximation to F.

Holland, Sahney and Tzimbalario [27] proved the following two theorems.

THEOREM 1. Let K be a finite dimensional subspace of a strictly convex normed space X. Then there exists a unique best simultaneous approximation from the elements of k to any given compact subset $F \subset X$.

Proof. The existence of a best simultaneous approximation is guaranteed by Lemma 2. Suppose k_1 and k_2 $(k_1 \neq k_2)$ are two best simultaneous approximations to F, i.e.

$$\inf_{k \in S} \sup_{f \in F} ||f - k||$$

$$= \sup_{f \in F} ||f - k_1||$$

$$= \sup_{f \in F} \|f - k_2\|$$

$$= d \quad \text{say}.$$

By Lemma 3, $\dfrac{k_1 + k_2}{2}$ is also a best simultaneous approximation, i.e.

$$\sup \left\| f - \frac{k_1 + k_2}{2} \right\| = d.$$

Since F is compact, there exists a f_0 such that

$$\left\| f_0 - \frac{k_1 + k_2}{2} \right\| = d.$$

Now, since X is strictly convex and

$$\|f_0 - k_1\| = d \quad \text{and} \quad \|f_0 - k_2\| = d,$$

so

$$\left\| f_0 - \frac{k_1 + k_2}{2} \right\| < d,$$

contradiction. So $k_1 = k_2$, a unique best simultaneous approximation.

THEOREM 2. Let K be a closed and convex subset of a uniformly convex Banach space X. For any compact subset $F \subset X$, there exists a unique best simultaneous approximation to F from the elements of K.

Proof. The outline of proof follows as given below. Let

$$d = \inf_{k \in K} \sup_{f \in F} \|f - k\|$$

and $\{k_n\}$ any sequence of elements in K such that

$$\lim_{n \to \infty} \sup_{f \in F} \|f - k_n\| = d.$$

By using uniform convexity of X and the compactness of F it is shown that $\{k_n\}$ is a Cauchy sequence; hence, it converges to some k in X. Since K is closed $k \in K$. The element k is a unique best simultaneous approximation follows by using strict convexity.

We give the following where uniform convexity condition is relaxed [45].

THEOREM 3. Let X be a strictly convex Banach space and C a weakly compact, convex subset of X. Then there exists a unique best simultaneous approximation from the elements of C to any given compact subset F of X.

DEFINITION. A bounded subset F of a normed linear space X is said to be remotal with respect to a subset $K \subset X$, if for each $k \in K$ there exists a point $f_0 \in F$ such that

$$\|k - f_0\| = \sup_{f \in F} \|k - f\| .$$

DEFINITION. A subset K of a normed linear space X is said to be proximal with respect to a subset $F \subset X$, if for each $f \in F$ there exists a point $k \in K$ such that

$$\|f - k\| = \inf_{x \in K} \|f - x\| .$$

In [46] Theorem 1 just gives existence and for uniqueness X has to be strictly convex. However, the following holds.

If X is strictly convex normed linear space and C is a compact, convex subset of X, $F \subset X$ is a remotal set with respect to C; then there exists a unique best simultaneous approximation to F from the elements of C.

Theorem 1 has been given for a finite dimensional space. A natural question is suggested by this theorem. Is the hypothesis of finite dimensional really necessary?

We prove the following where we do not have uniqueness [46]:

THEOREM 4. Let X be a normed linear space and C a reflexive subspace of X. Then for any nonempty bounded subset F of X, there exists a best simultaneous approximation in C.

Bosznay [6] gave the following in a more general setting:

THEOREM 5. Let X be a strictly convex normed linear space and let M be a reflexive subspace of X. Then for every nonempty compact set $F \subset X$, there exists a unique simultaneous best approximation point in M.

Sastry and Naidu [47] have proved the following:

THEOREM 6. If K is boundedly weakly sequentially compact subset of X, then there exists a best simultaneous approximation to a bounded subset $F \subset X$.

NOTE 1. If X is strictly convex, K is convex, F is a farthest point set with respect to K, then there exists at most one best simultaneous approximation to F from K.

NOTE 2. If X is strictly convex, K is boundedly weakly sequentially compact and convex, and F is a farthest point set with respect to K, then there exists a unique best simultaneous approximation to F from K.

THEOREM 7. If X is strictly convex and reflexive, K is closed and convex, and F is compact then there exists a unique best simultaneous approximation to F from K.

Proof. Since a closed, convex set in a reflexive Banach space is weakly compact and thus the proof follows from Note 1 and Note 2.

The following theorem suggested by Cheney is given in much more general setting.

LEMMA 4. If F is a bounded subset of a normed space X, then the function $\phi : X \to R$ defined by

$$\phi(x) = \sup_{y \in F} ||x - y||$$

is finite valued and lower semicontinuous.

<u>Proof</u>. The supremum of a family of continuous (or even lower semicontinuous) functions is lower semicontinuous (Rudin [44]).

NOTE. ϕ is even weakly lower semicontinuous because the map $x \to ||x - y||$ are weakly lower semicontinuous. The latter follows from the fact that

$$||x - y|| = \sup_{\substack{f \in x^* \\ ||f|| = 1}} f(x - y),$$

and each map $x \to f(x - y)$ is weakly lower semicontinuous. Thus, ϕ will assume its infimum on any weakly compact set. Hence the following.

THEOREM 8. If K is a weakly compact subset of a normed linear space X and if F is a bounded subset of X, then there exists a simultaneous approximant to F in the set K.

Ahuja and Narang [1] proved the following.

THEOREM 9. Let K be any subset of a normed linear space X. Then given any bounded subset F of X there exists a best simultaneous approximation to the set F by elements of K, if F is remotal with respect to K and K is proximal with respect to F.

If, in addition, K in above theorem is convex and the space is strictly convex, then the best simultaneous approximation is unique.

Uniqueness follows as in Theorem 1.

For the following terms see Franchetti and Cheney [20]. Let X be a Banach space and A a subset of X. For each $x \in X$, the distance from x to A is defined by

$$d(x, A) = \inf_{a \in A} |x - a|.$$

If a set of elements B is given in X, one might like to approximate all of the elements of B simultaneously by a single element in A. This type of problem arises when a function being approximated is not known precisely, but is known to belong to a set.

The Chebyshev radius of a set B with respect to A is defined by

$$r_A(B) = \inf_{a \in A} \sup_{b \in B} \|a - b\|.$$

In approximating simultaneously all elements of B by a single element of A, one identifies elements of the set

$$E_A(B) = \{a \in A \mid \sup_{b \in B} |a - b| = r_A(B)\}.$$

The elements of $E_A(B)$ are solutions to the simultaneous approximation problem. The set $E_A(B)$ is called the restricted center of B relative to A. The concept was first introduced by Garkavi [21]. In case $A = X$, $E_X(B)$ is called the Chebyshev center.

The elements of $E_A(B)$ are also called simultaneous approximants of B or global approximants of B or restricted centers of B. If $E_A(B) \neq \phi$ for every bounded set B in X, it is said that A is proximinal for the global approximation of bounded sets.

Mach [36] has shown that in certain subspaces A of a given Banach space X a best simultaneous approximation exists for every bounded set $F \subset X$. He states that by a simple compactness argument it is easy to show that every finite dimensional subspace of an arbitrary Banach space and every w*-closed subspace of a dual space have this property. He further has shown that every Weierstrass-Stone subspaces of $C(S, X)$, the space of all continuous X-valued functions x on a compact Hausdorff space S equipped with the norm $\|x\| = \sup_{t \in S} |x(t)|$, where $|\cdot|$ is the norm of X, has this property, if X is a uniformly convex Banach space, generalizing a result of Olech [19], and $E_A(F) \neq \phi$, if A is an M-ideal in a Lindenstrauss space and F is a compact set. Some further results are given by Mach in [36].

Amir and Ziegler [2, 3], Lambert and Milman [30], Smith and Ward [49, 50], Ward [51], Zamijatin [52, 53] and others have also given interesting results in this case.

A. Bacopoulos, G. Godini and I. Singer have considered best simultaneous approximations in the following way [4]. Let X be a linear space with real scalars endowed with two norms, say $\|\cdot\|_1$ and $\|\cdot\|_2$. A method of approximating an element $x \in X$ by elements of a subset G of X simultaneously in both norms $\|\cdot\|_1$ and $\|\cdot\|_2$, is to consider a norm $|||\cdot|||$ on R^2 and then to approximate x by the elements of G in the norm $\|\cdot\|$ on X defined by

$$\|x\| = |||(\|x\|_1, \|x\|_2)||| \qquad (x \in X). \tag{1}$$

Denote by $P_G(x)$ the set of all elements of best simultaneous approximation of x, out of G, in the norm (1), i.e.,

$$P_G(x) = \{g_0 \in G \mid \|x - g_0\| = \inf_{g \in G} \|x - g\|\}$$

$$= \{g_0 \in G \mid |||(\|x - g_0\|_1, \|x - g_0\|_2)|||$$

$$= \inf_{g \in G} \|\,|(\|x - g\|_1, \|x - g\|_2)\|\,|\}.$$

REMARK. $g_0 \in P_G(x)$ if and only if the point

$$p_0 = (\|x - g_0\|_1, \|x - g_0\|_2) \in R^2$$

satisfies $|||p_0||| = \text{dist}(0, A)$, where

$$A = \{(\|x - g\|_1, \|x - g\|_2) \mid g \in G\}. \tag{2}$$

This remark reduces the problem of best simultaneous approximation of x in X by the elements of G to the problem of best approximation (in the plane R^2) of 0 by the elements of the set A defined by (2). Therefore, using the results known for best approximation in R^2, one can draw conclusions on best simultaneous approximation of x in X. They have given several interesting results.

Bosznay [6] has proved the continuity of the best simultaneous operator, whereas Sastry and Naidu [48] have discussed the weaker form of continuity in a more general setting.

REFERENCES

1. Ahuja, G. C. and Narang, T. D., "On best simultaneous approximation", Nieuw Arch. voor Wiskunde (3) XXVII (1979), 255-261.

2. Amir, D. and Ziegler, Z, "Relative Chebyshev centers in normed linear spaces I", to appear J. Approx. Theory.

3. Amir, D. L., Ziegler, Z., "Characterizations of relative Chebyshev centers", Approximation Theory III, Ed. Cheney, E. W., Academic Press (1980) 157-162.

4. Bacopoulos, A., Godini, G., Singer, I., "On infima of sets in the plane and best approximation, simultaneous and vectorial in a linear space with two norms", Special Topics of Applied Math. Ed. Frehse et. al. North-Holland (1979), 219-239.

5. Blatt, H. P., "Nicht-linear Gleichmässige simultan-approximation", Dissertation, Univ. des Saarlandes, Saarbrücken (1970).

6. Bosznay, A. P., "A remark on simultaneous approximation", J. Approx. Theory 23 (1978), 296-298.

7. Bredendiek, E., "Simultan approximationen", Arch. Rat. Mech. & Anal. 33 (1969), 307-330.

8. Brøndsted, A., "A note on best simultaneous approximation in normed linear spaces", Canad. Math. Bull. 19 (1976), 359-360.

9. Carroll, M. P., "On simultaneous L_1 approximation, approximation theory", Edited by G. G. Lorentz, Academic Press, New York (1973).

10. Carroll, M. P., "Simultaneous L_1 approximation of a compact set of real valued functions", Numer. Math. 19 (1972), 110-115.

11. Cheney, E. W., Phillips, G. M., and McCabe, J. H., "A mixed-norm bivariate approximation problem with applications to Lewanowicz operators", Multivariate Approximations, Ed. D. C. Handscomb, Academic Press, New York, (1978), 315-323.

12. Chui, C. K., Rahman, Q. I., Sahney, B. N. and Smith, P. W., "On simultaneous Chebyshev approximation", J. Approx. Theory 22 (1978), 350-351.

13. Diaz, J. B. and McLaughlin, H. W., "On simultaneous Chebyshev approximation and Chebyshev approximation with an additive weight", J. Approx. Theory, 6 (1972), 68-71.

14. Diaz, J. B. and McLaughlin, H. W., "On simultaneous Chebyshev approximation of a set of bounded complex valued functions", J. Approx. Theory 2 (1969) 419-432.

15. Diaz, and McLaughlin, H. W., "On simultaneous Chebyshev approximation and Chebyshev approximation with an additive function", J. Approx. Theory 6 (1972), 68-71.

16. Diaz, J. B. and McLaughlin, H. W., "Simultaneous approximation of a set of bounded real functions", Math. Comp. 23 (1969), 583-594.

17. Dunham, C. B., "Simultaneous Chebyshev approximation of functions on an interval", Proc. Amer. Math. Soc. 18 (1967) 472-477.

18. Dunham, C. B., "Simultaneous approximation of a set", J. Approx. Theory, 21 (1977), 205-208.

19. Dunham, C. B., "Approximation with respect to Chebyshev norms", Aequationes Math. 8 (1972), 267-270.

20. Franchetti, C. and Cheney, E. W., "Simultaneous approximation and restricted Chebyshev centers in function spaces , preprint.

21. Garkavi, A. L., "The Chebyshev center and the convex hull of a set", Usp. Mat. Nauk. 18 (1964), 139-145.

22. Gillotte, M. J. and McLaughlin, H. W., "On additive weight approximation and simultaneous Chebyshev approximation using varisolvent families", J. Approx. Theory 17 (1976), 35-43.

23. Goel, D. S., Holland, A. S. B., Nasim, C. and Sahney, B. N., "Characterization of an element of best L_p-simultaneous approximation", S. R. Ramanujan Memorial Volume, Madras (1974), 10-14.

24. Goel, D. S., Holland, A. S. B., Nasim, C. and Sahney, B. N., "On best simultaneous approximation in normed linear spaces", Canad. Math. Bull. 17 (1974), 523-527.

25. Holland, A. S. B. and Sahney, B. N., "Some remarks on best simultaneous approximation", Theory of Approximation with Applications, Ed. Law & Sahney, Academic Press, New York, (1976), 332-337.

26. Holland, A. S. B., McCabe, J. H., Phillips, G. M. & Sahney, B. N., "Best simultaneous L_1 approximation", J. Approx. Theory, 24 (1978), 361-365.

27. Holland, A. S. B., Sahney, B. N. and Tzimbalario, J., "On best simultaneous approximation", J. Indian Math. Soc. 40 (1976), 69-73.

28. Holland, A. S. B., Sahney, B. N. and Tzimbalario, J., "On best simultaneous approximation", J. Approx. Theory 17 (1976), 187-188.

29. Johnson, D. J., "Nonuniqueness of simultaneous approximation by Trigonometric polynomials", Approximation Theory 11, Ed. G. G. Lorentz, C. K. Chui and L. L. Schumaker, Academic Press, New York (1976), 411-415.

30. Lambert, J. M. and Milman, P. D., "Restricted Chebyshev centers of bounded subsets in arbitrary Banach spaces", J. Approx. Theory 26 (1979), 71-78.

31. Laurent, P. J. and Tuan, P. D., "Global approximation of a compact set by elements of a convex set in a normed space", Numer. Math. 15 (1970), 137-150.

32. Lim, K. P., "Simultaneous approximation of a function and its derivatives", J. Approx. Theory 18 (1976), 346-349.

33. Ling, W. H., "On simultaneous Chebyshev approximations in the sum norm", Proc. Amer. Math. Soc. 48 (1975), 185-188.

34. Ling, W. H., McLaughlin, H. W. and Smith, M. L., "Approximation of random functions", J. Approx. Theory, 20 (1977), 10-22.

35. Lorentz, G. G., "Approximation of functions", Holt, Rinehart and Winston, New York, (1966).

36. Mach, J.. "On the existence of best simultaneous approximation", J. Approx. Theory 25 (1979), 258-265.

37. McCabe, J. H. and Phillips, G. M., "On the simultaneous approximation of the functions", preprint.

38. Meir, A. and Sharma, A., "Simultaneous approximation of a function and its derivatives", SIAM J. Numer. Anal. 3 (1966), 553-563.

39. Millman, P. D., "On best simultaneous approximation in normed linear spaces", J. Approx. Theory, 20 (1977), 223-238.

40. Olech, C., "Approximation of set valued functions by continuous functions", Colloq. Math. 19 (1968), 285-293.

41. Owens, R. W., "Characterization of an element of best simultaneous approximation", preprint.

42. Phillips, G. M., McCabe, J. H. and Cheney, E. W., "On simultaneous Chebyshev approximation", J. Approx. Theory 27 (1979), 93-98.

43. Phillips, G. M. and Sahney, B. N., "On best simultaneous approximation in the L_1 and L_2 norms", Theory of Approximations with Applications, Ed. Law and Sahney, Academic Press, New York, (1976), 213-219.

44. Rudin, W., "Real and complex analysis", McGraw-Hill, New York (1966).

45. Sahney, B. N. and Singh, S. P., "On best simultaneous approximation in Banach spaces", J. Approx. Theory (to appear).

46. Sahney, B. N. and Singh, S. P., "On best simultaneous approximation", Approximation Theory III, Ed. Cheney, E. W., Academic Press (1980), 782-789.

47. Sastry, K. P. R. and Naidu, S. V., "On best simultaneous approximation in normed linear spaces", Proc. Nat. Acad. Sci. India 48 (1978), 249-250.

48. Sastry, K. P. R. and Naidu, S. V., "Upper semi-continuity of the best simultaneous approximation operator", Pure and Applied Math. Sciences X(1979), 7-8.

49. Smith, P. W. and Ward, J. D., "Restricted centers in $C(\Omega)$", Proc. Amer. Math. Soc. 48 (1975), 165-172.

50. Smith, P. W. and Ward, J. D., "Restricted centers in subalgebras of C(X)", J. Approx. Theory 15 (1975), 54-59.

51. Ward, J. D., "Chebyshev centers in spaces of continuous functions", Pacif. J. Math. 52 (1974), 283-287.

52. Zamjatin, V. M., "Chebyshev centers in the spaces C(S)", First Scientific Conf. of Young Scholars of the Akygei (1971), 28-35, (Russian).

53. Zamjatin, V. M., "Relative Chebyshev centers in a space of continuous functions", Soviet Math. Dokl. 14 (1973), 610-614.

INDEX

N

O

P

Q

R

S